p-adic Geometry

Lectures from the
2007 Arizona Winter School

University
LECTURE
Series

Volume 45

p-adic Geometry

Lectures from the
2007 Arizona Winter School

Matthew Baker
Brian Conrad
Samit Dasgupta
Kiran S. Kedlaya
Jeremy Teitelbaum

Edited by
David Savitt
Dinesh S. Thakur

American Mathematical Society
Providence, Rhode Island

2000 *Mathematics Subject Classification*. Primary 14G22; Secondary 11F85, 14F30.

The 2007 Arizona Winter School was supported by NSF grant DMS-0602287.
The first author was supported by NSF grant DMS-0600027.
The second author was partially supported by NSF grant DMS-0600919.
The third author was partially supported by NSF grant DMS-0653023.
The fourth author was supported by NSF CAREER grant DMS-0545904,
and a Sloan Research Fellowship.
The fifth author was supported by NSF grant DMS-0245410.

For additional information and updates on this book, visit
www.ams.org/bookpages/ulect-45

Library of Congress Cataloging-in-Publication Data

Arizona Winter School (2007 : University of Arizona)
  p-adic geometry : lectures from the 2007 Arizona Winter School / Matthew Baker... [et al.] ;
edited by David Savitt, Dinesh S. Thakur.
    p. cm. — (University lecture series ; v. 45)
    Includes bibliographical references.
    ISBN 978-0-8218-4468-7 (alk. paper)
    1. Arithmetical algebraic geometry–Congresses. 2. p-adic analysis–Congresses 3. Geometry, Algebraic–Congresses I. Baker, Matthew, 1973– II. Savitt, David. III. Thakur, Dinesh S.
IV. Title.

QA242.5.A757   2007
516.3′5–dc22

2008023597

Contents

Preface

This volume contains notes which accompanied the lectures at the tenth annual Arizona Winter School, held from March 10 to 14, 2007 at the University of Arizona in Tucson. The Arizona Winter School is an intensive five-day meeting, each year organized around a different central topic in arithmetic geometry, featuring several courses by leading and emerging experts ("an annual pilgrimage," in the words of one participant). The Winter School is the main activity of the Southwest Center for Arithmetic Geometry, which was founded in 1997 by a group of seven mathematicians working in the southwest United States, and which has been supported since that time by the National Science Foundation.

The special character of the Arizona Winter School comes from its format. Each speaker proposes a project, and a month before the Winter School begins, the speaker is assigned a group of graduate students who work on the project. The speakers also provide lecture notes and a bibliography. During the actual school the speaker and and his or her group of students work every evening on the assigned project. On the last day of the workshop, the students from each group present their work to the whole school. The result is a particularly intense and focused five days of mathematical activity (for the students and speakers alike).

The topic of the Winter School in 2007 was p-adic geometry, and the speakers were Matthew Baker, Brian Conrad, Kiran Kedlaya, and Jeremy Teitelbaum. Samit Dasgupta joins Teitelbaum as a co-author. We thank the authors for their hard work before, during, and after the Winter School. We are also grateful to John Tate and Vladimir Berkovich, two pioneers of non-archimedean geometry, who agreed to describe the early history of some of their contributions to the subject. The anonymous reviewers made numerous valuable comments, and we thank them for their careful reading of this manuscript. Finally, we are indebted to the other members (past and present) of the Southwest Center; it is thanks to their efforts that the Winter School exists in its present form.

David Savitt
Dinesh Thakur

Foreword

JOHN TATE

This book contains the notes of the four lecture series of the 2007 Arizona Winter School on non-archimedean geometry, together with a short account by Berkovich of how he developed, in difficult circumstances, the remarkable theory of the spaces which bear his name.

Brian Conrad's talks are an excellent introduction to the general theory of the original rigid analytic spaces, of Raynaud's view of them as the generic fibers of formal schemes, and Berkovich spaces. The p-adic upper half plane as rigid analytic space and several applications of it are discussed in the notes of Samit Dasgupta and Jeremy Teitelbaum. Matt Baker's lectures offer a detailed discussion of the Berkovich projective line and p-adic potential theory on that and more general Berkovich curves. Finally, Kiran Kedlaya discusses cohomologies, deRham and rigid, the comparison between them, and how to compute them.

In this foreword I want to do two things: (1) tell more than is in the brief preface about the institution which gave rise to this book, the AWS, for I think it's a great concept which would be good to try in other subjects and places, and (2) give a bit of ancient history of the book's subject, p-adic geometry, especially the part I was involved in.

It's hard to describe the AWS adequately. "Intense", "interactive", "infectious enthusiasm", "flexibility in support" give some idea of the flavor. You have to experience it to believe it can work so well. The unique and novel feature of the AWS is that after the usual set of lectures during the day that all such schools have, there are working sessions until late in the night in which students work with the help of the speakers and their assistants on projects related to the lectures. In fact, even the lecture series part of the school is not so usual, because the speakers produce a complete set of notes for their lectures a month ahead of time, with background references, so that the participants can be better prepared (this book is essentially the notes for the 2007 school). The speakers also describe ahead of time the projects which their group of students will work on at the school. These projects can involve elaborations of the theory, working out special examples, computing some quantity in various cases, writing computer code to do that, background work on foundational aspects... anything to challenge the participants and help them learn by working with the material. The projects associated to the lectures in this book, and also the lectures and projects from several earlier Winter Schools, can be found on the Arizona Winter School website. On the last day of the school students from each group make a joint presentation of the results the group has obtained.

Another key feature of the AWS is that the participants come from a wide variety of programs from all over the country, with very different levels of preparation, pressure and expectations. The organizers try to preserve this diversity in each of the four working groups. Typically a group will have an advanced student working in an area close to the group's project, it will have students with a very strong, but also not so strong background, it will have students from many different universities, some elite, some not, it will have students of both sexes, and at least one student from the southwest, the region represented by the organizers. Preferences for projects expressed by students on their applications are taken into account, but not all can be satisfied. In fact, the AWS has become so popular that not all students are officially assigned to groups. But all are encouraged to hang out and informally participate anyway. The whole scheme is flexible.

The interactions between students from different backgrounds that the AWS makes possible has a positive impact on all attending. For all students it is an opportunity to feel the exhilaration, drive and power that the students from the top programs show. On the other hand, the very nature of the AWS is to create an environment where students with lower level backgrounds are encouraged to ask basic questions. Discussion of these can be very valuable to all; it may reveal that what seemed clear was not quite so clear after all. The cross-pollination of mathematical cultures which takes place at the AWS is of benefit to everybody. Ultimately participants get a clear picture of what it means to do mathematics in the real world and this can be a significant learning experience for students and postdocs of any background.

Almost all of the participants are housed in the same hotel, and the evening sessions are held there too. There is no escape. Everyone is constantly involved in small and large discussion groups, on the lectures, on the project topics and toward the joint presentations to be given on the last day.

The school is five days of very hard work — 16 hour days — for all participants. In spite of that, or perhaps just because of the intellectual intensity, most seem to thrive. As an informal participant in several of the schools I am aware of many testimonials, oral and written, that it is a rewarding experience, both scientifically and interpersonally. The topic changes every year and the frequency with which many students return year after year is another indication of the value of the school. Work at the school has been the germ of many Ph.D. theses.

The first Winter School was held in 1998, and it has been going strong since then. The organizers (which by now include a few ex-students from early schools) deserve great credit, for the original conception, for improving it in various ways over the years, and for the mostly excellent choices of topics and speakers they have made each year. The 2007 lectures were outstanding and unusually closely related. Hence the idea to collect them in a book.

Now I would like to switch gears and discuss some old (pre-1965) history. The basic problem in creating a global theory of analytic manifolds over a non-archimedean local field K is that analytic continuation in the usual sense does not work. We can agree that a function in a "closed" disc $|x - a| \leq r$ (which is also open), or in an "open" disc $|x - a| < r$ (which is also closed), is analytic if and only if it has a power series expansion convergent in the disc, and, in fact, this turns out ultimately to be the right idea. But this is a much stricter condition than to have a power series expansion in a neighborhood of every point of the disc, because in

the non-archimedean metric every disc is a *disjoint* union of smaller discs. So this
is not a local definition, and there's no obvious way to globalize it.

The first to overcome this difficulty was Marc Krasner. He made a good theory
of analytic functions of one variable. They were defined on certain subsets of
$\mathbb{P}^1(K) = K \cup \infty$ which he called "quasiconnected". For example, the set obtained
by removing from $\mathbb{P}^1(K)$ a finite set of discs of the form $|(ax + b)/(cx + d)| < r$
is quasiconnected and an analytic function on it is one which is a uniform limit
of rational functions with poles in its complement. Krasner's theory had little
influence on the later global higher dimensional developments described in this
volume, but was valued and further developed by p-adic analysts doing the theory
of p-adic differential equations, etc. I remember that Dwork was very upset that
there was no mention of Krasner in the introduction to the book [BGR] referred to
by Conrad in this volume, and I think he was right to be.

The earliest steps toward the subject of this book were mine. My motivation
was the isomorphism $K^*/q^{\mathbb{Z}} \xrightarrow{\sim} E_q(K)$, for $q \in K, 0 < |q| < 1$, where E_q is an
elliptic curve over K with invariant $j = q^{-1} + 744 + 196884q + \cdots$. I still remember
the thrill and amazement I felt when it occurred to me that the classical formulas
for such an isomorphism over $\mathbb{C}$ made sense p-adically when properly normalized.[1]

This uniformization of some elliptic curves made me wonder if there might not
be a general theory of p-adic manifolds. Two years later in the fall of 1961, very
much influenced by Grothendieck's theory of schemes, I was ready to make a serious
attempt to create such a theory. In contrast to the difficult circumstances Berkovich
faced as he developed his theory, my situation could not have been more favorable,
with a good job at Harvard and friends like Serre and Grothendieck to help me. I
recorded my progress in a series of letters to Serre. He wrote that he was keeping
them carefully, but not reading them carefully. But he did find and fix a gap in my
proof of the acyclicity theorem.[2] He was interested at the time in another aspect
of p-adic analysis, namely Dwork's spectacular proof of the rationality of the zeta
function of algebraic varieties over a finite field, and was developing a theory of
Fredholm determinants in p-adic Banach spaces in order to simplify Dwork's proof.
In his course that winter Serre discussed the curve E_q, p-adic Banach spaces, and
Dwork's proof and his own generalization of it to some L-functions. In Serre's
seminar, Houzel talked on my letters, which Serre had had typed at the IHES for
limited distribution. Ten years later they were published as a paper in Inventiones
math. 12 (1971).

Grothendieck was visiting Harvard at the time I was writing the letters, and
his presence was a great help. By then, in contrast to the bizarre negativism he had
shown in his letter to Serre quoted at the end of Brian Conrad's introduction, he had
become wildly optimistic, writing, again to Serre, in Oct.'61: "...Sooner or later it
will be necessary to subsume ordinary analytic spaces, rigid analytic spaces, formal
schemes, and maybe schemes themselves into a single kind of structure for which
all the usual theorems hold: Stein spaces, Grauert finiteness, Remmert-Grauert
GAGA, maybe also Rothstein type theorems...."

---

[1]The notes I wrote at the time (fall '59) are published in the book *Elliptic curves, modular
forms and Fermat's last theorem*, International Press, Cambridge MA, 1995. Though I published
nothing earlier, the curve E_q became known thanks to others.

[2]I had carelessly thought that the two complexes which are mentioned in the proof of
Lemma 8.5 of my Inventiones 1971 paper are the same. Serre explained to me that they are
not at all the same, but are, in fact, homotopic, so the lemma is OK.

The "affinoid" part of the theory I was trying to make certainly benefited from discussions with Grothendieck, but where his help was essential was in how to define a global rigid space by gluing affinoids together. In that section of the letters I say "we follow fully and faithfully a plan furnished by Grothendieck" and to the best of my recollection that is not much of an exaggeration. The plan, with its "h-structures" and "special coverings" is rather complicated and I'm not sure I ever really understood it well. The definition of rigid space given in the present book is certainly much simpler and more satisfying than the one Grothendieck and I arrived at. That they are equivalent is far from obvious.

That's the end of my involvement in the history. I regret that I am too ignorant to say much about the further development of the theory. It was soon substantially simplified by the introduction of rational subdomains by Gerritzen and Grauert. Grothendieck's vision of a grand unification quoted above did not take place, but what happened was much more interesting, especially the great idea of Berkovich. To understand how the theory has evolved, the present book is an excellent place to start.

Non-archimedean analytic geometry: first steps

Vladimir Berkovich

When Dinesh Thakur asked me to write an introduction to this volume, I carelessly agreed. Later I started thinking that a short description of my journey to non-archimedean analytic geometry and of some of the circumstances accompanying it might be an entertaining complement to the notes of Matthew Baker and Brian Conrad. Since I had no other ideas, I've written what is presented below.

I start by briefly telling about myself. I was very lucky to be accepted to Moscow State University for undergraduate and, especially, for graduate studies in spite of the well-known Soviet policy of that time towards Jewish citizens. I finished studying in 1976, and got a Ph.D. the next year. (My supervisor was Professor Yuri Manin.) Getting an academic position would be too much luck, and the best thing I could hope for was the job of a computer programmer at a factory of agricultural machines and, later, at the institute of information in agriculture. As a result, I practically stopped doing mathematics, did not produce papers, and was considered by my colleagues as an outsider. It took me several years to become an expert in computers and nearby fields, and to learn to control my time. Gradually I started doing mathematics again, and my love for it blazed up with new force and became independent of surrounding circumstances. By the time my story begins, I was hungry for mathematics as never before.

Thus, my story begins one July evening of 1985 in a train in which I was returning to Moscow after having visited my numerous relatives in Gomel, Belarus. Instead of talking to people near me — my usual occupation during long train trips — I opened a book on classical functional analysis by Yuri Lyubich which my eldest brother Yakov had given me a couple of days before. The basic material of the book was familiar to me. Nevertheless, I was thrilled to read about it again and, suddenly, asked myself: what is the analog of all this over a non-archimedean field k? In particular, what is the spectrum of a bounded linear operator acting on a Banach space over k?

It did not take much time to find that, if one defines the spectrum in the same way as in the classical situation, it may be empty even if k is algebraically closed. Indeed, if K is a non-archimedean field larger than k, then the multiplication by any element of K which does not lie in k is an operator with empty spectrum. That such a larger field always exists is easily seen: one can take the completion of the field of rational functions in one variable over k with respect to the Gauss norm. I was very intrigued, and decided to understand what all this meant.

I knew that my fellow Manin student Misha Vishik had written a paper on non-archimedean spectral theory. The next day, I found the paper and started reading it. It turned out Vishik was studying bounded linear operators on a Banach space over k with the property that their resolvents are analytic on the complement of the spectrum defined in the usual way as a subset of k (the field was assumed to be algebraically closed). When I understood that, a very natural idea came to me. In the classical situation, the spectrum of an operator coincides with the complement of the analyticity set of its resolvent. Can one find out what the spectrum of a non-archimedean operator is by investigating a similar analyticity set of its resolvent? That such a resolvent takes values in the Banach algebra of bounded linear operators was not a problem. It was the notion of analyticity set that was not clear. But at least, one could try to investigate sets from a reasonable class at which the resolvent is analytic. For example, the resolvent is analytic at the complement of a closed (or open) disc with center at zero of a big enough radius.

At the beginning, I considered the so-called quasiconnected (and infraconnected) sets introduced by M. Krasner in 1940s, and I found a curious phenomenon whose slightly weakened form states the following. If the resolvent of a bounded linear operator is analytic at a standard set (i.e., the complement of a nonempty finite disjoint union of open discs in the projective line), then it is analytic at a strictly bigger standard set (i.e., all of the radii of the corresponding discs are strictly bigger or smaller). Of course, in the light of our present knowledge this phenomenon is completely clear since the standard sets being defined by non-strict inequalities are *closed* subsets of the *compact* projective line. But the analyticity set of the resolvent being the complement of the compact spectrum is an *open* set.

At that time I was not so smart to see the above. I considered the analyticity sets as strictly increasing families of finite disjoint unions of standard sets, and the spectra as strictly decreasing families of complementary sets of the same type. (A precise definition of complementary sets is given on p. 141 of my book.) The latter families can be viewed as filters of finite unions of standard sets. It turned out that one can easily describe the maximal elements in the family of filters, i.e., ultrafilters, and there are four types of them.

First of all, every element $a \in k$ defines an ultrafilter which is formed by the sets (finite unions of standard sets) that contain the point a, i.e., a base of this ultrafilter is formed by closed discs with center at a. Furthermore, every closed disc $E(a; r)$ of radius $r > 0$ with center at $a \in k$ defines an ultrafilter. If $r \in |k^*|$, a base of the corresponding ultrafilter $p(E(a; r))$ is formed by standard sets of the form $E(a; r') \backslash \bigcup_{i=1}^{n} D(a_i; r_i)$ with $r_i < r < r'$, $|a_i - a| \leq r$ and $|a_i - a_j| = r$ for $1 \leq i \neq j \leq n$, where $D(a_i; r_i)$ is the open disc of radius r_i with center at a_i. If $r \notin |k^*|$, a base of the corresponding ultrafilter $p(E(a; r))$ is formed by the closed annuli $E(a; r') \backslash D(a; r'')$ with $r'' < r < r'$. Finally, if the field k is not maximally complete, then every family of nested discs with empty intersection is a base of an ultrafilter. (By the way, it is easy to see that there is a natural bijection between the set of ultrafilters and the set of nested families of closed discs.) The above four types of ultrafilters correspond to what are now known as points of types (1)-(4) of the affine line, and elements of the ultrafilters are precisely affinoid neighborhoods of those points.

In fact, as soon as I found the above description, I knew that the space of all ultrafilters must be considered as the affine line $\mathbf{A}^1$ over k. This space is endowed with a natural topology with respect to which it is locally compact: its basis consists of sets of ultrafilters which contain a given standard set. It is also endowed with a natural sheaf of local rings, the sheaf of analytic functions $\mathcal{O}_{\mathbf{A}^1}$. But my main reason to view the space $\mathbf{A}^1$ as the affine line was the fact that it provided an answer to the question on the spectrum of a bounded linear operator I posed to myself at the very beginning. Namely, the spectrum of such an operator is a nonempty compact subset of $\mathbf{A}^1$, and it coincides with the complement of the analyticity set of its resolvent. The field k is naturally embedded in the affine line as a dense subset, and the operators studied by Misha Vishik were precisely those with spectrum contained in k.

It was a pleasant exercise to extend Vishik's results to arbitrary operators, and it helped me to understand better the topological tree-like structure of the non-archimedean affine and projective lines, to get used to them, and to accept them as reality. During this work I met with Misha several times to tell him about the progress. At that time, he was the only person (besides my wife) who shared my excitement about all this. The usual reaction of my colleagues was simple indifference at best, and the quite understandable reason for that was nicely expressed by Professor Manin. When I told him about what I was doing, he observed that it is worthwhile to develop a general theory only having in mind a concrete problem. Of course, understanding what the spectrum of a non-archimedean operator should be was not a concrete problem. Had I followed this wise advice, I would have turned back to concrete problems I had in abundance in the area of computers, and would probably have become rich during the present age of the high tech boom since I was a really good programmer. Fortunately, I was already stupid enough to miss such an attractive opportunity, and I continued my exploration of the unknown new world revealed to me by a fluke.

My job occupied me five days per week from 8am till 5pm. It took me several years to learn to devote an hour or two to mathematics during working hours. Time free from my job belonged to my family, and when I was completely hooked on mathematics and an hour or two per day was not enough for it, I discovered an additional source of time. I learned to get up every day very early (often as early as at 2am), and thus extended the time for doing mathematics. At this time of the day, the world around me was quiet and fresh, nobody and nothing disturbed me, my head was clear, and I could plunge into another world to explore and describe it.

When I had finished writing everything I had in mind, I could look at it quietly and listen to an inner feeling that something was not satisfactory. I thought about this from time to time more and more often, but could not even express what tormented me. One day at the very end of 1985 all this obsessed me. I could not stop thinking about it at my job, and later at home. I did not go to bed early as usual. The right question and an immediate answer to it came early the next morning.

As I mentioned above, the affine line $\mathbf{A}^1$ is provided with a sheaf of rings $\mathcal{O}_{\mathbf{A}^1}$. Its stalk $\mathcal{O}_{\mathbf{A}^1,x}$ at a point x is a local ring with residue field $\kappa(x)$ provided with a valuation that extends that on k. If x is of type (1) (i.e., corresponds to an element of k), then $\mathcal{O}_{\mathbf{A}^1,x}$ is the algebra of convergent power series at that point, and so $\kappa(x) = k$. Otherwise, it is a field of infinite degree over k, and so it coincides with the non-complete field $\kappa(x)$. If $\mathcal{H}(x)$ denotes the completion of $\kappa(x)$, one gets a character $k[T] \to \mathcal{H}(x)$ over k. The question that came to me on that early morning was the following. What are all possible characters $k[T] \to K$ to non-archimedean fields K over k?

After the above question had been formulated, I already knew the answer: every such character goes through a character $k[T] \to \mathcal{H}(x)$ for a unique point $x \in \mathbf{A}^1$. The proof is very easy. Indeed, given a character $\chi : k[T] \to K$, consider the family of closed discs of the form $E(a; |\chi(T - a)|)$ with $a \in k$ and $\chi(T - a) \neq 0$. It is easy to see that it is a nested family of discs and, if x is the corresponding point of $\mathbf{A}^1$, the character χ goes through the character $k[T] \to \mathcal{H}(x)$.

Thus, the affine line $\mathbf{A}^1$ can be defined as the set of equivalence classes of characters $k[T] \to K$ to non-archimedean fields K over k or, equivalently, as the set of all multiplicative seminorms on $k[T]$ that extend the valuation on k. Wow, this definition was so simple and easily seen to be applicable in a much more general setting (e.g., for defining affine spaces of higher dimension). It also gave a clear idea how to define the non-archimedean analog of the Gelfand spectrum of a complex commutative Banach algebra. But the main thing I was struck by was the fact that this definition was also applicable in the classical situation and gave the corresponding classical objects. In this way, I was thrown into the new (for me) area of analytic geometry. It took me several days to calm down and to quietly look at what all that meant.

The above observation made it clear how to define analytic spaces over an arbitrary field k complete with respect to a nontrivial valuation (archimedean or not). First of all, one should start one step earlier and define the affine space $\mathbf{A}^n$ as the set of multiplicative seminorms on the ring of polynomials $k[T_1, \ldots, T_n]$ that extend the valuation on k. The space $\mathbf{A}^n$ is endowed with the evident topology, and each point $x \in \mathbf{A}^n$ defines a character $k[T_1, \ldots, T_n] \to \mathcal{H}(x) : f \mapsto f(x)$ to a complete valuation field $\mathcal{H}(x)$ over k so that the corresponding seminorm is the function $f \mapsto |f(x)|$. Furthermore, as we were taught by Krasner, an analytic function f on an open subset $\mathcal{U} \subset \mathbf{A}^n$ should be defined as a local limit of rational functions. The latter means that f is a map that takes each point $x \in \mathcal{U}$ to an element $f(x) \in \mathcal{H}(x)$ with the following property: one can find an open neighborhood $x \in \mathcal{U}' \subset \mathcal{U}$ such that, for every $\varepsilon > 0$, there exist polynomials $g, h \in k[T_1, \ldots, T_n]$ with $h(x') \neq 0$ and $\left| f(x') - \frac{g(x')}{h(x')} \right| < \varepsilon$ for all points $x' \in \mathcal{U}'$. Finally, arbitrary analytic spaces are those locally ringed spaces which are locally isomorphic to a local model of the form $(X, \mathcal{O}_X)$, where X is the set of common zeros of a finite system of analytic functions $f_1, \ldots, f_m$ on an open subset $\mathcal{U} \subset \mathbf{A}^n$ and $\mathcal{O}_X$ is the restriction of the quotient $\mathcal{O}_\mathcal{U}/\mathcal{J}$ by the subsheaf of ideals $\mathcal{J} \subset \mathcal{O}_\mathcal{U}$ generated by $f_1, \ldots, f_m$. By the way, the spectrum $\mathcal{M}(\mathcal{A})$ of a commutative Banach k-algebra $\mathcal{A}$ should be defined as the space of all bounded multiplicative seminorms on $\mathcal{A}$.

If $k = \mathbf{C}$, the affine space $\mathbf{A}^n$ is the maximal spectrum of the ring of polynomials $\mathbf{C}[T_1, \ldots, T_n]$ (i.e., the vector space $\mathbf{C}^n$), analytic functions are local limits of polynomials, and the spectrum of a complex commutative Banach algebra is the Gelfand space of its maximal ideals. If $k = \mathbf{R}$, the above construction gives a new object: the real analytic affine space $\mathbf{A}^n$ is the maximal spectrum of the ring of polynomials $\mathbf{R}[T_1, \ldots, T_n]$ (i.e., the quotient of $\mathbf{C}^n$ by the complex conjugation), and local limits of polynomials with real coefficients are not enough to define analytic functions.

The above definition of an analytic space was a lodestar in my journey, but I was unable to work with it directly. The difficulty was in establishing functional analytic properties of the analytic spaces, whereas establishing their geometric properties was much easier. Fortunately, the fundamental paper by John Tate on rigid analytic

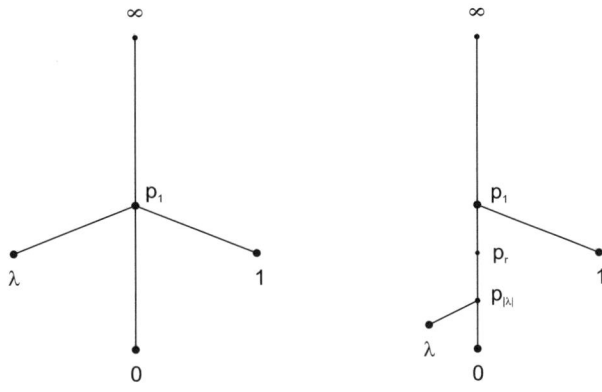

FIGURE 1

spaces was available since it was translated and published in the Soviet Union (even before it was published in the West). It was Tate's theory of affinoid algebras and affinoid domains that compensated for the lack of the usual complex analytic tools in the non-archimedean world. I studied Tate's work intensively and adjusted to the new framework, introducing a category of analytic spaces which eventually coincided with that given above. In the present framework, it is precisely the full subcategory of the category of analytic spaces consisting of the spaces without boundary. They are glued from the interiors of affinoid spaces (the interior of $\mathcal{M}(\mathcal{A})$ consists of the points x for which the corresponding character $\mathcal{A} \to \mathcal{H}(x)$ is a completely continuous operator), and include the analytifications of algebraic varieties. The new analytic spaces were applied to define the common spectrum of a finite family of commuting bounded linear operators, to develop holomorphic functional calculus, and to prove the Shilov idempotent theorem. The latter states that for any open and closed subset of the spectrum $\mathcal{M}(\mathcal{A})$ of a commutative Banach algebra $\mathcal{A}$ there exists a unique idempotent $e \in \mathcal{A}$ which is equal to 1 precisely at that subset.

At that time I found that the process of writing down of what I was starting to understand was very enjoyable and extremely helpful for better understanding. The need to express an idea forced me to concentrate on each small object or detail of reasoning. This concentration helped me see hidden and refined nuances which could change the whole picture, or to discover again and again a deep-rooted prejudice or wrong vision or simple stupidity.

The first typewritten text was finished in April of 1986, and I succeeded in passing it to Professor Barry Mazur, who knew me from my previous work. Later on, I was surprised to learn that my text had been accepted by the American Mathematical Society for publication as a book. But I was actually lucky that everybody at AMS immediately forgot about me finishing that book, and so I could continue to rewrite the text infinitely many times, gradually extending the framework of the new analytic spaces and investigating their amazing properties.

Tate's paper was still the only source of my knowledge in rigid analytic geometry, when I considered the following situation. Assume that the ground field k is algebraically closed and the characteristic of its residue field is not 2, and let $\mathcal{E}$ be an elliptic curve over k defined by the affine equation $y^2 = x(x-1)(x-\lambda)$ with

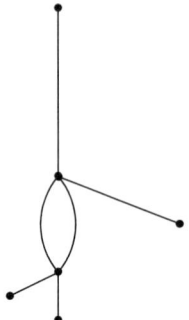

FIGURE 2

$0 < |\lambda| \leq 1$ and $|\lambda - 1| = 1$. (The latter can be always achieved after a change of coordinates.) Recall that the projective line $\mathbf{P}^1$ has the property that any two different points can be connected by a unique path. Let us connect the points 0, 1, λ and ∞ in $\mathbf{P}^1$. We get one of the two graphs Γ_λ presented in Figure 1, which correspond to the cases $|\lambda| = 1$ and $|\lambda| < 1$. (For brevity the point $p(E(0;r))$ is denoted by p_r.) The complement of Γ_λ in $\mathbf{P}^1$ is a disjoint union of open discs of the form $D(a; r_a)$ with $a \in k \backslash \{0, 1, \lambda\}$ and $r_a = \min\{|a|, |a - 1|, |a - \lambda|\}$. Every such disc is glued to its boundary which is a point of Γ_λ, and Γ_λ is a strong deformation retraction of the whole projective line $\mathbf{P}^1$. Consider now the x-projection $\pi : \mathcal{E}^{an} \to \mathbf{P}^1$ from the analytification $\mathcal{E}^{an}$ of $\mathcal{E}$. Since the characteristic of the residue field of k is not two, the square root of each of the linear factors of $x(x - 1)(x - \lambda)$ can be extracted at $D(a; r_a)$ and, therefore, the preimage of $\pi^{-1}(D(a; r_a))$ is a disjoint union of two open discs which are glued to their boundaries at the preimage $\pi^{-1}(\Gamma_\lambda)$. Thus, the latter is a strong deformation retraction of $\mathcal{E}^{an}$. If $0 < r < |\lambda|$, then the square roots of $x - \lambda$ and $x - 1$ are extracted at the open annulus $D(0; r + \varepsilon) \backslash E(0; r - \varepsilon)$ with $0 < r - \varepsilon < r + \varepsilon < |\lambda|$, but the square root of x is not. This means that each point of the interval that connects 0 with $p_{|\lambda|}$ has a unique preimage in $\mathcal{E}^{an}$. Similarly, each point from the intervals that connect 1 with p_1, λ with $p_{|\lambda|}$, and ∞ with p_1, has a unique preimage in $\mathcal{E}^{an}$. In particular, if $|\lambda| = 1$, then $\pi^{-1}(\Gamma_\lambda) \xrightarrow{\sim} \Gamma_\lambda$. If now $|\lambda| < 1$, then the square roots of $x - 1$ and of the product $x(x - \lambda)$ are extracted at the open annulus $D(0; r + \varepsilon) \backslash E(0; r - \varepsilon)$ with $|\lambda| < r - \varepsilon < r + \varepsilon < 1$. This means that each point p_r with $|\lambda| < r < 1$ has two preimages in $\mathcal{E}^{an}$, and the graph $\pi^{-1}(\Gamma_\lambda)$ has the form presented in Figure 2.

Thus, the analytic curve $\mathcal{E}^{an}$ is contractible if $|\lambda| = 1$, and homotopy equivalent to a circle if $|\lambda| < 1$. It is well known that these two cases correspond to those when the modular invariant $j(\mathcal{E})$ is integral or not. But the latter case $|j(\mathcal{E})| > 1$ is precisely that of a Tate elliptic curve. Wow, such a curve is homotopy equivalent to a circle! I was always fascinated by Tate elliptic curves, but never understood the reason for which they admit uniformization. And here I had a very elementary explanation of this astonishing phenomenon discovered by Tate; it reminded me of the classical construction of the Riemann surface of an algebraic function.

Of course, all this strongly lifted up my spirit and eagerness in exploration of the new spaces. This was very timely since it distracted me from a serious health problem I had at that time, not to mention my job in a dull institution and the reality of a country in an advanced stage of decaying.

The great day of liberation for me and my family came in August, 1987, when we got out of the cesspool of the Soviet Union and arrived in the wonderful, sunny and crazy State of Israel. Although I was not so young, I was given an opportunity to renew my scientific career without also having to earn my family's living in some other way. I was again hungry for mathematics as never before, and ready for the next steps in my journey.

CHAPTER 1

Several approaches to non-archimedean geometry

BRIAN CONRAD

Introduction

Let k be a *non-archimedean field*: a field that is complete with respect to a specified nontrivial non-archimedean absolute value $|\cdot|$. There is a classical theory of k-analytic manifolds (often used in the theory of algebraic groups with k a local field), and it rests upon versions of the inverse and implicit function theorems that can be proved for convergent power series over k by adapting the traditional proofs over $\mathbf{R}$ and $\mathbf{C}$. Serre's Harvard lectures [**S**] on Lie groups and Lie algebras develop this point of view, for example. However, these kinds of spaces have limited geometric interest because they are totally disconnected. For global geometric applications (such as uniformization questions, as first arose in Tate's study of elliptic curves with split multiplicative reduction over a non-archimedean field), it is desirable to have a much richer theory, one in which there is a meaningful way to say that the closed unit ball is "connected". More generally, we want a satisfactory theory of coherent sheaves (and hence a theory of "analytic continuation"). Such a theory was first introduced by Tate in the early 1960's, and then systematically developed (building on Tate's remarkable results) by a number of mathematicians. Though it was initially a subject of specialized interest, in recent years the importance and power of Tate's theory of *rigid-analytic spaces* (and its variants, due especially to the work of Raynaud, Berkovich, and Huber) has become ever more apparent. To name but a few striking applications, the proof of the local Langlands conjecture for GL_n by Harris–Taylor uses étale cohomology on non-archimedean analytic spaces (in the sense of Berkovich) to construct the required Galois representations over local fields, the solution by Raynaud and Harbater of Abyhankar's conjecture concerning fundamental groups of curves in positive characteristic uses the rigid-analytic GAGA theorems (whose proofs are very similar to Serre's proofs in the complex-analytic case), and recent work of Kisin on modularity of Galois representations makes creative use of rigid-analytic spaces associated to Galois deformation rings.

The aim of these lectures is to explain some basic ideas, results, and examples in Tate's theory and its refinements. In view of time and space constraints, we have omitted most proofs in favor of examples to illustrate the main ideas. To become a serious user of the theory it is best to closely study a more systematic development. In particular, we recommend [**BGR**] for the "classical" theory due

This work was partially supported by NSF grants DMS-0600919 and DMS-0602287. The author is grateful to the participants of the Arizona Winter School, especially Michael Temkin and Antoine Ducros, for their comments and suggestions.

to Tate, [**BL1**] and [**BL2**] for Raynaud's approach based on formal schemes, and [**Ber1**] and [**Ber2**] for Berkovich's theory of k-analytic spaces. Some recent lecture notes by Bosch [**B**] explain both rigid geometry and Raynaud's theory with complete proofs, and a recent Bourbaki survey by Ducros [**D**] treats Berkovich's theory in greater depth. There are other points of view as well (most notably the work of Huber [**H**] and Fujiwara-Kato [**FK**]), but we will pass over these in silence (except for a few comments on how Huber's adic spaces relate to Berkovich spaces).

Before we begin, it is perhaps best to tell a story that illustrates how truly amazing it is that there can be a theory of the sort that Tate created. In 1959, Tate showed Grothendieck some *ad hoc* calculations that he had worked out with p-adic theta functions in order to uniformize certain p-adic elliptic curves by a multiplicative group, similarly to the complex-analytic case. Tate wondered if his computations could have deeper meaning within a theory of global p-adic analytic spaces, but Grothendieck was doubtful. In fact, in an August 18, 1959 letter to Serre, Grothendieck expressed serious pessimism that such a global theory could possibly exist: *"Tate has written to me about his elliptic curves stuff, and has asked me if I had any ideas for a global definition of analytic varieties over complete valuation fields. I must admit that I have absolutely not understood why his results might suggest the existence of such a definition, and I remain skeptical. Nor do I have the impression of having understood his theorem at all; it does nothing more than exhibit, via brute formulas, a certain isomorphism of analytic groups; one could conceive that other equally explicit formulas might give another one which would be no worse than his (until proof to the contrary!)"*

1. Affinoid algebras

1.1. Tate algebras. In this first lecture, we discuss the commutative algebra that forms the foundation for the local theory of rigid-analytic spaces, much as the theory of polynomial rings over a field is the basis for classical algebraic geometry. (The primary reference for this lecture and the next one is [**BGR**].) The replacement for polynomial rings over a field will be Tate algebras.

Unless we say to the contrary, throughout this lecture and all subsequent ones we shall fix a non-archimedean field k, and we write R to denote its *valuation ring* and $\tilde{k}$ its *residue field*:

$$R = \{t \in k \mid |t| \le 1\}, \ \ \tilde{k} = R/\mathfrak{m}$$

where $\mathfrak{m} = \{t \in R \mid |t| < 1\}$ is the unique maximal ideal of R.

EXERCISE 1.1.1. The *value group* of k is $|k^\times| \subseteq \mathbf{R}^\times_{>0}$. Prove that R is noetherian if and only if $|k^\times|$ is a discrete subgroup of $\mathbf{R}^\times_{>0}$, in which case R is a discrete valuation ring.

It is a basic fact that every finite extension k'/k admits a unique absolute value $|\cdot|'$ (necessarily non-archimedean) extending the given one on k, and that k' is complete with respect to this absolute value. Explicitly, if $x' \in k'$ then $|x'|' = |\mathrm{N}_{k'/k}(x')|^{1/[k':k]}$, but it is not obvious that this latter definition satisfies the non-archimedean triangle inequality (though it is clearly multiplicative). The absolute value on k therefore extends uniquely to any algebraic extension of k (using that it extends to every finite subextension, necessarily compatibly on overlaps due to uniqueness), though if k'/k is not finite then k' may not be complete. In view of the uniqueness, this extended absolute value is invariant under all k-automorphisms of k'. It is an important fact that if $\overline{k}$ is an algebraic closure of k then its completion $\overline{k}^{\wedge}$ is still algebraically closed and its residue field is an algebraic closure of $\widetilde{k}$. In the special case $k = \mathbf{Q}_p$, this completed algebraic closure is usually denoted $\mathbf{C}_p$.

DEFINITION 1.1.2. For $n \geq 1$, the n-variable *Tate algebra* over k is

$$T_n = T_n(k) = \left\{ \sum a_J X^J \,|\, |a_J| \to 0 \text{ as } \|J\| \to \infty \right\},$$

where for a multi-index $J = \{j_1, \ldots, j_n\}$ we write X^J to denote $\prod X_i^{j_i}$ and $\|J\|$ to denote $\sum_i j_i$. In other words, $T_n(k)$ is the subring of formal power series in $k[\![X_1, \ldots, X_n]\!]$ that converge on R^n. This k-algebra is also denoted $k\langle X_1, \ldots, X_n \rangle$.

The *Gauss norm* (or *sup norm*, for reasons to become clear shortly) on T_n is

$$\left\| \sum a_J X^J \right\| = \max_J |a_J| \geq 0.$$

Obviously $\|f\| = 0$ if and only if $f = 0$.

EXERCISE 1.1.3. This exercise develops properties of the Gauss norm on T_n. This gives T_n a topological structure that goes beyond its mere algebraic structure.

(1) Prove that the Gauss norm is a k-Banach algebra norm on T_n. That is, $\|f_1 + f_2\| \leq \max(\|f_1\|, \|f_2\|)$ for all $f_1, f_2 \in T_n$, $\|cf\| = |c| \cdot \|f\|$ for all $c \in k$ and $f \in T_n$, $\|f_1 f_2\| \leq \|f_1\| \|f_2\|$ for all $f_1, f_2 \in T_n$, and T_n is complete for the metric $\|f_1 - f_2\|$.

(2) By using $k^{\times}$-scaling to reduce to the case of unit vectors, show that $\|f_1 f_2\| = \|f_1\| \cdot \|f_2\|$ for all $f_1, f_2 \in T_n$. That is, the Gauss norm is multiplicative.

(3) Let $\overline{k}$ be an algebraic closure of k, endowed with the unique absolute value (again denoted $|\cdot|$) extending the given one on k. Using $k^{\times}$-scaling to reduce to the case of unit vectors, prove that the Gauss norm computes a supremum of magnitudes over the closed unit n-ball over $\overline{k}$:

$$\|f\| = \sup_x |f(x)| = \max_x |f(x)|,$$

where $x = (x_1, \ldots, x_n)$ varies with $x_j \in \overline{k}$ and $|x_j| \leq 1$. In particular, this supremum and maximum are finite.

(4) Show that the use of $\overline{k}$ in the previous part is essential: give an example of $f \in \mathbf{Q}_p\langle X \rangle$ such that $\|f\| > \sup_{x \in \mathbf{Z}_p} |f(x)|$.

We want $T_n(k)$ to be the "coordinate ring" of the closed unit n-ball over k, but as with algebraic geometry over a field that may not be algebraically closed, we have to expect to work with points whose coordinates are not all in k. That is, the underlying space for the closed unit n-ball over k should admit points with coordinates in finite extensions of k. Let's now see that T_n admits many k-algebra maps to finite extensions of k.

EXERCISE 1.1.4. Let k'/k be a finite extension, and choose $c'_1, \ldots, c'_n \in k'$ with $|c'_j| \leq 1$. Prove that there exists a unique continuous k-algebra map $T_n \to k'$ (using the Gauss norm on T_n) such that $X_j \mapsto c'_j$ for all j. Conversely, prove that every continuous k-algebra map $T_n \to k'$ arises in this way. (Hint for converse: $c \in k'$ satisfies $|c| \leq 1$ if and only if the sequence $\{c^m\}_{m \geq 1}$ in k' is bounded.)

The basic properties of T_n are summarized in the following result that is analogous to properties of polynomial rings over a field. The proofs of these properties are inspired by the local study of complex-analytic spaces (via Weierstrass Preparation techniques to carry out induction on n).

THEOREM 1.1.5. *The Tate algebra $T_n = T_n(k)$ satisfies the following properties:*

(1) *The domain T_n is noetherian, regular, and a unique factorization domain. For every maximal ideal $\mathfrak{m}$ of T_n the local ring $(T_n)_{\mathfrak{m}}$ has dimension n and residue class field $T_n/\mathfrak{m}$ that has finite degree over k.*

(2) *The ring T_n is Jacobson: every prime ideal $\mathfrak{p}$ of T_n is the intersection of the maximal ideals containing it. In particular, if I is an ideal of T_n then an element of T_n/I is nilpotent if and only if it lies in every maximal ideal of T_n/I.*

(3) *Every ideal in T_n is closed with respect to the Gauss norm.*

As a consequence of this theorem, we can reinterpret Exercise 1.1.3(3) and Exercise 1.1.4 in a more geometric manner, as follows. Consider the set $\mathrm{MaxSpec}(T_n)$ of maximal ideals of T_n. A point in this set will usually be denoted as x, though if we want to emphasize its nature as a maximal ideal we may denote it as $\mathfrak{m}_x$. To each such point there is associated the residue class field $k(x) = T_n/\mathfrak{m}_x$ of finite degree over k, and this field is equipped with the unique absolute value (which we also denote as $|\cdot|$) that extends the given one on k. For any $f \in T_n$ we write $f(x)$ to denote the image of f in $k(x)$. We can combine Exercise 1.1.3(3) and Exercise 1.1.4 to say that for all $f \in T_n$,

$$\|f\| = \sup_x |f(x)| = \max_x |f(x)|,$$

where now x varies through $\mathrm{MaxSpec}(T_n)$; there is no intervention of the auxiliary $\bar{k}$ here. In particular, the function $x \mapsto |f(x)|$ on $\mathrm{MaxSpec}(T_n)$ is bounded and attains a maximal value. (It is "as if" $\mathrm{MaxSpec}(T_n)$ were a compact topological space, an idea that becomes a reality within the framework of Berkovich spaces, as we shall see later.) One curious consequence of this formula for the Gauss norm in terms of $\mathrm{MaxSpec}(T_n)$ and the intrinsic k-algebra structure of T_n is that the Gauss norm is intrinsic to the k-algebra T_n and does not depend on its "coordinates" $X_j \in T_n$; in particular, it is invariant under all k-algebra automorphisms of T_n (which is not obvious from the initial definition of the Gauss norm).

1.2. Affinoid algebras. Much as affine algebraic schemes over a field can be obtained from quotients of polynomial rings, and these in turn are the local model spaces from which more general algebraic schemes are constructed via gluing, the building blocks for rigid-analytic spaces will be obtained from quotients of Tate algebras. This distinguished class of k-algebras is given a special name, as follows.

DEFINITION 1.2.1. A *k-affinoid algebra* is a k-algebra A admitting an isomorphism $A \simeq T_n/I$ as k-algebras, for some ideal $I \subseteq T_n$. The set $\mathrm{MaxSpec}(A)$ of maximal ideals of A is denoted $M(A)$.

EXAMPLE 1.2.2. We have $R^n \subseteq M(T_n)$ in an evident manner, but if k is not algebraically closed (e.g., $k = \mathbf{Q}_p$) then $M(T_n)$ has many more points than just those coming from R^n. This underlies the enormous difference between rigid-analytic spaces over k and the more classical notion of a k-analytic manifold.

By Theorem 1.1.5, every k-affinoid algebra A is noetherian and Jacobson with finite Krull dimension, and $A/\mathfrak{m}$ is a finite extension of k for every $\mathfrak{m} \in M(A)$. For a point $x \in M(A)$ we write $k(x)$ to denote this associated finite extension of k and we write $a(x) \in k(x)$ to denote the image of $a \in A$ in $k(x)$. By the Jacobson property of A, $a \in A$ is nilpotent if and only if $a(x) = 0$ for all $x \in M(A)$. Obviously $a \in A^{\times}$ if and only if $a(x) \neq 0$ for all $x \in M(A)$. In this respect, we can view elements of A as "functions" on $M(A)$ (valued in varying fields $k(x)$) much like we do for coordinate rings of affine algebraic schemes over a field, and the function $x \mapsto a(x)$ determines a up to nilpotents.

EXERCISE 1.2.3. Recall that any domain of finite dimension over a field is itself a field. Using this, prove that $M(A)$ is functorial via pullback. That is, if $f : A' \to A$ is a map of k-affinoid algebras then the prime ideal $f^{-1}(\mathfrak{m}) \subseteq A'$ is a maximal ideal of A' for every maximal ideal $\mathfrak{m}$ of A.

Geometrically, if we choose an isomorphism $A \simeq T_n/I$ and we let $\{f_1, \ldots, f_m\}$ be generators of I then functoriality provides an injection $M(A) \hookrightarrow M(T_n)$ onto the subset of points

$$\{x \in M(T_n) \mid \text{ all } f_j(x) = 0\} = \{x \in M(T_n) \mid f(x) = 0 \text{ for all } f \in I\}.$$

In this sense, we want to think of $M(A)$ as being the underlying set of the "space" of points in the closed unit n-ball over k where the f_j's simultaneously vanish. (Keep in mind that, just as for $M(T_n)$, if k is not algebraically closed then there are generally many points $x \in M(A)$ with $k(x) \neq k$, which is to say that $M(A)$ usually has many points that are not k-rational. This abundance of non-rational points over the base field is a fundamental distinction between rigid-analytic spaces and the more classical concept of a k-analytic manifold. In Berkovich's theory there will nearly always be even more points than these, and in particular lots of non-rational points even if k is algebraically closed! This is analogous to the fact that an algebraic scheme over an algebraically closed field nearly always has many non-rational points.) In order to give geometric substance to the sets $M(A)$, we need to endow them with a good function theory, and this in turn requires an understanding of the topological structure of A. Thus, we now turn to this aspect of k-affinoid algebras.

A k-*Banach space* is a k-vector space V equipped with a function $\|\cdot\| : V \to \mathbf{R}_{\geq 0}$ such that $\|v\| = 0$ if and only if $v = 0$,

$$\|v + v'\| \leq \max(\|v\|, \|v'\|), \quad \|cv\| = |c| \cdot \|v\|$$

(for all $v, v' \in V$ and $c \in k$), and V is complete for the metric $\|v - v'\|$. Likewise, a k-*Banach algebra* (always understood to be commutative) is a k-algebra $\mathscr{A}$ equipped with a k-Banach space structure $\|\cdot\|$ that is submultiplicative with respect to the multiplication law on $\mathscr{A}$: $\|a_1 a_2\| \leq \|a_1\| \cdot \|a_2\|$ for all $a_1, a_2 \in \mathscr{A}$. For example, we have seen that T_n with the Gauss norm is a k-Banach algebra, and in fact any k-affinoid algebra A admits a k-Banach algebra structure. To see this, we choose an isomorphism $A \simeq T_n/I$ as k-algebras, and since I is closed in T_n we may use the residue norm from T_n to define a k-Banach structure on the quotient T_n/I (and hence on A), as is explained in the next exercise.

EXERCISE 1.2.4. Let $(V, \|\cdot\|)$ be a k-Banach space and W a closed subspace. For $\overline{v} \in V/W$ define the *residue norm* on $\overline{v}$ to be

$$\|\overline{v}\|' = \inf_{v \bmod W = \overline{v}} \|v\|,$$

the infimum of the norms of all representatives of $\overline{v}$ in V. Using that W is closed, prove that this is a k-Banach space structure on V/W; what goes wrong if W is not closed in V? In the special case that $V = \mathscr{A}$ is a k-Banach algebra and $W = I$ is a closed ideal, prove that the residue norm is a k-Banach algebra structure on $\mathscr{A}/I$.

EXERCISE 1.2.5. A linear map $L : V \to V'$ between k-Banach spaces is *bounded* if there exists $C > 0$ such that $\|L(v)\|' \leq C\|v\|$ for all $v \in V$. Prove that L is bounded if and only if it is continuous. (Here you will need to use that $|k^\times| \neq \{1\}$.) Also prove the *Banach Open Mapping Theorem*: a bijective bounded linear map $L : V \to V'$ between k-Banach spaces has bounded inverse. (Hint: copy the classical proof over $\mathbf{R}$.) This theorem is fundamental in non-archimedean analysis and geometry, and it fails if the absolute value on k is trivial.

If we choose two different presentations $T_n/I \simeq A$ and $T_m/J \simeq A$ of a k-affinoid algebra A as a quotient of a Tate algebra, then the resulting residue norms on A are generally not the same. In this sense, A usually has no canonical k-Banach structure (in contrast with T_n). However, it turns out that any two k-Banach algebra structures on A (even those perhaps not arising from a presentation of A as a quotient of a Tate algebra) are bounded by positive multiples of each other, and hence the resulting k-Banach *topology* and concepts such as "boundedness" are in fact intrinsic to A. In particular, for this intrinsic k-Banach topology all ideals of A are closed (since the "residue norm" construction via an isomorphism $A \simeq T_n/I$ reduces this to the known case of Tate algebras). These and further remarkable features of the k-Banach algebra structures on k-affinoid algebras are summarized in the next result.

THEOREM 1.2.6. *Let A be a k-affinoid algebra.*

(1) *If $\|\cdot\|$ and $\|\cdot\|'$ are k-Banach algebra norms on A then there exist $C \geq c > 0$ such that*

$$c\|\cdot\| \leq \|\cdot\|' \leq C\|\cdot\|,$$

so both norms define the same topology and the same concept of boundedness. In particular, for $a \in A$ the property that the sequence $\{a^n\}_{n \geq 1}$ is bounded (i.e., a is power-bounded) is independent of the choice of k-Banach algebra structure.

(2) *Any k-algebra map $A' \to A$ between k-affinoid algebras is continuous for the intrinsic k-Banach topologies, or equivalently is a bounded linear map with respect to any choices of k-Banach algebra norms.*

(3) *Any A-algebra A' with module-finite structure map $A \to A'$ is necessarily a k-affinoid algebra.*

(4) *(Noether normalization theorem) If $d = \dim A \geq 0$ then there is a module-finite k-algebra injection $T_d(k) \hookrightarrow A$. In particular, if A is a domain then all of its maximal ideals have height d.*

(5) *(Maximum Modulus Principle) For any $f \in A$ we have the equality*

$$\|f\|_{\sup} := \sup_{x \in M(A)} |f(x)| = \max_{x \in M(A)} |f(x)| < \infty.$$

In particular, the function $x \mapsto |f(x)|$ on $M(A)$ is bounded and attains a maximal value. If A is reduced (i.e., has no nonzero nilpotent elements) then this is a k-Banach algebra structure on A.

The final part of this theorem provides a canonical k-Banach algebra structure on any reduced k-affinoid algebra, recovering the Gauss norm in the special case of Tate algebras. This k-Banach algebra structure may not be multiplicative, but it is clearly power-multiplicative: $\|a^n\|_{\sup} = \|a\|_{\sup}^n$ for all $a \in A$ and $n \geq 1$. In particular, for a reduced k-affinoid algebra A we deduce the important consequence that $a \in A$ is power-bounded if and only if $\|a\|_{\sup}^n$ is bounded for $n \geq 1$, which is to say $\|a\|_{\sup} \leq 1$, or in other words that $|a(x)| \leq 1$ for all $x \in M(A)$. In fact, it can be shown that this characterization of power-boundedness in k-affinoid algebras is valid without assuming reducedness. That is, if A is any k-affinoid algebra then $a \in A$ is power-bounded if and only if $|a(x)| \leq 1$ for all $x \in M(A)$.

REMARK 1.2.7. For units in k-affinoid algebras A (i.e., $u \in A$ such that $u(x) \neq 0$ for all $x \in M(A)$!) there is a "minimum modulus principle": for $u \in A^\times$, $\inf_{x \in M(A)} |u(x)| = \min_{x \in M(A)} |u(x)| > 0$. Indeed, this is a reformulation of the Maximum Modulus Principle for $1/u$.

We conclude this lecture with an exercise that provides a universal mapping property for Tate algebras within the category of k-affinoid algebras (and even k-Banach algebras), reminiscent of the universal mapping property of polynomial rings.

EXERCISE 1.2.8. Let $\mathscr{A}$ be a k-Banach algebra, and let $\mathscr{A}^0$ be the subset of power-bounded elements: $a \in \mathscr{A}$ such that the sequence $\{a^n\}_{n \geq 1}$ is bounded with respect to the k-Banach norm on $\mathscr{A}$.

(1) Prove that $\mathscr{A}^0$ is a subring of $\mathscr{A}$, and in fact is a subalgebra over the valuation ring R of k.

(2) Prove that $\mathscr{A}^0$ is functorial in $\mathscr{A}$ within the category of k-Banach algebras (using continuous maps). In particular, any k-Banach algebra map $T_n = T_n(k) \to \mathscr{A}$ carries each X_j to an element $a_j \in \mathscr{A}^0$.

(3) Show that the map of sets $\operatorname{Hom}(T_n, \mathscr{A}) \to (\mathscr{A}^0)^n$ defined by
$$\phi \mapsto (\phi(X_1), \ldots, \phi(X_n))$$
is bijective. This is the universal mapping property of T_n within the category of k-Banach algebras, and in particular within the full subcategory of k-affinoid algebras.

(4) As an application of the universal property, we can "recenter the polydisc" at any k-rational point. That is, for $c_1, \ldots, c_n \in k$ with $|c_j| \leq 1$, prove that there is a unique automorphism of T_n satisfying $X_j \mapsto X_j - c_j$.

2. Global rigid-analytic spaces

2.1. Topological preparations. In the first lecture we studied some basic algebraic and topological properties of k-affinoid algebras, and in particular for any such algebra A we introduced the set $M(A)$ of maximal ideals of A. We wish to impose a suitable topology (really a mild Grothendieck topology) on $M(A)$ with respect to which notions such as connectedness will have a good meaning. But before doing that we want to explain how $M(A)$ has a Hausdorff "canonical topology" that is closer in spirit to the totally disconnected topology that arises in the classical theory of k-analytic manifolds. This canonical topology is not especially useful, but it is psychologically satisfying to know that it exists; the subtle issue is that $M(A)$ usually has many points that are not k-rational and it is also not a set of $\overline{k}$-points either (unless $k = \overline{k}$). For this reason, it requires some thought to define the canonical topology. The motivation for the definition comes from the following concrete description of $M(A)$.

EXERCISE 2.1.1. Let A be a k-affinoid algebra, and $\overline{k}$ an algebraic closure of k. For each $x \in M(A)$ if we choose a k-embedding $i : k(x) \hookrightarrow \overline{k}$ then we get a k-algebra map $A \to \overline{k}$ whose image lies in a subextension of finite degree over k. Let $A(\overline{k})$ denote the set of k-algebra maps $A \to \overline{k}$ with image contained in a subfield of finite degree over k; this set has contravariant functorial dependence on A. Observe that $\mathrm{Aut}(\overline{k}/k)$ acts on this set via composition.

(1) Show that if we change the choice of i then the resulting map in $A(\overline{k})$ changes by the action of $\mathrm{Aut}(\overline{k}/k)$. Hence, we get a well-defined map of sets $M(A) \to A(\overline{k})/\mathrm{Aut}(\overline{k}/k)$ into the space of orbits of $\mathrm{Aut}(\overline{k}/k)$ on $A(\overline{k})$.

(2) Prove that the map $M(A) \to A(\overline{k})/\mathrm{Aut}(\overline{k}/k)$ is functorial in A, and that it is a bijection.

(3) For any $x \in A(\overline{k})$ and $f \in A$ we get a well-defined element $f(x) \in \overline{k}$ and hence a number $|f(x)|$. Show that the loci

$$\{x \in A(\overline{k}) \,|\, |f_i(x)| \geq \varepsilon_i, |g_j(x)| \leq \eta_j \text{ for all } i, j\}$$

for $f_1, \ldots, g_m \in A$ and $\varepsilon_1, \ldots, \eta_m > 0$ are a basis of open sets for a topology on $A(\overline{k})$. Give $M(A)$ the resulting quotient topology. Prove that this topology on $M(A)$ is Hausdorff and totally disconnected, and that it is functorial in A (in the sense that the pullback map $M(A) \to M(A')$ induced by a k-algebra map $A' \to A$ of k-affinoid algebras is continuous). This is the *canonical* topology on $M(A)$.

(4) Show that $M(T_n)$ with its canonical topology is the disjoint union of two open sets, $\{|x_1| = \cdots = |x_n| = 1\}$ and its complement.

Having introduced the canonical topology, we now prepare to build up the Tate topology that will replace it. The basic idea is to artfully restrict both the open sets and the coverings of one open set by others that we permit ourselves to consider. (The restriction on coverings is more fundamental than the restriction on open sets.) In this way, disconnectedness will be eliminated where it is not desired. For example, the decomposition in Exercise 2.1.1(4) will be eliminated in Tate's theory, and in fact $M(T_n)$ will (in an appropriate sense) wind up becoming connected. To construct Tate's theory, we need to introduce several important classes of open subsets of $M(A)$: *Weierstrass domains*, *Laurent domains*, and *rational domains*. These are analogues of basic affine opens as used in algebraic geometry, but the main difference is that we can consider loci defined by (non-strict!) inequalities of the type $|f_1| \leq |f_2|$ on absolute values, whereas in algebraic geometry (with the Zariski topology) we can only use conditions of the type $f_1 \neq f_2$. It should be noted that later in the theory we will permit strict inequalities of the type $|f_1| < |f_2|$, but at the beginning it is non-strict inequalities that are more convenient to use as the building blocks. (Roughly what is happening is that non-strict inequalities define loci that will behave "as if" they are compact, which is in fact what happens within the framework of Berkovich's theory, whereas strict inequalities define loci that lack a kind of compactness property.)

In order to define interesting open domains within $M(A)$, it will be useful to first introduce a relative version of Tate algebras, much as we do with polynomial rings over a general commutative ring.

DEFINITION 2.1.2. Let $(\mathscr{A}, \|\cdot\|)$ be a k-Banach algebra. The *Tate algebra over* $\mathscr{A}$ (in n variables) is

$$\mathscr{A}\langle Y_1, \ldots, Y_n \rangle = \left\{ \sum a_J Y^J \in \mathscr{A}[\![Y_1, \ldots, Y_n]\!] \mid a_J \to 0 \right\};$$

this is also denoted $\mathscr{A}\langle \underline{Y} \rangle$ if n is understood from context. We define a norm on this ring as follows:

$$\left\| \sum a_J Y^J \right\| = \max_J \|a_J\|.$$

EXERCISE 2.1.3. Let $\mathscr{A}$ be a k-Banach algebra.

(1) Check that in the preceding definition, the norm is a k-Banach algebra structure on $\mathscr{A}\langle \underline{Y} \rangle$, and that if the k-Banach algebra structure on $\mathscr{A}$ is replaced with an equivalent such norm (i.e., one bounded above and below by a positive constant multiple of the given one) then the resulting norm on the Tate algebra over $\mathscr{A}$ is also replaced with an equivalent one. In particular, if $\mathscr{A}$ is k-affinoid then all of its k-Banach algebra structures define equivalent norms on the Tate algebras over $\mathscr{A}$.

(2) If $\mathscr{A} = A$ is k-affinoid, and say $T_m/I \simeq A$ is an isomorphism, show that the resulting natural map $T_{n+m} \to A\langle Y_1, \ldots, Y_n \rangle$ is surjective, so the Tate algebras over A are also k-affinoid.

(3) In the category of k-Banach algebras over $\mathscr{A}$ (i.e., the category of k-Banach algebras equipped with a continuous map from $\mathscr{A}$, and morphisms are as $\mathscr{A}$-algebras), state and prove a universal mapping property similar to that for $T_n(k)$ in the category of k-Banach algebras. Using this, construct a "transitivity" isomorphism of the type $(\mathscr{A}\langle \underline{X} \rangle)\langle \underline{Y} \rangle \simeq \mathscr{A}\langle \underline{X}, \underline{Y} \rangle$.

Let A be a k-affinoid algebra. The following class of rings will wind up being the coordinate rings of subsets of $M(A)$ to be called *Laurent domains*. Let A be a k-affinoid algebra. For $\underline{a} = (a_1, \ldots, a_n) \in A^n$ and $\underline{a}' = (a'_1, \ldots, a'_m) \in A^m$, define

$$A\langle \underline{a}, \underline{a}'^{-1} \rangle = A\langle \underline{X}, \underline{Y} \rangle / (X_1 - a_1, \ldots, X_n - a_n, a'_1 Y_1 - 1, \ldots, a'_m Y_m - 1).$$

REMARK 2.1.4. Beware that relative Tate algebras cannot be treated as easily as polynomial rings. For example, $A\langle a \rangle = A\langle X \rangle / (X - a)$ is generally *not* the same as A; geometrically what happens is that we are "forcing" a to become power-bounded, which may not be the case in A at the outset. We will see that the natural map $M(A\langle a \rangle) \to M(A)$ is an injection onto the set of $x \in M(A)$ such that $|a(x)| \le 1$.

In view of the universal property of relative Tate algebras, the A-algebra $A\langle \underline{a}, \underline{a}'^{-1} \rangle$ has the following universal property: for any map of k-Banach algebras

$\phi : A \to B$, we can fill in a commutative diagram

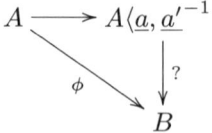

in at most one way, and such a diagram exists if and only if $\phi(a_i) \in B^0$ for all i (i.e., all $\phi(a_i)$ are power-bounded) and $\phi(a'_j) \in B^{\times}$ with $\phi(a'_j)^{-1} \in B^0$ for all j. In more geometric language, if B is k-affinoid then we can say that the structure map $\phi : A \to B$ factors through $A\langle \underline{a}, \underline{a}'^{-1} \rangle$ if and only if the map of sets $M(\phi) : M(B) \to M(A)$ factors through the subset

$$\{x \in M(A) \mid |a_i(x)| \leq 1, |a'_j(x)| \geq 1 \text{ for all } i, j\} \subseteq M(A).$$

EXERCISE 2.1.5. By taking B to vary through finite extensions of k, use the above universal property to deduce that the map of k-affinoid algebras $A \to A\langle \underline{a}, \underline{a}'^{-1} \rangle$ induces a bijection

$$M(A\langle \underline{a}, \underline{a}'^{-1} \rangle) \to \{x \in M(A) \mid |a_i(x)| \leq 1, |a'_j(x)| \geq 1 \text{ for all } i, j\}.$$

The significance of the conclusion of Exercise 2.1.5 is that the purely algebraic condition that a map of k-affinoid algebras $\phi : A \to B$ factors through the canonical map $A \to A\langle \underline{a}, \underline{a}'^{-1} \rangle$ is equivalent to the set-theoretic condition that $M(\phi) : M(B) \to M(A)$ factors through the locus of points of $M(A)$ defined by the pointwise conditions $|a_i| \leq 1$ and $|a'_j| \geq 1$. Subsets of $M(A)$ defined by such conditions are called *Laurent domains*, and if there are no a'_j's then we call the subset a *Weierstrass domain*. In particular, by Yoneda's Lemma, a Laurent domain in $M(A)$ functorially determines the k-affinoid A-algebra $A\langle \underline{a}, \underline{a}'^{-1} \rangle$ that gives rise to it, so this latter algebra is intrinsic to the image of its MaxSpec in $M(A)$, and hence it enjoys some independence of the choice of the a_i's and a'_j's. (An analogue in algebraic geometry is that the localizations $A[1/a]$ and $A[1/a']$ are isomorphic as A-algebras if and only if there is the set-theoretic equality of the non-vanishing loci of a and a' in $\operatorname{Spec} A$, in which case such an isomorphism is unique.)

This characterization of an algebra by means of a set-theoretic condition is reminiscent of the situation for affine open subschemes of an affine scheme in algebraic geometry: if $\operatorname{Spec}(A')$ is an open subscheme of $\operatorname{Spec}(A)$, then a map of schemes $\operatorname{Spec}(B) \to \operatorname{Spec}(A)$ factors through $\operatorname{Spec}(A')$ as schemes if and only if it does so on underlying sets. Note that closed subschemes rarely have such a set-theoretic characterization (unless they are also open), since we can replace the defining ideal by its square without changing the underlying set but this nearly always changes the closed subscheme. This set-theoretic mapping property suggests that we ought to consider a Laurent domain as an "open subset" of $M(A)$ with associated coordinate ring given by its canonically associated A-algebra as above.

EXERCISE 2.1.6. Let us work out an example of a Laurent domain (explaining the reason for the name "Laurent domain"). Pick $c \in k$ with $0 < |c| \leq 1$, and consider the Laurent domain in $M(T_1)$ defined by the conditions $|c| \leq |t| \leq 1$, where $T_1 = k\langle t \rangle$; this is an "annulus". The associated coordinate ring is

$$k\langle t, X, Y \rangle / (X - t, c^{-1}tY - 1) = k\langle t, X, Y \rangle (X - t, tY - c).$$

Prove that the natural map $k\langle t\rangle \to k\langle t, X\rangle/(X-t)$ is an isomorphism by considering universal mapping properties, and deduce that the annulus has associated k-affinoid algebra $k\langle t, Y\rangle/(tY - c)$.

Prove that this is a domain. (Hint: show that every element of $k\langle t, Y\rangle/(tY - c)$ can be represented by a unique series of the form $c_0 + \sum_{j\geq 1} c_j t^j + \sum_{j\geq 1} c_{-j} Y^j$ with $c_j \to 0$ as $j \to \infty$ and $c_{-j} \to 0$ as $j \to \infty$. Show that this defines an injection into a k-algebra of doubly infinite Laurent series in t satisfying certain convergence properties.)

EXERCISE 2.1.7. If you are familiar with étale maps of schemes, then to put in perspective the role of open subschemes in set-theoretic mapping properties, consider the following problem. Let $i : U \to X$ be a locally finitely presented map of schemes with the property that a map of schemes $X' \to X$ factors through i if and only if its image is contained in $i(U)$, in which case such a factorization is unique. Prove that i is an open immersion. (Hint: Show that i is étale via the functorial criterion, and that it is set-theoretically injective and induces purely inseparable residue field extensions. Thus, it is an open immersion by [**EGA**, IV$_4$, 17.9.1].)

Now we introduce another important class of subsets of $M(A)$, the *rational domains*. For these subsets, the relevant input is a collection of elements $a_1, \ldots, a_n, a' \in A$ with *no common zero*. Given such data, we define

$$A\langle\frac{a_1}{a'}, \ldots, \frac{a_n}{a'}\rangle = A\langle X_1, \ldots, X_n\rangle/(a'X_1 - a_1, \ldots, a'X_n - a_n).$$

What is happening in this A-algebra is that we are forcing $|a_j| \leq |a'|$ at all points (easier to remember as the imprecise condition $|a_j/a'| \leq 1$). To make this precise, we state and prove a universal mapping property.

LEMMA 2.1.8. *Let A be a k-affinoid algebra, and $a_1, \ldots, a_n, a' \in A$ be elements with no common zero. For any map of k-affinoid algebras $\phi : A \to B$, there is at most one way to fill in the commutative diagram*

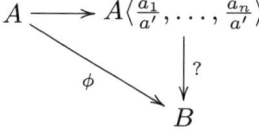

and such a diagram exists if and only if $M(\phi) : M(B) \to M(A)$ factors through the subset of $x \in M(A)$ such that $|a_j(x)| \leq |a'(x)|$ for all j, or in other words $|\phi(a_j)(y)| \leq |\phi(a')(y)|$ for all $y \in M(B)$.

PROOF. By the universal property of relative Tate algebras, to give such a diagram is to give power-bounded elements $b_1, \ldots, b_n \in B$ such that $\phi(a')b_j = \phi(a_j)$ for all j. This implies that $\phi(a')$ must be a unit in B because at any $y \in M(B)$ where it vanishes we get that all $\phi(a_j)$ also vanish, so the point $M(\phi)(y) \in M(A)$ is a common zero of the a_j's and a', contrary to hypothesis. Hence, the b_j's are uniquely determined if they exist, so we get the uniqueness of the diagram if it exists. Moreover, power-boundedness of such b_j's forces $|\phi(a_j)(y)|/|\phi(a')(y)| = |b_j(y)| \leq 1$ for all $y \in M(B)$, which is to say that $M(\phi)$ factors through the desired subset of $M(A)$. Conversely, if this set-theoretic condition holds then $|\phi(a_j)| \leq |\phi(a')|$ pointwise on $M(B)$, so $\phi(a') \in B$ has to be a unit because if it is not a unit then there would exist some $y \in M(B)$ at which $\phi(a')$ vanishes, and hence all $\phi(a_j)$

vanish, yielding the point $M(\phi)(y) \in M(A)$ as a common zero of the a_j's and a' (contrary to hypothesis). But with $\phi(a')$ a unit in B (even if a' is not a unit in A) it makes sense to consider $b_j = \phi(a_j)/\phi(a') \in B$. To construct the desired commutative diagram, the problem is to prove that b_j is power-bounded in B, or equivalently that $|b_j(y)| \leq 1$ for all $y \in M(B)$. This is exactly the assumed system of inequalities $|\phi(a_j)(y)| \leq |\phi(a')(y)|$ for all $y \in M(B)$ since $|\phi(a')(y)| \neq 0$ for all such y. $\qquad\square$

We call a subset in $M(A)$ of the form

$$\{x \in M(A) \mid |a_j(x)| \leq |a'(x)|\}$$

for $a_1, \dots, a_n, a' \in A$ with no common zero a *rational domain*. The universal property in the preceding lemma shows that such a subset canonically determines the A-algebra $A\langle a_1/a', \dots, a_n/a' \rangle$.

EXAMPLE 2.1.9. We write $\sqrt{|k^\times|} \subseteq \mathbf{R}_{>0}^\times$ to denote the divisible subgroup that is generated by $|k^\times|$, which is to say the set of positive real numbers α such that $\alpha^N \in |k^\times|$ for some integer $N > 0$. Note that this is a dense subgroup of $\mathbf{R}_{>0}^\times$. If $\alpha \in \sqrt{|k^\times|}$ and $\alpha^N = |c|$ with $c \in k^\times$, then for k-affinoid A and $f \in A$ the inequality $|f(x)| \leq \alpha$ for $x \in M(A)$ is equivalent to the inequality $|c^{-1}f^N(x)| \leq 1$ Thus, in the definitions of Weierstrass, Laurent, and rational domains it is no more general to permit real scaling factors from $\sqrt{|k^\times|}$ in the inequalities.

For example, in the closed unit disc over $k = \mathbf{Q}_p$, the locus $|t| \leq 1/\sqrt{p}$ is a Weierstrass domain: it is the same as the condition $|pt^2| \leq 1$, and so has associated "coordinate ring" $k\langle t, X\rangle/(X - pt^2)$. In contrast, it is true and unsurprising (but perhaps not obvious how to prove) that for $r \notin \sqrt{|\mathbf{Q}_p^\times|} = p^{\mathbf{Q}}$ the locus $\{|t| \leq r\}$ in the closed unit disc $M(\mathbf{Q}_p\langle t\rangle)$ over $\mathbf{Q}_p$ is *not* a Weierstrass domain.

2.2. Affinoid subdomains and admissible opens. The domains of Weierstrass, Laurent, and rational type are the most important examples of the following general concept:

DEFINITION 2.2.1. Let A be a k-affinoid algebra. A subset $U \subseteq M(A)$ is an *affinoid subdomain* if there exists a map $i : A \to A'$ of k-affinoids such that $M(i) : M(A') \to M(A)$ lands in U and is universal for this condition in the following sense: for any map of k-affinoid algebras $\phi : A \to B$, there is a commutative diagram

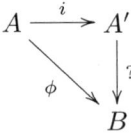

if and only if $M(\phi)$ carries $M(B)$ into U, in which case such a diagram is unique.

By Yoneda's Lemma, if $U \subseteq M(A)$ is an affinoid subdomain then the k-affinoid A-algebra A' as in the preceding definition is unique up to unique A-algebra isomorphism. It is therefore legitimate to denote this A-algebra as A_U: it is functorially determined by U. We call A_U the *coordinate ring* of U (with respect to A). For example, the universal property of the domains of Weierstrass, Laurent, and rational types shows that each is an affinoid subdomain and provides an explicit description of A_U in such cases. By chasing points valued in finite extensions of k, it is not hard to show that the natural map $M(A_U) \to M(A)$ is an injection onto

$U \subseteq M(A)$. By the universal property we likewise see that if $V \subseteq U$ is an inclusion of affinoid subdomains of $M(A)$ then there is a unique A-algebra map of coordinate rings $\rho_V^U : A_U \to A_V$, and by uniqueness this is transitive with respect to another inclusion $W \subseteq V$ of affinoid subdomains in $M(A)$ (in the sense that $\rho_W^V \circ \rho_V^U = \rho_W^U$). This is to be considered as analogous to restriction maps for the structure sheaf of a scheme, so for $f \in A_U$ we usually write $f|_V$ to denote $\rho_V^U(f) \in A_V$. Akin to the case of schemes, it can be shown (by a method entirely different from the case of schemes) that A_U is A-flat for any affinoid subdomain $U \subseteq M(A)$.

EXERCISE 2.2.2. Prove that Weierstrass, Laurent, and rational domains in $M(A)$ are all open for the canonical topology. If the condition "no common zero" is dropped from the definition of a rational domain then it still makes sense to consider the underlying set in $M(A)$ defined by the simultaneous conditions $|a_j(x)| \leq |a'(x)|$. Show by example that this locus can fail to be open if there is a common zero.

EXERCISE 2.2.3. This exercise develops two kinds of completed tensor product operations that arise in rigid-analytic geometry. The theory of the completed tensor product can be developed in greater generality than we shall do, but for our limited purposes we adopt a more utilitarian approach.

(1) Let A and A' be k-affinoid algebras. We wish to construct a k-affinoid "completed tensor product" $A \widehat{\otimes}_k A'$. To do this, first choose presentations $A \simeq T_n/I$ and $A' \simeq T_{n'}/I'$. Using the natural maps $T_n \to T_{n+n'}$ and $T_{n'} \to T_{n+n'}$ onto the first n and last n' variables, it makes sense to let $J, J' \subseteq T_{n+n'}$ be the ideals generated by I and I' respectively. Consider the k-affinoid algebra $T_{n+n'}/(J+J')$. There are evident k-algebra maps $\iota : A \to T_{n+n'}/(J+J')$ and $\iota' : A' \to T_{n+n'}/(J+J')$. Prove that this pair of maps is universal in the following sense: for any k-Banach algebra B and any k-Banach algebra maps $\phi : A \to B$ and $\phi' : A' \to B$, there is a unique k-Banach algebra map $h : T_{n+n'}/J \to B$ so that $h \circ \iota = \phi$ and $h \circ \iota' = \phi'$. In view of this universal property, the triple $(T_{n+n'}/(J+J'), \iota, \iota')$ is unique up to unique isomorphism, so we may denote $T_{n+n'}/(J+J')$ as $A \widehat{\otimes}_k A'$. The product $\iota(a)\iota'(a')$ is usually denoted $a \widehat{\otimes} a'$.

(2) Let $j : A'' \to A$ and $j' : A'' \to A'$ be a pair of maps of k-affinoid algebras. Define the k-affinoid algebra

$$A \widehat{\otimes}_{A''} A' := (A \widehat{\otimes}_k A')/(j(a'') \widehat{\otimes} 1 - 1 \widehat{\otimes} j'(a'') \,|\, a'' \in A'').$$

Formulate and prove a universal property for this in the category of k-Banach A''-algebras (analogous to the universal property of tensor products of rings).

(3) Let A be a k-affinoid algebra, and let K/k be an analytic extension field (i.e., a non-archimedean field K equipped with a structure of extension of k respecting the absolute values). Beginning with the case of Tate algebras, define a K-affinoid algebra $K \widehat{\otimes}_k A$ as a solution to a universal mapping problem for continuous maps (over $k \to K$) from A to K-Banach algebras.

EXERCISE 2.2.4. Let A be a k-affinoid algebra.

(1) If $U, U' \subseteq M(A)$ are affinoid subdomains, then prove that $U \cap U'$ is one as well: the k-affinoid A-algebra $A_{U \cap U'}$ is $A_U \widehat{\otimes}_A A_{U'}$. Check this via universal mapping properties.

(2) Let $\phi : A \to B$ be a map of k-affinoid algebras, and let $U \subseteq M(A)$ be an affinoid subdomain. Show that $M(\phi)^{-1}(U) \subseteq M(B)$ is also an affinoid subdomain: its coordinate ring is $A_U \widehat{\otimes}_A B$. Check this via universal mapping properties. Is there a similar result for Weierstrass, Laurent, and rational domains?

(3) Let $U \subseteq M(A)$ be an affinoid subdomain, with corresponding coordinate ring A_U. Using the natural bijection $U = M(A_U)$, prove that a subset $U' \subseteq U = M(A_U)$ is an affinoid subdomain of $M(A_U)$ if and only if it is an affinoid subdomain of $M(A)$.

The introduction of the concept of affinoid subdomains was a genuine advance beyond Tate's original work, in which he got by with just Weierstrass, Laurent, and rational subdomains. In order to make affinoid subdomains easy to handle (e.g., are they open?), the crucial result required is the *Gerritzen–Grauert theorem* that describes them in terms of rational domains:

THEOREM 2.2.5. *Let A be a k-affinoid algebra. Every affinoid subdomain $U \subseteq M(A)$ is a finite union of rational domains. In particular, affinoid subdomains are open with respect to the canonical topology.*

It is very hard to determine when a given finite union of rational domains (let alone affinoid subdomains) is an affinoid subdomain. This is analogous to the difficulty of detecting when a finite union of affine open subschemes of a scheme is again affine. Since Laurent domains are a basis for the canonical topology, in order to get a good theory of non-archimedean analytic spaces we cannot permit ourselves to work with arbitrary unions of affinoid subdomains (or else we will encounter the total disconnectedness problem). Tate's idea is to restrict attention to a class of open subsets (for the canonical topology) and a *restricted* collection of coverings of these opens by such opens so as to "force" affinoid subdomains to appear to be compact. The key definition in the theory is as follows.

DEFINITION 2.2.6. Let A be a k-affinoid algebra. A subset $U \subseteq M(A)$ is an *admissible open* subset if it has a set-theoretic covering $\{U_i\}$ by affinoid subdomains $U_i \subseteq M(A)$ with the following finiteness property under affinoid pullback: for any map of k-affinoid algebras $\phi : A \to B$ such that $M(\phi) : M(B) \to M(A)$ has image contained in U, there are finitely many U_i's that cover this image; equivalently, the open covering $\{M(\phi)^{-1}(U_i)\}$ of $M(B)$ by affinoid subdomains has a finite subcovering.

A collection $\{V_j\}$ of admissible open subsets of $M(A)$ is an *admissible cover* of its union V if, for any k-affinoid algebra map $\phi : A \to B$ with $M(\phi)(M(B)) \subseteq V$, the covering $\{M(\phi)^{-1}(V_j)\}$ of $M(B)$ has a refinement by a covering consisting of finitely many affinoid subdomains. (This forces V to be admissible open, by using the affinoid subdomain covering $\{V_{jk}\}_{j,k\in K_j}$ where $\{V_{jk}\}_{k\in K_j}$ is a covering of each V_j by affinoid subdomains as in the definition of admissibility of each V_j.)

Note that the set-theoretic covering $\{U_i\}$ of U in the definition of an admissible open subset of $M(A)$ is necessarily an admissible cover.

EXAMPLE 2.2.7. Let $U_1, \ldots, U_n$ be affinoid subdomains of $M(A)$ for a k-affinoid algebra A. Then $U = \cup U_j$ is an admissible open subset, with $\{U_j\}$ an admissible covering of U. (For example, for $a \in A$ the Laurent domains $M(A\langle a \rangle) = \{|a| \leq 1\}$ and $M(A\langle 1/a \rangle) = \{|a| \geq 1\}$ constitute an admissible covering of $M(A)$.) The content here is that the pullback of each U_j under $M(\phi)$ for a k-affinoid algebra map $\phi : A \to B$ is an affinoid subdomain of $M(B)$. It is difficult to determine if such a U is an affinoid subdomain.

Now we come to a key example that shows the significance of the finiteness requirement in the definition of admissibility.

EXAMPLE 2.2.8. Let $T_1 = k\langle t\rangle$ be the Tate algebra in one variable. Within the closed unit ball $M(T_1)$, the locus $V = \{|t| = 1\}$ is a Laurent domain. The subset $U = \{|t| < 1\}$ is open for the canonical topology, and more importantly it is an admissible open. Indeed, it is covered by the Weierstrass domains $U_n = \{|t| \le |c|^{1/n}\}$ for a fixed $c \in k$ with $0 < |c| < 1$ and $n \ge 1$, and these satisfy the admissibility condition due to the Maximum Modulus Principle: if $\phi : T_1 \to B$ is a map to a k-affinoid algebra such that $M(B) \to M(T_1)$ lands in U, then the function $\phi(t) \in B$ has absolute value < 1 pointwise on $M(B)$, and so the Maximum Modulus Principle on $M(B)$ provides $0 < \alpha < 1$ such that $|\phi(t)(y)| \le \alpha$ for all $y \in M(B)$. Hence, for n_0 so large that $\alpha < |c|^{1/n_0} < 1$ we have $M(\phi)(M(B)) \subseteq U_{n_0}$. Thus, the required finite subcover property is satisfied.

The pair of admissible opens $V = \{|t| = 1\}$ and $U = \{|t| < 1\}$ covers $M(T_1)$ set-theoretically, and these are disjoint. However, this is not an *admissible* covering. Indeed, by the definition of admissibility of a covering (applied to the identity map of $M(T_1)$) it would follow that $\{U, V\}$ has as a refinement a finite covering of $M(T_1)$ by affinoid subdomains. But by the Maximum Modulus Principle, any affinoid subdomain of $M(T_1)$ contained in U is contained in some $U_n = \{|t| \le |c|^{1/n}\}$, and hence if there were a refinement of $\{U, V\}$ by a finite collection of affinoid subdomains then we would get that $M(T_1)$ is covered by V and by U_{n_0} for some large n_0. By using a suitable finite extension of k we can certainly find a point in $M(T_1)$ lying in the locus $\{|t| = |c|^{1/(n_0+1)}\}$ that is disjoint from $U_{n_0} \cup V$. This gives a contradiction, so $\{U, V\}$ is not an admissible covering of $M(T_1)$. (Note that here it is essential that our spaces have points that are not necessarily k-rational.)

EXERCISE 2.2.9. Choose $0 < r < 1$ with $r \notin \sqrt{|k^\times|}$. Prove that $\{|t| \le r\} = \{|t| < r\}$ is an admissible open subset of $M(k\langle t\rangle)$, and give an admissible covering by Weierstrass domains. Prove that this locus does not have an admissible covering by finitely many affinoid subdomains. (Hint: use the Maximum Modulus Principle.)

EXERCISE 2.2.10. Generalize the method of Example 2.2.8 to show that for any k-affinoid algebra A and $a, a' \in A$, the set

$$U = \{x \in M(A) \,|\, |a(x)| < |a'(x)|\}$$

is an admissible open subset; give an admissible open affinoid covering of U (and be careful about points in $M(A)$ where a' vanishes).

EXERCISE 2.2.11. Let A be a k-affinoid algebra. Prove that if an admissible open $U \subseteq M(A)$ is covered set-theoretically by some admissible opens U_i then $\{U_i\}$ is an admissible covering of U if and only if it admits an admissible refinement. Also show that admissibility for subsets is a "local" property in the following sense: if $U \subseteq M(A)$ is an admissible open and $\{U_i\}$ is an admissible covering of U by admissible opens then a subset $V \subseteq U$ is admissible open in $M(A)$ if and only if $V \cap U_i$ is admissible open in $M(A)$ for all i.

EXAMPLE 2.2.12. Let $A = k\langle x, y\rangle$. Inside $M(A)$ we will construct a subset U that is open for the canonical topology but is not an admissible open subset. Fix $c \in k^{\times}$ with $0 < |c| < 1$, and for $i \geq 1$ let $U_i = \{|x| \leq |c|^i, |y| \leq |c|^{1/i}\} \subseteq M(A)$. Let $V = \{|y| = 1\} \subseteq M(A)$. Each of V and the U_i's is an affinoid subdomain, and in particular is obviously open for the canonical topology. Hence, the union U of V and the U_i's is open for the canonical topology. We claim that U is not an admissible open locus in $M(A)$. The covering of U by V and the U_i's is not an admissible cover because intersecting with $\{x = 0\} = M(k\langle y\rangle) = M(T_1)$ gives the covering that we proved is non-admissible in Example 2.2.8. However, this does not prove that U is not an admissible open subset of $M(A)$, since it must be proved that there is *no* set-theoretic covering of U by affinoid domains in $M(A)$ for which the admissibility requirement is satisfied. To rigorously prove that U is not admissible, we have to use deeper properties of admissible open sets. More specifically, since U contains the locus $\{x = 0\}$ in $M(A)$, it follows from the discussion at the end of §5.2 that if U were admissible then it would have to contain some "tube" $\{|x| \leq |c|^{i_0}\} \subseteq M(A)$ for some $i_0 > 0$. But for $b \in \overline{k}$ such that $|c|^{1/i_0} < |b| < 1$ the point $(c^{i_0}, b) \in M(A)$ (i.e., the maximal ideal kernel of the map $A \to k(b)$ defined by $x \mapsto c^{i_0}$, $y \mapsto b$) lies in this tube but does not lie in U. Hence, U is not admissible.

2.3. The Tate topology. The admissible opens and their admissible coverings within $M(A)$ lead to the definition of a mild Grothendieck topology (in the sense that it only involves subsets of the ambient spaces, which is not a requirement in the general theory of Grothendieck topologies, such as the étale topology on a scheme):

DEFINITION 2.3.1. The *Tate topology* (or *G-topology*) on $M(A)$ has as objects the admissible open subsets and as coverings the admissible open coverings.

EXERCISE 2.3.2. Let A be a k-affinoid algebra. For $a \in A$ let $V(a)$ be the locus of $x \in M(A)$ for which $a(x) \neq 0$. Prove that this is an admissible open subset (give an admissible Laurent covering), and show that the $V(a)$'s are a base of opens for a topology on $M(A)$; this is called the *analytic Zariski topology*. Show that the closed sets for this topology are the subsets $M(A/I)$ for ideals I of A (these are called *analytic sets* in $M(A)$), and that all Zariski-opens and Zariski-open covers of Zariski-opens are admissible. (Hint for admissibility: if B is k-affinoid with $\|\cdot\|$ a k-Banach algebra norm on B and $b_1, \ldots, b_n \in B$ some elements that generate 1, say $\sum \beta_j b_j = 1$ with $\beta_j \in B$, then show that $M := \max(\|\beta_1\|, \ldots, \|\beta_n\|) > 0$ and the Laurent domains $\{|b_j| \geq 1/M\}$ cover $M(B)$.)

The Tate topology is generally not a topology on $M(A)$ in the usual sense, but is instead a Grothendieck topology. It is not crucial (for our purposes) to delve into the general formalism of Grothendieck topologies. The main point that matters for working with the Tate topology is to keep in mind that we do not consider general unions of admissible opens, and when doing sheaf theory we only consider admissible coverings of admissible opens. It generally does not make sense to evaluate a sheaf for the Tate topology on a general open set for the canonical topology, and even for evaluation on admissible opens we cannot expect a sheaf for the Tate topology to satisfy the sheaf axioms for a non-admissible covering of an admissible open by admissible opens (e.g., the pair $\{U, V\}$ in the closed unit disc in Example 2.2.8). When we define disconnectedness later, it will be expressed in terms of an *admissible* covering by a pair of disjoint non-empty admissible opens.

Example 2.2.8 shows that this rules out many classical sources of disconnectedness of the canonical topology. The fundamental result that gets the theory off the ground is the existence of a "structure sheaf" with respect to the Tate topology. This is Tate's *Acylicity Theorem*:

THEOREM 2.3.3. *Let A be a k-affinoid algebra. The assignment $U \mapsto A_U$ of the coordinate ring to every affinoid subdomain of $M(A)$ uniquely extends to a sheaf $\mathscr{O}_A$ with respect to the Tate topology on $M(A)$. In particular, if $\{U_i\}$ is a finite collection of affinoid subdomains with $U = \cup U_i$ also an affinoid subdomain of $M(A)$ then the evident sequence*

$$0 \to A_U \to \prod A_{U_i} \to \prod A_{U_i \cap U_j}$$

is exact.

Tate proved this theorem by heavy use of Čech-theoretic methods to reduce to the special case of a Laurent covering of $M(A)$ by the pair $\{M(A\langle a \rangle), M(A\langle 1/a \rangle)\}$ for $a \in A$. In this special case he could carry out a direct calculation. The next exercise gives the simplest instance of this calculation.

EXERCISE 2.3.4. Choose $c \in k$ such that $0 < |c| < 1$. In $M(T_1)$, let $U = \{|t| \leq |c|\}$ and $V = \{|t| \geq |c|\}$, so $U \cap V = \{|t| = |c|\}$. By calculating with convergent Laurent series in t (using Exercise 2.1.6 to describe $A_V \simeq k\langle t, Y \rangle/(tY - c)$ as a $k\langle t \rangle$-algebra of certain Laurent series $\sum_{n \in \mathbf{Z}} c_n t^n$), show that if $f \in A_U$ and $g \in A_V$ satisfy $f|_{U \cap V} = g|_{U \cap V}$ in $A_{U \cap V}$ then there is a unique $h \in T_1$ such that $h|_U = f$ and $h|_V = g$.

DEFINITION 2.3.5. Let A be a k-affinoid algebra. The *affinoid space* $\mathrm{Sp}(A)$ is the pair $(M(A), \mathscr{O}_A)$ consisting of the set $M(A)$ endowed with its Tate topology and sheaf of k-algebras $\mathscr{O}_A$ with respect to the Tate topology. If $A = T_n = T_n(k)$ then this is denoted $\mathbf{B}^n = \mathbf{B}_k^n$. Usually we write $\mathscr{O}_X$ rather than $\mathscr{O}_A$ (with $X = \mathrm{Sp}(A)$).

EXERCISE 2.3.6. Prove that for $x \in X = \mathrm{Sp}(A)$, the stalk

$$\mathscr{O}_{X,x} = \varinjlim_{x \in U} \mathscr{O}_X(U)$$

(limit over admissible opens, or equivalently affinoid subdomains, containing x) is a local ring. Describe $\mathscr{O}_{\mathbf{B}^1, 0}$ as an intermediate ring strictly between the algebraic local ring $k[t]_{(t)}$ and the completion $k[\![t]\!]$.

In general, $\mathscr{O}_{X,x}$ is a noetherian ring that is faithfully flat over the algebraic local ring $A_{\mathfrak{m}_x}$, and in fact it has the same completion, but this requires more work to prove. This is the key to developing a good *dimension theory* for rigid-analytic spaces.

In order to make global definitions, we have to first define the category in which we will be working. A *G-topologized space* is a set X equipped with a set $\mathfrak{U}$ of subsets $U \subseteq X$ (to be called the "open subsets") and a set of set-theoretic coverings $\mathrm{Cov}(U)$ of each $U \in \mathfrak{U}$ by collections of members of $\mathfrak{U}$ such that certain natural locality and transitivity properties from ordinary topology are satisfied (see [**BGR**, Ch. 9] for a precise discussion): $\mathfrak{U}$ is stable under finite intersections, $\emptyset \in \mathfrak{U}$, $\{U\} \in \mathrm{Cov}(U)$ for all $U \in \mathfrak{U}$, if $\{U_i\} \in \mathrm{Cov}(U)$ and $V \subseteq U$ then $V \in \mathfrak{U}$ if and only if $V \cap U_i \in \mathfrak{U}$ for all i, and if $\{V_{ij}\}_{j \in J_i} \in \mathrm{Cov}(U_i)$ for $\{U_i\} \in \mathrm{Cov}(U)$ then $\{V_{ij}\}_{i,j} \in \mathrm{Cov}(U)$. In particular, we *omit* the requirement that $\mathfrak{U}$ is stable under arbitrary (or even finite) unions.

One very important construction for G-topologized subspaces is the analogue of the "open subspace" topology. To be precise, if X is a G-topologized space with associated collection $\mathfrak{U}$ of open subsets, then for any $U \in \mathfrak{U}$ we endow U with a structure of G-topologized space by using $\mathfrak{U}_U = \{V \in \mathfrak{U} \mid V \subseteq U\}$ as the collection of "open subsets" of U and using the same collection of coverings; that is, for each $V \in \mathfrak{U}_U$ its associated collection $\mathrm{Cov}_U(V)$ of set-theoretic coverings is $\mathrm{Cov}(V)$. This construction does satisfy all of the axioms to be a G-topologized space, and it is called the *open subspace structure* on U.

A *sheaf* (of sets, groups, etc.) on a G-topologized space X is a contravariant assignment $U \mapsto \mathscr{F}(U)$ of a set (or group, etc.) to each $U \in \mathfrak{U}$ such that the usual sheaf axioms are satisfied for coverings in $\mathrm{Cov}(U)$ for all $U \in \mathfrak{U}$. In this respect, we are restricting both the concept of openness and the concept of open covering from ordinary topology. Note that if $U \in \mathfrak{U}$ and $\mathscr{F}$ is a sheaf on X then the functor $\mathscr{F}|_U : V \mapsto \mathscr{F}(V)$ on $\mathfrak{U}_U$ is easily seen to be a sheaf on U with respect to its open subspace structure. We leave it to the reader's imagination (or see [**BGR**, Ch. 9]) to formulate how one glues G-topologized spaces or sheafifies presheaves on them (this latter issue requires some care).

EXAMPLE 2.3.7. Consider a pair $(X, \mathscr{O}_X)$ consisting of a G-topologized space X and a sheaf of k-algebras $\mathscr{O}_X$ on this space. If $U \in \mathfrak{U}$ then with the open subspace structure on U we get another such pair $(U, \mathscr{O}_U)$ where $\mathscr{O}_U$ is the sheaf of k-algebras $\mathscr{O}_X|_U$. This is called an *open subspace* of $(X, \mathscr{O}_X)$.

For a second such pair $(X', \mathscr{O}_{X'})$, a *morphism* $(X', \mathscr{O}_{X'}) \to (X, \mathscr{O}_X)$ is a pair $f : X' \to X$ and $f^\sharp : \mathscr{O}_X \to f_*(\mathscr{O}_{X'})$ where f is continuous (in the sense that pullback under f respects the class of "opens" and their "coverings") and $f^\sharp$ is a map of sheaves of k-algebras (with f_* defined in the usual way; f_* carries sheaves to sheaves because f is continuous). Composition of morphisms is defined exactly as in the theory of ringed spaces. In case the stalks of the structure sheaves are local rings, we can also define the more restrictive notion of a morphism of locally ringed G-topologized spaces.

As a fundamental example, for any k-affinoid algebra A we have constructed such a space $\mathrm{Sp}(A)$, and if $U \subseteq \mathrm{Sp}(A)$ is an affinoid subdomain then the corresponding open subspace $(U, \mathscr{O}_A|_U)$ is naturally identified with $\mathrm{Sp}(A_U)$ due to Exercise 2.2.4(3). More importantly, if $\phi : A \to B$ is a map of k-affinoid algebras then we get a morphism of locally ringed G-topologized spaces $\mathrm{Sp}(\phi) : \mathrm{Sp}(B) \to \mathrm{Sp}(A)$ as follows: on underlying spaces we use the map $f = M(\phi) : M(B) \to M(A)$, and the map $\mathscr{O}_A \to f_*(\mathscr{O}_B)$ that is defined on affinoid subdomains $U \subseteq \mathrm{Sp}(A)$ via the k-algebra map $A_U \to A_U \widehat{\otimes}_A B = B_{f^{-1}(U)}$ and is uniquely extended to general admissible opens via the sheaf axioms. One checks readily that this assignment $A \rightsquigarrow \mathrm{Sp}(A)$ is thereby a contravariant functor, and one shows by copying the proof in the case of affine schemes that this is a *fully faithful* functor. In this way, the (opposite category of the) category of k-affinoid algebras is identified with a full subcategory of the category of locally ringed G-topologized spaces with k-algebra structure sheaf (and maps that respect this k-structure).

An unfortunate fact of life is that the concept of stalk (at a point of X) is not as useful as in ordinary topology. For example, even for the G-topologized spaces $X = M(A)$ it can and does happen that there exist abelian sheaves $\mathscr{F}$ on X and nonzero $s \in \mathscr{F}(X)$ such that $s_x \in \mathscr{F}_x$ vanishes for all $x \in X$. The reason that this does not violate the sheaf axiom is that the vanishing in the stalk merely provides $U_x \in \mathfrak{U}$ containing $x \in X$ so that $s|_{U_x} \in \mathscr{F}(U_x)$ vanishes, but perhaps $\{U_x\}_{x \in X}$ is not in $\mathrm{Cov}(X)$! Hence, one cannot conclude $s = 0$. For the purposes of coherent sheaf theory in rigid-analytic geometry, this pathology will not intervene. However, it is a very serious issue when working with more general abelian sheaves, and so it arises in any attempt to set up a good theory of étale cohomology on such spaces. The work of Berkovich and Huber enlarges the underlying sets of affinoid spaces (and their global counterparts) to have "enough points" so as to permit stalk arguments to work as in classical sheaf theory. This is one of the technical merits of these other approaches to non-archimedean geometry.

2.4. Globalization. Having introduced the notion of G-topologized spaces and sheaves on them, we can now make the key global definition.

DEFINITION 2.4.1. A *rigid-analytic space* over k is a pair $(X, \mathscr{O}_X)$ consisting of a locally ringed G-topologized space whose structure sheaf is a sheaf of k-algebras such that there is a covering $\{U_i\} \in \mathrm{Cov}(X)$ with each open subspace $(U_i, \mathscr{O}_X|_{U_i})$ isomorphic to an affinoid space $\mathrm{Sp}(A_i)$ for a k-affinoid algebra A_i. Morphisms are taken in the sense of locally ringed G-topologized spaces with k-algebra structure sheaf (and maps respecting this k-structure), as in Example 2.3.7.

EXAMPLE 2.4.2. Let $(X, \mathscr{O}_X)$ be a rigid-analytic space and let $U \subseteq X$ be an admissible open subset. The pair $(U, \mathscr{O}_U)$ is a rigid-analytic space. Indeed, let $\{U_i\}$ be an admissible open covering of X such that each $(U_i, \mathscr{O}_{U_i})$ is an affinoid space. By the axioms for a G-topologized space, $\{U_i \cap U\}$ is an admissible open covering of U. Thus, if we can find an admissible open covering of each $U_i \cap U$ by affinoid spaces V_{ij} ($j \in J_i$) then the entire collection $\{V_{ij}\}_{i,j}$ is an admissible covering of U (by the G-topology axioms and the definition of the open subspace structure), so $(U, \mathscr{O}_U)$ thereby has an admissible covering by affinoid spaces, as required. Hence, we can rename U_i as X to reduce to the case when $X = \mathrm{Sp}(A)$ is an affinoid space. But then by definition of the G-topology on $M(A)$, any admissible open $U \subseteq \mathrm{Sp}(A)$ has an admissible covering by affinoid domains.

For any rigid-analytic space X and $x \in X$, the stalk $\mathscr{O}_{X,x}$ can be computed using any affinoid open around x, so its residue field $k(x)$ is a finite extension of k. For any $f \in \mathscr{O}_X(X)$ we therefore get a value $f(x) \in k(x)$ and hence a number $|f(x)| \geq 0$. We can imitate the definitions of Weierstrass, Laurent, and rational domains by imposing non-strict inequalities on the $|f(x)|$'s, and these are admissible opens in X (since they meet each admissible affinoid open in an admissible open subset). One can also construct rigid-analytic spaces by gluing procedures that are similar to the case of gluing ringed spaces. Rather than delve into general details, we illustrate with some examples.

EXAMPLE 2.4.3. We first construct rigid-analytic affine n-space over k. This rigid space, $\mathbf{A}_k^{n,\mathrm{an}}$, is defined by choosing $c \in k$ with $0 < |c| < 1$ and gluing a rising chain of closed balls centered at the origin with polyradius $|c|^{-j}$ for $j \geq 1$. More precisely, let $D_j = \mathbf{B}_k^n$ be the closed unit n-ball over k, with coordinates $\xi_{1,j}, \ldots, \xi_{n,j}$. We define the map $D_j \to D_{j+1}$ to correspond to the k-affinoid algebra map $\xi_{i,j+1} \mapsto c\xi_{i,j}$ (which makes sense since $|c| \leq 1$). This identifies D_j with an affinoid subdomain $\{|\xi_{i,j}| \leq |c|\}$ in D_{j+1}, and in particular for each $1 \leq i \leq n$ the analytic functions $\xi_{i,j}/c^j$ on D_j are compatible with change in j. Thus, on the gluing of the D_j's we get unique global sections ξ_i of the structure sheaf such that $\xi_i|_{D_j} = \xi_{i,j}/c^j$ for all j. This gluing is denoted $\mathbf{A}_k^{n,\mathrm{an}}$, and the locus $\{|\xi_1| \leq |c|^{-j}, \dots, |\xi_n| \leq |c|^{-j}\}$ is the open subspace D_j (with $c^j \xi_i|_{D_j} = \xi_{i,j}$).

It is instructive to see that this deserves to be called an affine n-space by showing that it has the right universal property. Namely, for any rigid space X and any morphism $f : X \to \mathbf{A}_k^{n,\mathrm{an}}$ we get pullback functions $f^\sharp(\xi_i) \in \mathscr{O}_X(X)$, and hence a natural map of sets

$$\mathrm{Hom}(X, \mathbf{A}_k^{n,\mathrm{an}}) \to \mathscr{O}_X(X)^n$$

given by $f \mapsto (f^\sharp(\xi_1), \dots, f^\sharp(\xi_n))$. We claim that this is bijective, so it provides a universal property: affine n-space is the universal rigid space equipped with an ordered n-tuple of global functions. This gives a viewpoint that is independent of the auxiliary choice of c. To prove the bijectivity, by naturality and gluing for morphisms it suffices to treat the case when X is affinoid. But then by the Maximum Modulus Principle, the k-affinoid ring A of global functions on X is the rising union of its subsets A_j of elements with sup-norm at most $|c|^{-j}$ for $j \geq 1$ (since $0 < |c| < 1$). Since a map $f : X \to \mathbf{A}_k^{n,\mathrm{an}}$ lands in D_j if and only if each $f^\sharp(\xi_i) \in A$ has sup-norm at most $|c|^{-j}$ (why?), we reduce ourselves to a problem for the D_j's separately. Upon working with $\xi_{i,j} = c^j \xi_i|_{D_j}$, the problem for each D_j viewed as a closed unit n-ball becomes exactly the universal property of the n-variable Tate algebra!

EXAMPLE 2.4.4. By imitating the gluing procedures used to make projective spaces as a union of affine spaces, we can construct rigid-analytic projective spaces. These satisfy the usual universal property in terms of line bundles, by the same method of proof as in algebraic geometry, once the theory of coherent sheaves (to be discussed later) is fully developed. It is rather crucial for applications (e.g., see Examples 3.2.5 and 3.3.9) that rigid-analytic projective spaces can also be constructed by gluing closed unit polydiscs rather than affine spaces. More precisely, let $\Delta_0, \dots, \Delta_n$ be such polydiscs with Δ_j having coordinates $t_{1j}, \dots, t_{nj}$, and for $j' \neq j$ glue $\Delta_{j'}$ to Δ_j along the isomorphism $\Delta_{j'}\langle t_{jj'}^{-1}\rangle \simeq \Delta_j\langle t_{j'j}^{-1}\rangle$ defined by the habitual transition formulas. One checks that this satisfies the requirements to make a gluing, and that the natural map from the resulting glued space to the rigid-analytic projective n-space built by gluing rigid-analytic affine n-spaces is an isomorphism.

EXAMPLE 2.4.5. In Exercise 2.2.3 we saw how to define the completed tensor product $A \widehat{\otimes}_{A''} A'$ for a pair of maps of k-affinoid algebras $A'' \to A$ and $A'' \to A'$. Via its universal property, one readily checks (much like in the case of affine schemes with tensor products) that $\mathrm{Sp}(A \widehat{\otimes}_{A''} A')$ equipped with its evident morphisms to $\mathrm{Sp}(A)$ and $\mathrm{Sp}(A')$ (agreeing upon composition to $\mathrm{Sp}(A'')$) is a fiber product $\mathrm{Sp}(A) \times_{\mathrm{Sp}(A'')} \mathrm{Sp}(A')$ in the category of affinoid spaces, and then (by gluing maps) that it is such a fiber product in the category of rigid-analytic spaces. One can then copy the same gluing method as for schemes to globalize this construction to obtain the existence of fibers products $X \times_{X''} X'$ for any pair of maps $X \to X''$ and $X \to X''$ of rigid-analytic spaces.

EXAMPLE 2.4.6. In complex-analytic geometry, a very useful tool is the procedure of analytification for both algebraic $\mathbf{C}$-schemes and coherent sheaves on them. The resulting functors $\mathscr{X} \rightsquigarrow \mathscr{X}^{\mathrm{an}}$ and $\mathscr{F} \rightsquigarrow \mathscr{F}^{\mathrm{an}}$ from algebraic $\mathbf{C}$-schemes to complex-analytic spaces (and on their categories of sheaves of modules) satisfy a number of nice properties that we will not list here. The one aspect we note is that there is a natural map $i_{\mathscr{X}} : \mathscr{X}^{\mathrm{an}} \to \mathscr{X}$ of locally ringed spaces with $\mathbf{C}$-algebra structure sheaves such that $i_{\mathscr{X}}$ carries $\mathscr{X}^{\mathrm{an}}$ bijectively onto $\mathscr{X}(\mathbf{C})$, it induces isomorphisms on completed local rings, and it is final among maps from complex-analytic spaces to $\mathscr{X}$. For any $\mathscr{O}_X$-module $\mathscr{F}$ on X, one defines $\mathscr{F}^{\mathrm{an}} = i_{\mathscr{X}}^*(\mathscr{F})$. The GAGA theorems of Serre (as extended by Grothendieck from the projective to the proper case) concern three aspects: the equivalence of categories of coherent sheaves on $\mathscr{X}$ and $\mathscr{X}^{\mathrm{an}}$ when $\mathscr{X}$ is proper, the comparison isomorphisms of cohomology for $\mathscr{F}$ and $\mathscr{F}^{\mathrm{an}}$ when $\mathscr{X}$ is proper and $\mathscr{F}$ is coherent, and the full faithfulness of the functor $\mathscr{X} \rightsquigarrow \mathscr{X}^{\mathrm{an}}$ when $\mathscr{X}$ is proper.

A similar procedure works in the rigid-analytic setting as follows. First of all, algebraic affine n-space $\mathbf{A}_k^n$ equipped with its standard ordered n-tuple of global functions can be shown to be a final object in the category of locally ringed G-topologized spaces equipped with a k-algebra structure sheaf and an ordered n-tuple of global functions. In particular, we get a unique morphism

$$\mathbf{A}_k^{n,\mathrm{an}} \to \mathbf{A}_k^n$$

compatible with the standard ordered n-tuple of global functions on each space. By the universal property of the source and target, this is final among all rigid-analytic spaces over k equipped with a morphism to $\mathbf{A}_k^n$. In general we define an *analytification* of a locally finite type k-scheme $\mathscr{X}$ to be a map $i_{\mathscr{X}} : \mathscr{X}^{\mathrm{an}} \to \mathscr{X}$ that is final among all maps from rigid spaces over k to $\mathscr{X}$ (as locally ringed G-topologized spaces with a k-algebra structure sheaf, and maps that respect this k-structure). The preceding shows that for $\mathscr{X} = \mathbf{A}_k^n$ an analytification exists: rigid-analytic affine n-space. Then one uses arguments with coherent ideal sheaves (discussed in the next lecture!) to pass from this case to all affine algebraic k-schemes, and finally to the general case by gluing arguments. We omit the details, except to remark that analytification is naturally compatible with the formation of fiber products and that $i_{\mathscr{X}}$ carries $\mathscr{X}^{\mathrm{an}}$ bijectively onto the set of closed points in $\mathscr{X}$ (inducing an isomorphism on completed local rings).

EXERCISE 2.4.7. Let X be a rigid-analytic space and $\{U_i\}$ a collection of admissible opens that cover X set-theoretically. Show that this is an admissible covering if and only if for every morphism $f : \mathrm{Sp}(A) \to X$ from an affinoid space, the set-theoretic covering $\{f^{-1}(U_i)\}$ by admissible opens has a finite affinoid open refinement. (This is an easy exercise in unwinding definitions, with "morphism" defined as in Example 2.3.7. Keep in mind that by definition X has an admissible covering by open affinoid subspaces, and that the functor $A \rightsquigarrow \mathrm{Sp}(A)$ is a fully faithful functor in the sense discussed near the end of Example 2.3.7.)

EXERCISE 2.4.8. A rigid-analytic space X is said to be *quasi-compact* if it has an *admissible* covering consisting of finitely many affinoid opens. A morphism $f : X' \to X$ of rigid-analytic spaces is *quasi-compact* if there is an admissible covering of X by affinoid opens U_i such that each admissible open $f^{-1}(U_i)$ in X' (endowed with its open subspace structure) is quasi-compact for all i.

(1) Prove that if $X = \mathrm{Sp}(A)$ is affinoid and $f : X' \to X$ is quasi-compact, then for every affinoid subdomain $V = \mathrm{Sp}(B) \subseteq \mathrm{Sp}(A)$ the preimage $f^{-1}(V)$ is quasi-compact. In particular, X' is quasi-compact. (Hint: Use the Gerritzen-Grauert Theorem (Theorem 2.2.5) to reduce to the case when V is a rational domain.)

(2) Prove that if $f : X' \to X$ is quasi-compact in the sense of the above definition then for every quasi-compact admissible open $U \subseteq X$ the admissible open preimage $f^{-1}(U) \subseteq X'$ is quasi-compact. (Hint: Show that it suffices to treat the case when U is affinoid.)

(3) Assume that $f : X' \to X$ is a local isomorphism in the sense that there is an admissible open covering $\{U_i'\}$ of X' such that f maps U_i' isomorphically onto an admissible open $U_i \subseteq X$. If f is bijective and quasi-compact then prove that it is an isomorphism. Also give an example of such an f that is bijective but not an isomorphism.

We conclude with an exercise that demonstrates the power of Tate's theory by rescuing connectedness.

EXERCISE 2.4.9. A rigid-analytic space X is *disconnected* if there exists an admissible open covering $\{U, V\}$ of X with $U, V \neq \emptyset$ and $U \cap V = \emptyset$. Otherwise we say that X is *connected*. (Under this definition, $X = \emptyset$ is connected.)

(1) Using that the k-algebra of global functions on $\mathrm{Sp}(A)$ is A, prove that $\mathrm{Sp}(A)$ is connected if and only if A has no nontrivial idempotents. (This includes $A = 0$ as an uninteresting special case.) Equivalently, $\mathrm{Sp}(A)$ is connected if and only if $\mathrm{Spec}(A)$ is connected. Also show that if $A = A_1 \times A_2$ is a product of nonzero k-affinoid algebras then $\mathrm{Sp}(A_1)$ and $\mathrm{Sp}(A_2)$ are affinoid subdomains of $\mathrm{Sp}(A)$ (hint: impose inequalities on pointwise absolute values of idempotents) with $\{\mathrm{Sp}(A_1), \mathrm{Sp}(A_2)\}$ an admissible covering of $\mathrm{Sp}(A)$.

(2) Let X be a rigid-analytic space. For any $x \in X$, let U_x be the set of points $x' \in X$ that can be linked to x by a connected chain of finitely many connected admissible affinoid opens. That is, there exist connected admissible affinoid opens $U_1, \dots, U_n$ in X such that $x \in U_1$, $x' \in U_n$, and $U_i \cap U_{i+1} \neq \emptyset$ for all $1 \leq i < n$. Prove that the U_x's are admissible open in X and that for any $x_1, x_2 \in X$ either $U_{x_1} = U_{x_2}$ or $U_{x_1} \cap U_{x_2} = \emptyset$. Prove that the collection of U_x's (without repetition) is an admissible cover of X.

(3) Building on the previous part, prove that X is connected if and only if $\mathcal{O}_X(X)$ has no nontrivial idempotents (just like for locally ringed spaces). Moreover, in the context of the previous part, show that the U_x's are connected and that any connected admissible open in X is contained in some U_x. For this reason, we call the U_x's the *connected components* of X.

3. Coherent sheaves and Raynaud's theory

3.1. Coherent sheaves. To go further in the theory (e.g., to define closed immersions, separatedness, etc.), we need to discuss coherent sheaves. Kiehl extended Tate's methods to prove the following basic result (in which the module-finiteness hypothesis can be omitted).

THEOREM 3.1.1. *Let $X = \mathrm{Sp}(A)$ be an affinoid space over k, and let M be a finite A-module. The assignment $U \mapsto A_U \otimes_A M$ for affinoid subdomains $U \subseteq X$ uniquely extends to an $\mathcal{O}_X$-module $\widetilde{M}$. In particular, $M \simeq \widetilde{M}(X)$ and the natural map*

$$\mathrm{Hom}_{\mathcal{O}_X}(\widetilde{M}, \mathscr{F}) \to \mathrm{Hom}_A(M, \mathscr{F}(X))$$

induced by the global sections functor is bijective for any $\mathcal{O}_X$-module $\mathcal{F}$.

Kiehl also proved the following globalization.

THEOREM 3.1.2. *Let X be a rigid space, and $\{U_i\}$ an admissible affinoid cover. Let $\mathcal{F}$ be an $\mathcal{O}_X$-module. The following properties are equivalent:*

(1) *For every admissible affinoid open $V \subseteq X$, $\mathcal{F}|_V \simeq \widetilde{M_V}$ for a finite $\mathcal{O}_X(V)$-module M_V.*
(2) *For every i, $\mathcal{F}|_{U_i} \simeq \widetilde{M_i}$ for a finite $\mathcal{O}_X(U_i)$-module M_i.*

An $\mathcal{O}_X$-module $\mathcal{F}$ that satisfies these equivalent conditions is called *coherent*. As for locally noetherian schemes, coherence is inherited by kernels, cokernels, tensor products, and extensions. There is a naive approach to trying to define quasi-coherence, but it is not satisfactory, as we now explain. Motivated by the case of locally noetherian schemes, one may consider to define a quasi-coherent sheaf on X to be an $\mathcal{O}_X$-module that, locally on the space (i.e., on the constituents of an admissible covering) can be expressed as a direct limit of coherent sheaves. (This is the definition suggested in [**FvP**, Exer. 4.6.7].) It can be shown that this property is preserved under the formation of kernels, cokernels, extensions, tensor products, and direct limits, and that it suffices to work with coherent *subsheaves* in the local direct limit process used in the definition. However, it is generally not true that on an arbitrary admissible affinoid open in the space such a sheaf is a direct limit of coherent sheaves (thereby answering in the negative the "open problem" mentioned in [**FvP**, Exer. 4.6.7]). More specifically, Gabber has given an example of a sheaf of modules $\mathcal{F}$ on the closed unit disk $\mathbf{B}^1$ such that $\mathcal{F}$ is locally a direct limit of coherent sheaves but with nonzero degree-1 sheaf cohomology, so $\mathcal{F}$ cannot be expressed as a direct limit of coherent sheaves over the entire affinoid space (because the formation of sheaf cohomology commutes with the formation of direct limits on an affinoid rigid space; this is left as an exercise for readers who are familiar with the Čech to derived functor cohomology spectral sequence, which has to be carried over to the rigid-analytic case). There may be a better definition of quasi-coherence that enjoys the stability properties under basic operations as in algebraic geometry and is equivalent over an affinoid space to some module-theoretic data (perhaps with topological structure) over the coordinate ring, but I do not know what such a definition should be.

EXERCISE 3.1.3. If $X = \mathrm{Sp}(A)$ is affinoid and M is a finite A-module, prove that $\widetilde{M}_x \simeq M \otimes_A \mathcal{O}_{X,x}$ for all $x \in X$. Using that $\mathcal{O}_{X,x}$ is a local noetherian ring with the same completion as $A_{\mathfrak{m}_x}$, deduce that if $m \in M$ vanishes in $\widetilde{M}_x$ for all $x \in X$ then $m = 0$. More globally, deduce that a global section of a coherent sheaf on any rigid-analytic space vanishes if and only if it vanishes in all stalks on the space. (This property is special to coherent sheaves; it fails for general abelian sheaves.)

EXAMPLE 3.1.4. Let $X = \mathrm{Sp}(B) \to \mathrm{Sp}(A) = Y$ be a map of k-affinoid spaces and let I be the kernel of the natural map $B\widehat{\otimes}_A B \to B$ induced by multiplication. Then I/I^2 is a finite module over $(B\widehat{\otimes}_A B)/I = B$. This is generally not the same as the module of relative algebraic Kähler differentials (which is typically huge, since the usual tensor product $B \otimes_A B$ may not be noetherian). This finite B-module gives rise to a coherent sheaf $\Omega^1_{X/Y}$ on X. There is an $\mathcal{O}_Y$-linear derivation

$\mathscr{O}_X \to \Omega^1_{X/Y}$ that can be globalized in accordance with local formulas similar to the situation in algebraic geometry.

EXAMPLE 3.1.5. A map of rigid spaces $f : X' \to X$ is a *closed immersion* if there exists an admissible affinoid covering $\{U_i\}$ of X such that $U'_i = f^{-1}(U_i)$ is affinoid and the map of affinoids $U'_i \to U_i$ corresponds to a surjection on coordinate rings. In that case it can be proved that for every admissible affinoid open $U \subseteq X$ the preimage $f^{-1}(U)$ is affinoid and $f^{-1}(U) \to U$ corresponds to a surjection on coordinate rings, and moreover that $\mathscr{O}_{X'} \to f_*(\mathscr{O}_X)$ is surjective with coherent kernel $\mathscr{I}$. In fact, this coherent ideal sheaf $\mathscr{I}$ determines the map $f : X' \hookrightarrow X$ up to unique X-isomorphism, and conversely every coherent ideal sheaf $\mathscr{I} \subseteq \mathscr{O}_X$ arises in this way (by gluing $\mathrm{Sp}(A/I)$'s for admissible affinoid open $\mathrm{Sp}(A) \subseteq X$ with $I = \mathscr{I}(\mathrm{Sp}(A)) \subseteq \mathscr{O}_X(\mathrm{Sp}(A)) = A$).

Using closed immersions and quasi-compactness, we can carry over some notions from algebraic geometry involving diagonal maps:

DEFINITION 3.1.6. A map $f : X \to Y$ of rigid spaces is *separated* if the diagonal map $\Delta_f : X \to X \times_Y X$ is a closed immersion. In case $Y = \mathrm{Sp}(k)$, we say that X is *separated*. If Δ_f is merely quasi-compact (equivalently, the overlap of any two affinoid opens in X over a common affinoid open in Y is quasi-compact) then f is *quasi-separated*.

A map of rigid spaces $f : X' \to X$ is *finite* if there exists an admissible affinoid covering $\{U_i\}$ of X such that each $U'_i = f^{-1}(U_i)$ is affinoid and the map of coordinate rings $\mathscr{O}_X(U_i) \to \mathscr{O}_{X'}(U'_i)$ is module-finite. (In this case it can then be shown that $f^{-1}(U)$ is affinoid with coordinate ring finite over that of U for any affinoid open $U \subseteq X$.)

REMARK 3.1.7. There is no rigid-analytic notion of "affinoid morphism" akin to the concept of affine morphism in algebraic geometry. The problem is that there is no good analogue of Serre's cohomological criterion for affineness; see [**Liu**] for counterexamples (i.e., quasi-compact and separated non-affinoid spaces whose coherent sheaves all have vanishing higher cohomology). Interestingly, [**Liu**] also gives an example of a quasi-compact and separated non-affinoid space admitting a finite surjection from an affinoid space, in contrast with a theorem of Chevalley in the case of schemes (a separated scheme admitting a finite surjection from an affine scheme is necessarily affine).

EXERCISE 3.1.8. Copy the proof from the case of schemes to show that if $f : X \to Y$ is a map of rigid spaces and Y is separated then for any admissible affinoid opens $U \subseteq Y$ and $V \subseteq X$, the overlap $V \cap f^{-1}(U)$ is affinoid. (Hint: Show that the graph map $(1, f) : X \to X \times Y$ is a closed immersion and consider $U \times V \subseteq X \times Y$.) Taking f to be the identity map, deduce that an overlap of finitely many admissible affinoid opens in a separated rigid space is again affinoid.

EXERCISE 3.1.9. Let X be a rigid space over k. Prove that X is quasi-separated if and only if it has an admissible covering by affinoid opens U_i such that each overlap $U_i \cap U_j$ is quasi-compact, in which case $U \cap V$ is quasi-compact for any quasi-compact admissible opens U and V in X. Prove that if X is a quasi-separated rigid space then for any finite collection $\{U_i\}$ of quasi-compact admissible opens in X, the union $U = \cup U_i$ is an admissible open in X for which the U_i's are an admissible covering.

Give an example of a quasi-compact rigid space that is not quasi-separated. Can you find such an example for which there is a finite union of admissible affinoid opens that is not an admissible open subset?

REMARK 3.1.10. On the category of quasi-separated rigid spaces over k one can use a gluing procedure to define change of base field functors $X \rightsquigarrow X_K$ from rigid spaces over k to ones over K for any analytic extension field K/k, compatibly with fiber products. This is very useful, such as with $k = \mathbf{Q}_p$ and $K = \mathbf{C}_p$, or more generally $K = \overline{k}^{\wedge}$ for any k. (It is a general fact that if $[\overline{k} : k]$ is infinite then $\overline{k}$ is not complete, so $\overline{k}^{\wedge}$ contains elements that are transcendental over k in such cases; a notable such example is $k = \mathbf{Q}_p$.) The idea underlying the definition of this functor is to first define it in the affinoid case via the operation $A \rightsquigarrow K\widehat{\otimes}_k A$ from k-affinoid algebras to K-affinoid algebras compatibly with completed tensor products, and to then globalize by gluing in the separated case (since an overlap of affinoids is affinoid). The quasi-separated case is obtained by another repetition of this process, using that an overlap of admissible affinoid opens in a quasi-separated space may fail to be affinoid but is at least quasi-compact and separated.

We omit the details, except to remark that this is merely a "construction" and it is not really a fiber product or characterized by an abstract universal property as in the case of schemes if K/k has infinite degree because in such cases rigid spaces over K cannot be mapped to rigid spaces over k in any reasonable way: MaxSpec is not functorial (with respect to pullback of prime ideals) between k-affinoid algebras and K-affinoid algebras! This lack of functoriality is a real nuisance, but in the approaches of Berkovich and Huber there are many more points in the underlying spaces and one can view change of base field functors as actual fiber products (and more specifically one can consider analytic spaces over k and K as part of a common category). For purposes of analogy, consider the classical concepts of a variety over $\overline{\mathbf{Q}}$ and over $\mathbf{C}$ (using only "closed points" from the scheme perspective; no "universal domain") and try to formulate the idea of the map $V_{\mathbf{C}} \to V$ for a $\overline{\mathbf{Q}}$-variety V: the trascendental points in $V_{\mathbf{C}}$ have nowhere to go in V since the variety V does not have generic points. This is due to the use of MaxSpec in classical algebraic geometry, and exactly the same problem arises in rigid geometry (i.e., rigid spaces lack "enough points").

EXERCISE 3.1.11. Since elements of $k[X_1, \ldots, X_n]$ have only finitely many nonzero coefficients whereas elements of $k\langle X_1, \ldots, X_n \rangle$ can have infinitely many nonzero coefficients, change of the base field in rigid geometry exhibits some features that may be surprising from the viewpoint of algebraic geometry. For example, it can happen that an affinoid space $X = \mathrm{Sp}(A)$ over k remains reduced after any finite extension on k but not after some infinite-degree analytic extension on k. (This never happens for algebraic schemes over a field.) Indeed, consider the following example. Let k be a non-archimedean field of characteristic $p > 0$ such that $[k : k^p]$ is infinite.

(1) Show that an example of such a k is $F((y))$ with the y-adic absolute value, where F is a field of characteristic p such that $[F : F^p]$ is infinite. Find such an F.

(2) Show that there exists an infinite sequence $\{a_n\}$ in k tending to 0 such that $|a_n| \leq 1$ for all n and the $a_n^{1/p}$ generate an infinite-degree extension of k.

(3) Choose such a sequence $\{a_n\}$, and let $f = \sum a_n X^{np} \in T_1$ and $A = T_1[Y]/(Y^p - f)$ (which is T_1-finite, hence affinoid). Prove that $A \simeq T_1\langle Y\rangle/(Y^p - f)$ and that $A \otimes_k k'$ is a domain for any k'/k of finite degree, but $A\widehat{\otimes}_k k^{p^{-1}}$ is not reduced. (Also show that $k^{p^{-1}}$ is complete!)

(4) Assume $p \neq 2$ and let $B = k\langle X, Y, t\rangle/(t^2 - (Y^p - f))$. Prove that $B \otimes_k k'$ is a normal domain for any finite-degree extension k'/k but that $B\widehat{\otimes}_k k^{p^{-1}}$ is reduced and not normal. It can be shown that $B\widehat{\otimes}_k K$ is reduced for any analytic extension field K/k.

EXERCISE 3.1.12. This exercise addresses some subtle features of the general concept of an admissible open subset, even within an affinoid space. Let X be a quasi-separated rigid space over k and let K/k be an analytic extension field. Let $i : U \to X$ be the natural inclusion from an admissible open $U \subseteq X$. We get an induced map of rigid spaces $i_K : U_K \to X_K$ over K. Is this an isomorphism onto an admissible open? To appreciate where the difficulties lie, consider some special cases as follows.

(1) Assume that $X = \mathrm{Sp}(A)$ is affinoid and that U is an affinoid subdomain. Prove that $U_K \to X_K$ is an isomorphism onto an affinoid subdomain. Do this by first showing that U_K has admissible open image in X_K via the Gerritzen–Grauert theorem (Theorem 2.2.5), and then work with the scalar extension of the coordinate ring A_U of U. (Exercise 2.4.8(2) will be useful here.) The reason that this special case requires serious input (Theorem 2.2.5) is that the universal property for $U \subseteq X$, even when formulated in purely algebraic terms via k-affinoid algebras, only involves maps with k-affinoid spaces, and not also K-affinoid spaces (if $[K : k]$ is infinite).

(2) If X is affinoid and U is a quasi-compact admissible open, use the previous part to show that i_K is an isomorphism onto a quasi-compact admissible open in X_K.

(3) Prove an affirmative answer if i is a quasi-compact map. The key difficulty in the general case appears to be to determine if i_K has admissible open image in X_K.

3.2. Cohomology, properness, and flatness. No discussion of coherent sheaves would be complete without addressing their cohomology, especially in the proper case. We first make some general observations concerning how to define sheaf cohomology on rigid spaces. Despite the problematic nature of stalks at points on rigid spaces when working with general abelian sheaves, one can adapt some methods of Grothendieck to prove that the category of abelian sheaves and the category of $\mathscr{O}_X$-modules on a rigid space X each have enough injectives, so we may (and do) define sheaf cohomology via derived functors in both cases. The concept of flasque sheaf as traditionally used on ringed spaces is a bit problematic in the general rigid-analytic case, but nonetheless the following result can be proved (and it is left as an exercise just for those whose taste is inclined toward such questions!):

EXERCISE 3.2.1. Let X be a quasi-separated rigid space. Prove that for an $\mathscr{O}_X$-module $\mathscr{F}$, the natural map from its sheaf cohomology in the sense of $\mathscr{O}_X$-modules to its sheaf cohomology in the sense of abelian sheaves is an isomorphism. More specifically, prove that an injective $\mathscr{O}_X$-module has vanishing higher sheaf cohomology in the sense of abelian sheaves, and that restriction to an admissible open preserves the property of being an injective sheaf of modules (resp. an injective abelian sheaf).

Thus, on quasi-separated spaces the theory of sheaf cohomology via derived functors presents no ambiguities. Kiehl showed that there is a good cohomology theory for coherent sheaves on rigid spaces. His work, coupled with some auxiliary arguments, gives the following result.

THEOREM 3.2.2. *Let X be a rigid space and $\mathscr{F}$ a coherent sheaf on X.*

(1) (*Acyclicity theorem for coherent sheaves*). *If X is affinoid and $\mathfrak{U}$ is a finite covering of X by admissible affinoid opens then the Čech cohomology $\mathrm{H}^i(\mathfrak{U}, \mathscr{F})$ vanishes for all $i > 0$. Moreover, the sheaf cohomology (defined via derived functors) $\mathrm{H}^i(X, \mathscr{F})$ vanishes for all $i > 0$.*

(2) *If X is a quasi-compact and separated rigid space and $\mathfrak{U}$ is a finite admissible affinoid open covering of X then the natural map $\mathrm{H}^i(\mathfrak{U}, \mathscr{F}) \to \mathrm{H}^i(X, \mathscr{F})$ is an isomorphism for all i.*

We next turn to the definition of properness. In view of the nature of the Tate topology, the condition of universal closedness that is used for schemes is not the right one to use in rigid geometry. Instead, we adapt a formulation similar to one that works in the complex-analytic case.

DEFINITION 3.2.3. A map $f : X \to Y$ of rigid spaces is *proper* if it is separated and quasi-compact and there exists an admissible affinoid open covering $\{U_i\}$ of Y and a pair of finite (necessarily admissible) affinoid open coverings $\{V_{ij}\}_{j \in J_i}$ and $\{V'_{ij}\}_{j \in J_i}$ (same index set J_i of j's!) of $f^{-1}(U_i)$ such that two conditions hold: $V_{ij} \subseteq V'_{ij}$ for all j, and for all $j \in J_i$ there is an $n \geq 1$ and a closed immersion $V'_{ij} \hookrightarrow U_i \times \mathbf{B}^n$ over U_i such that $V_{ij} \subseteq U_i \times \{|t_1|, \ldots, |t_{n_i}| \leq r\}$ for some $0 < r < 1$ with $r \in \sqrt{|k^\times|}$. (Equivalently, by the Maximum Modulus Principle, we can replace "$\leq r$" with "< 1".)

The condition on the inclusion $V_{ij} \subseteq V'_{ij}$ over U_i in Definition 3.2.3 is called *relative compactness* of V_{ij} in V'_{ij} over U_i. It is a replacement for saying that V_{ij} has U_i-proper closure in V'_{ij} in ordinary topology. Algebraically, if the coordinate rings of U_i, V_{ij}, and V'_{ij} are A_i, B_{ij}, and B'_{ij} then the condition is that the k-affinoid A_i-algebra B'_{ij} can be expressed as a quotient of a relative n-variable Tate algebra over A_i such that the images $b'_1, \ldots, b'_n$ in B'_{ij} of the standard variables $X_1, \ldots, X_n$ from the Tate algebra have images $b_1, \ldots, b_n \in B_{ij}$ with sup-norms less than 1. (That is, all $b'_r|_{\mathrm{Sp}(B_{ij})}$ have sup-norm less than 1 for all $j \in J_i$.)

EXERCISE 3.2.4. In the case of a submersion of complex manifolds, use the implicit function theorem to describe the condition of properness of the underlying map of topological spaces in terms similar to the definition of properness in the rigid-analytic case.

EXAMPLE 3.2.5. Let $X = \mathbf{P}_k^{n,\mathrm{an}}$ be rigid-analytic projective n-space. This can be constructed by gluing $n + 1$ affine n-spaces $U_0, \ldots, U_n$ exactly as in algebraic geometry so as to establish a universal property much like in algebraic geometry. It is therefore also identified with the analytification of the algebraic scheme $\mathbf{P}_{\mathrm{Spec}(k)}^n$. But this rigid-analytic space can likewise be constructed by analogous gluing using closed unit polydiscs $\mathbf{B}_k^n$! More precisely, let V_j be the closed unit polydisc centered at the origin in the jth standard affine n-space U_j and let V'_j be the closed polydisc centered at the origin with radius $1/|c|$ for some $c \in k^\times$ with $|c| < 1$. We have that V_j is relatively compact in V_j' for all j (over $\mathrm{Sp}(k)$), since identifying V_j' with a closed unit polydisc via scaling its coordinates by $|c|$ carries the affinoid subdomain $V_j \subseteq V_j'$ isomorphically onto the closed subdisc of radius $|c| < 1$ centered at the origin. But $\{V_j\}$ and $\{V_j'\}$ are each admissible covers of X, so X is proper.

The definition of properness is tricky to work with because the requirement on the U_i's does not obviously apply to all open affinoids in Y. For example, if Y is affinoid then it is not clear if one can find a pair of finite affinoid open coverings $\{V_j\}$ and $\{V_j'\}$ of X such that V_j is relatively compact in V_j' over Y for all j. To appreciate the difficulties of the situation, consider the following exercise.

EXERCISE 3.2.6. Prove that properness of a morphism $X \to Y$ is preserved by base change, as well as any change of the base field functor (assuming in the latter case that Y, and hence X, is quasi-separated, so the change of base field functors can be constructed by gluing across affinoids). Also show that if $Z \to X$ is a closed immersion and X is proper over a rigid-analytic space Y then Z is also proper over Y. (In particular, any projective rigid-analytic space is proper, by Example 3.2.5.) But try to prove that properness is preserved under composition. It is not easy (and was not proved in general until quite recently, by Temkin via Berkovich spaces)!

In algebraic geometry, proper maps enjoy some important cohomological properties, such as the theorem on coherence of higher direct images. Likewise, one has Grothendieck's theory of cohomology and base change, as well as his theorem on formal functions, that relates fibral cohomology to the structure of higher direct images. Results along these lines in the rigid-analytic case were proved by Kiehl, adapting both analytic techniques used to prove such results in the complex-analytic case as well as algebraic techniques used by Grothendieck. We just record the first of these results:

THEOREM 3.2.7. *If $f : X \to Y$ is a proper map of rigid spaces and $\mathscr{F}$ is a coherent sheaf on X then the higher direct image sheaves $\mathrm{R}^i(f_*)(\mathscr{F})$ on Y are coherent. In particular, if X is proper over $\mathrm{Sp}(k)$ then $\mathrm{H}^i(X, \mathscr{F})$ is finite-dimensional over k for all coherent sheaves $\mathscr{F}$ on X and all i.*

REMARK 3.2.8. As for ringed spaces, higher direct images may be computed via sheafified cohomology. However, it is worth noting that one does not see a priori (as one does for schemes) that for affinoid Y and a proper map $f : X \to Y$ the natural map $\delta_{i,\mathscr{F}} : \mathrm{H}^i(X, \mathscr{F}) \to \Gamma(Y, \mathrm{R}^i(f_*)(\mathscr{F}))$ is an isomorphism for coherent $\mathscr{F}$ on X. In fact, Kiehl's proof does not directly show this; the proof establishes such a result only over affinoids in an admissible covering of Y as in the definition of properness for f. But a posteriori such an isomorphism claim is true over the entire affinoid base Y. The point is that once Kiehl's coherence theorem for higher direct images is known, then for affinoid Y it follows from Kiehl's acyclicity theorem for coherent sheaves on Y that the Leray spectral sequence

$$E_2^{p,q} = \mathrm{H}^p(Y, \mathrm{R}^q(f_*)(\mathscr{F})) \Rightarrow \mathrm{H}^{p+q}(X, \mathscr{F})$$

degenerates (with $E_2^{p,q} = 0$ for $p > 0$). All that survives are the maps $\delta_{p,\mathscr{F}}$ as edge maps, so these are isomorphisms. In particular, if $Y = \mathrm{Sp}(A)$ and X is Y-proper then $\mathrm{H}^p(X, \mathscr{F})$ is a finite A-module for every coherent sheaf $\mathscr{F}$ on X and every $p \geq 0$.

EXAMPLE 3.2.9. The theory of proper maps provides the framework for Serre's GAGA theorems. We now explain this just in the case of proper objects over a field, though there are variants for proper morphisms to any rigid-analytic space. Let X be a proper k-scheme. It is a non-obvious fact (proved in general only recently, by Temkin) that the associated rigid-analytic space X^{an} is proper over $\mathrm{Sp}(k)$ in the sense defined above; in the special case that X is projective this is trivial because analytification carries closed immersions to closed immersions, and we have seen that projective rigid-analytic spaces are proper.

The GAGA theorems assert that the functor $X \rightsquigarrow X^{\mathrm{an}}$ from proper k-schemes to proper rigid-analytic spaces is fully faithful, and that for a fixed such X the analytification functor $\mathscr{F} \rightsquigarrow \mathscr{F}^{\mathrm{an}}$ on $\mathscr{O}_X$-modules is an equivalence of categories from the category of coherent sheaves on X to the category of coherent sheaves on X^{an} and the natural comparison morphism $\mathrm{H}^i(X, \mathscr{F}) \to \mathrm{H}^i(X^{\mathrm{an}}, \mathscr{F}^{\mathrm{an}})$ is an isomorphism for all i and all coherent $\mathscr{F}$ on X. The proof for projective k-schemes goes almost exactly as in Serre's original arguments over $\mathbf{C}$, except that one has to be careful about the presence of non-rational points (if k is not algebraically closed) and about admissibility issues (since Serre uses some pointwise compactness arguments). Also, Grothendieck's generalization to the proper case via Chow's Lemma carries over to the rigid-analytic case essentially verbatim as well.

Let us now pose some exercises that illustrate shortcomings of Tate's theory, and whose only known solutions are via later approaches developed by Raynaud, Berkovich, and others.

EXERCISE 3.2.10. Let $f : X' \to X$ be a map of rigid spaces over k. Say that f is *flat* if the local map $\mathscr{O}_{X,f(x')} \to \mathscr{O}_{X',x'}$ is flat for every $x' \in X'$. If in addition f is surjective then say it is *faithfully flat*. Prove that if $X' = \mathrm{Sp}(A')$ and $X = \mathrm{Sp}(A)$ are affinoid then f is flat (resp. faithfully flat) if and only if the map of coordinate rings $A \to A'$ is flat (resp. faithfully flat).

In algebraic geometry, it is a basic fact that a flat map between algebraic k-schemes is an open map. So it is natural to ask if a flat map $f : X' \to X$ has admissible open image. This is too much to expect to be true in general (due to the subtle nature of admissible opens), but if X and X' are quasi-compact and quasi-separated then it is a more reasonable question (since at least finite unions of admissible open affinoids in such spaces are necessarily admissible open). This problem has an affirmative answer, but it seems hopeless to attack this by the methods of rigid geometry. Raynaud's theory of formal scheme models provides the right techniques (to ultimately reduce this openness problem to the known analogue in algebraic geometry for schemes of finite type over the residue field $\widetilde{k}$!).

EXERCISE 3.2.11. If $f : X' \to X$ is a flat map between separated rigid spaces over k and if K/k is an analytic extension field, then is $f_K : X'_K \to X_K$ flat? Reduce this to the problem of showing that if $A \to A'$ is a flat map of k-affinoid algebras then $K \widehat{\otimes}_k A \to K \widehat{\otimes}_k A'$ is flat. This affinoid special case appears to be beyond the reach of the methods of rigid geometry. Once again, Raynaud's theory of formal scheme models (which in this case requires using *non-affine* formal schemes, even in the case of affinoid rigid spaces) provides the methods needed to give an affirmative answer (by reducing the problem to the known case of preservation of flatness under base change for schemes over suitable quotients of the valuation ring R of k).

EXERCISE 3.2.12. If you are familiar with descent theory for schemes, scheme maps, and quasi-coherent sheaves, try to formulate an analogue of faithfully flat and quasi-compact descent theory for rigid spaces, their maps, and coherent sheaves on them. How much can you prove in this direction? Without Raynaud's theory, you will probably find yourself not able to prove very much!

EXERCISE 3.2.13. (if familiar with étale cohomology for schemes) How would you define the étale site on a rigid space? (An *étale map* $f : X' \to X$ of rigid spaces can be defined in several equivalent ways, just like in algebraic geometry, one definition being flatness and the requirement that for all $x \in X$ the fiber $f^{-1}(x)$ over $k(x)$ is a disjoint (admissible) union of $\mathrm{Sp}(k')$ for various finite separable extensions $k'/k(x)$.) Admissibility makes things more subtle than one may have expected. The works of Huber and Berkovich were inspired in part by the aim of developing a good theory of étale cohomology in non-archimedean geometry. Their theories are not the same, but enjoy certain compatibilities and each has its own merits for various purposes.

3.3. Raynaud's formal models. The remainder of this lecture will be concerned with explaining some basic aspects of Raynaud's theory of formal scheme models for rigid spaces. In the final two lectures we will discuss Berkovich's theory. Roughly speaking, Raynaud's theory is very useful for "algebraic" questions (flatness, fiber dimension, behavior of properties with respect to base change, etc.) whereas Berkovich's theory is useful for deeper cohomological questions and for carrying out "pointwise" intuition in a manner that sometimes cannot be done within the framework of usual rigid geometry. (We will see some striking examples of this in later lectures.)

Recall that we let R denote the valuation ring of k. (This is noetherian if and only if k is discretely-valued.) A *topologically finitely presented* (tfp) *R-algebra* is an R-algebra of the form

$$\mathscr{A} = R\{X_1, \ldots, X_n\}/\mathscr{I}$$

where $\mathscr{I}$ is a finitely generated ideal in the ring $R\{X_1, \ldots, X_n\}$ of restricted powers series in n variables over R: power series $\sum a_J X^J$ with coefficients $a_J \in R$ such that $a_J \to 0$ as $\|J\| \to \infty$. If it is R-flat as well then it is called an *admissible R-algebra*.

EXERCISE 3.3.1. For $\pi \in R$ with $0 < |\pi| < 1$, show that $R\{X_1, \ldots, X_n\}$ is the π-adic completion of $R[X_1, \ldots, X_n]$. If k is discretely-valued, so R is a discrete valuation ring, this is the same as the $\mathfrak{m}$-adic completion, with $\mathfrak{m}$ the maximal ideal of R. But show that if k is algebraically closed (e.g., $k = \mathbf{C}_p$) then $\mathfrak{m}^2 = \mathfrak{m}$, and deduce that in such cases the $\mathfrak{m}$-adic completion is $\tilde{k}[X_1, \ldots, X_n]$!

In general (any k), show that

$$k \otimes_R R\{X_1, \ldots, X_n\} = R\{X_1, \ldots, X_n\}[1/\pi] \simeq T_n(k)$$

is the n-variable Tate algebra over k. Deduce that if $\mathscr{A}$ is a tfp R-algebra then $A = k \otimes_R \mathscr{A}$ is a k-affinoid algebra, and if $\mathscr{M}$ is a finitely generated $\mathscr{A}$-module then $k \otimes_R \mathscr{M}$ is a finite A-module.

Using special properties of the valuation rings R (even in the non-noetherian case), one can prove the following useful lemma.

LEMMA 3.3.2. *Let $\mathscr{A}$ be a tfp R-algebra. It is π-adically separated and complete, and the ideal $\mathscr{A}[\pi^\infty]$ of π-power torsion elements is finitely generated. Also, if I is any ideal in T_n then its intersection with $R\{X_1, \ldots, X_n\}$ is finitely generated.*

EXERCISE 3.3.3. Prove that an R-module is flat if and only if it is torsion-free (even in the non-noetherian case). Deduce from Lemma 3.3.2 that every k-affinoid algebra A has the form $k \otimes_R \mathscr{A}$ with $\mathscr{A}$ an admissible R-algebra. (In particular, the natural map $\mathscr{A} \to A$ is injective.)

The preceding exercise shows that every k-affinoid algebra A admits a flat integral model in the sense of admissible R-algebras (i.e., $A \simeq k \otimes_R \mathscr{A}$ with $\mathscr{A}$ admissible over R.) Rather less evident is how this can globalize to non-affinoid rigid spaces over k. In fact, even in the affinoid setting there are further non-obvious questions concerning integral models: if M is a finite A-module, can we choose the admissible R-algebra $\mathscr{A}$ so that M arises from a finitely presented $\mathscr{A}$-module $\mathscr{M}$? And if M is A-flat then can we arrange that $\mathscr{M}$ is $\mathscr{A}$-flat? The key to these latter affinoid questions is to attack them within a more global theory of formal scheme models for rigid spaces. This is a remarkable achievement of Raynaud, as we shall now see. First we develop the required notions from the theory of formal schemes.

EXERCISE 3.3.4. For a tfp R-algebra $\mathscr{A}$, consider $\overline{X} = \mathrm{Spec}(\mathscr{A}/\mathfrak{m}\mathscr{A})$ as a topological space. For any $\pi \in R$ with $0 < |\pi| < 1$ and any $n \geq 1$, explain how to naturally identify $\overline{X}$ with $X_n = \mathrm{Spec}(\mathscr{A}/(\pi^n))$ as topological spaces. For any $f \in \mathscr{A}$, the non-vanishing locus $\overline{X}_f$ of f in X_n has coordinate ring $\mathscr{A}[1/f]/(\pi^n)$. Show that this ring only depends on n and f mod $\mathfrak{m}$; the inverse limit of these coordinate rings is the π-adic completion $\mathscr{A}_{\{f\}}$ of $\mathscr{A}[1/f]$. Prove that this completion is isomorphic to $\mathscr{A}\{X\}/(1 - fX)$, and so is a tfp R-algebra.

Explain why this completion only depends on the non-vanishing locus $\overline{X}_f$ of f in $\overline{X}$ (in particular, it is naturally independent of π and of the choice of f giving rise to it). Thus, it is well-defined to assign $\mathscr{A}_{\{f\}}$ to this open subset of $\overline{X}$. Prove that this assignment satisfies the sheaf axioms for the covering of one such open subset by others of this type. (Hint: use the existence of structure sheaves on the schemes X_n.) Deduce that this uniquely extends to a sheaf of R-algebras $\mathscr{O}_{\mathscr{A}}$ on $\overline{X}$, and that the stalks of this sheaf are local rings.

DEFINITION 3.3.5. With notation as in the preceding exercise, the *tfp affine formal scheme* $\mathrm{Spf}(\mathscr{A})$ over R is the locally ringed space $(\overline{X}, \mathscr{O}_{\mathscr{A}})$. (Note that its R-algebra of global functions is $\mathscr{A}$.) A *tfp formal scheme* over R is a quasi-compact locally ringed space $(\mathfrak{X}, \mathscr{O}_{\mathfrak{X}})$ with R-algebra structure sheaf such that it is locally isomorphic (respecting the R-structure) to a tfp affine formal scheme over R. *Morphisms* of tfp formal schemes are morphisms of locally ringed spaces that respect the R-structure.

An *admissible formal R-scheme* is a tfp formal R-scheme $\mathfrak{X}$ whose local rings are R-flat. (It is equivalent, but not obviously so in the non-noetherian case, to require that the coordinate rings of some open affine covering to be R-flat, or for the coordinate ring of every open affine to be R-flat.)

Observe that if $\mathfrak{X}$ is a tfp formal R-scheme then its underlying topological space is noetherian (even if R is not noetherian!), since it is identified with the underlying space of a scheme of finite type over the residue field $\widetilde{k}$.

EXERCISE 3.3.6. Prove that the assignment $\mathscr{A} \rightsquigarrow \mathrm{Spf}(\mathscr{A})$ is a fully faithful contravariant functor from the category of tfp R-algebras to the category of tfp formal schemes over R. For any $f \in \mathscr{A}$, show that the set $\{f \neq 0\}$ in $\mathrm{Spf}(\mathscr{A})$ is an open subset whose induced open formal subscheme structure is naturally isomorphic to $\mathrm{Spf}(\mathscr{A}_{\{f\}})$; these are called *basic open affines* in $\mathrm{Spf}(\mathscr{A})$.

If $\mathfrak{X}$ is a tfp formal R-scheme then a *formal open affine* in $\mathfrak{X}$ is an open subspace $(\mathfrak{U}, \mathscr{O}_{\mathfrak{X}}|_{\mathfrak{U}})$ that is a tfp formal affine scheme. Prove that any such $\mathfrak{X}$ has a base of open subsets that are formal open affines (when endowed with their open subspace structure as locally ringed spaces).

By using finitely presented modules over formal open affines, one can create a good theory of coherent sheaves on such formal schemes (requiring some work in the non-noetherian case). Globally, a coherent sheaf on such an $\mathfrak{X}$ is simply an $\mathscr{O}_{\mathfrak{X}}$-module that is locally finitely presented. (The hard part in the non-noetherian case is to justify that this latter definition is equivalent to a concrete one using modules in the affine case.)

One may be annoyed by the non-noetherian case, but in fact it is a great feature of the theory: if we want to study base change from $\mathbf{Q}_p$ to $\mathbf{C}_p$ for rigid spaces and prove results about preservation of flatness and other kinds of properties (fiber dimension, etc.), it is very useful to have the theory of these formal schemes available over both $\mathbf{Z}_p$ and the (non-noetherian) valuation ring of $\mathbf{C}_p$.

EXERCISE 3.3.7. Let $f : X' \to X$ be a map between quasi-compact and quasi-separated rigid spaces over k. Let K/k be an analytic extension field. If all fibers of f have dimension $\leq d$, prove the same is true for $f_K : X'_K \to X_K$. (This exercise is meant to make you appreciate how hard the problem is from a classical point of view when $[K : k]$ is not finite; without using further deep results in Raynaud's theory, it is probably too hard to solve.)

The interest in tfp formal R-schemes is revealed in the next exercise.

EXERCISE 3.3.8. Explain how the assignment $\mathrm{Spf}(\mathscr{A}) \rightsquigarrow \mathrm{Sp}(k \otimes_R \mathscr{A})$ from tfp formal affine schemes over R to affinoid rigid spaces over k is functorial, and prove that it carries Zariski-open immersions to quasi-compact admissible opens. (Hint: First show that basic open affines go over to Laurent domains!)

Once you have handled the affine case, prove that this construction uniquely extends to a functor $\mathfrak{X} \rightsquigarrow \mathfrak{X}_k$ from tfp formal R-schemes to rigid spaces over k such that open immersions go over to admissible opens and fiber products are preserved (in particular, there is compatibility with overlaps of opens). This is Raynaud's *generic fiber* functor.

EXAMPLE 3.3.9. Let X be an R-scheme of finite presentation. A rigid-analytic space can be associated to X in *two ways*. First, we can form the analytification X_k^{an} of the generic fiber $X_k = X \otimes_R k$. Second, if we let $\mathfrak{X}$ be its π-adic completion (a tfp formal R-scheme, admissible if X is R-flat), then we can form the Raynaud generic fiber $\mathfrak{X}_k$. For example, if $X = \mathbf{A}_R^n$ then $X_k^{\mathrm{an}} = \mathbf{A}_k^{n,\mathrm{an}}$ is rigid-analytic affine n-space whereas $\mathfrak{X} = \mathrm{Spf}(R\{t_1, \ldots, t_n\})$ is formal affine n-space over $\mathrm{Spf}(R)$, so $\mathfrak{X}_k = \mathbf{B}_k^n$, which is much smaller than X_k^{an}. The natural quasi-compact open immersion $\mathbf{B}_k^n \hookrightarrow \mathbf{A}_k^{n,\mathrm{an}}$ can be uniquely generalized for any such X to a functorial morphism of rigid spaces $i_X : \mathfrak{X}_k \to X_k^{\mathrm{an}}$ that is compatible with fiber products and recovers the canonical map in case $X = \mathbf{A}_R^1$. Moreover, as long as X is R-separated, it follows from the valuative criterion for separatedness (and auxiliary arguments) that this map is a quasi-compact open immersion, and (via the valuative criterion for properness) that it is an isomorphism when X is R-proper. For example, if $X = \mathbf{P}_R^n$ then X_k^{an} is rigid-analytic projective n-space from the viewpoint of gluing $n + 1$ affine n-spaces whereas $\mathfrak{X}_k$ is projective n-space from the viewpoint of gluing $n + 1$ closed unit polydiscs of dimension n (and i_X is the canonical isomorphism between these two gluing constructions).

EXERCISE 3.3.10. Show that if $\mathfrak{X}$ is quasi-compact then the associated generic fiber rigid space $\mathfrak{X}_k$ is quasi-compact and quasi-separated. Also show that on the category of admissible formal $\mathfrak{X}$'s the generic-fiber functor is faithful (i.e., if two maps between such formal schemes agree on the associated rigid spaces then they must be the same map). For this faithfulness aspect you may wish to just treat the case of R-separated admissible formal schemes over R (with R-separatedness defined in an evident manner); the general case requires some notions (rig-points) that we have not developed.

The preceding exercise naturally raises the question of whether every quasi-compact and quasi-separated rigid space X over k admits a *formal model*: does there exist an admissible formal R-scheme $\mathfrak{X}$ whose generic fiber is X? If so, what relations are there between the various formal models of X? More specifically, given a map $f : X' \to X$ between such rigid spaces, can we choose the formal models $\mathfrak{X}'$ and $\mathfrak{X}$ so that f is induced by a map $\mathfrak{f} : \mathfrak{X}' \to \mathfrak{X}$ between such formal schemes over R? We also want to know if $\mathfrak{f}$ can be arranged to satisfy analogues of properties of f (such as separatedness, flatness, having fiber dimension at most d, having geometrically reduced fibers, etc., when such notions are appropriately defined for morphisms between tfp formal schemes over R).

Raynaud gave definitive (and largely affirmative) answers to such questions. Before we explain part of his answer, it is instructive to see an example of how a formal model can be changed without affecting the associated rigid space. This operation intuitively corresponds to the idea of making a blow-up in the "closed fiber" (over $\widetilde{k}$), and so it should not affect the "generic fiber" (over k).

EXAMPLE 3.3.11. Let $\mathfrak{X} = \mathrm{Spf}(\mathscr{A})$ be an affine admissible formal R-scheme (so $\mathscr{A}$ if tfp and flat over R). Let $\mathscr{I}$ be a finitely generated ideal in $\mathscr{A}$, say with generators $f_1, \ldots, f_n$. Pick $\pi \in R$ with $0 < |\pi| < 1$. Assume that some power of π lies in $\mathscr{I}$. (Intuitively, this means that this ideal cuts out a closed subscheme "supported in the closed fiber" over R.) Imitating the charts in a blow-up, consider the R-algebra A_i that is the quotient of $\mathscr{A}[T_{i1}, \ldots, T_{in}]/(f_i T_{ij} - f_j)$ modulo its f_i-power torsion. Finally, let $\mathscr{A}_i$ denote the π-adic completion of A_i. This is an admissible R-algebra (in particular, it is R-flat). We can glue the $\mathrm{Spf}(\mathscr{A}_i)$'s much as we glue the charts in a blow-up. The resulting glued admissible formal scheme over R is called the *admissible formal blow-up* of $\mathfrak{X}$ along $\mathscr{I}$. It can be characterized by a universal property so as to show it is independent of the choice of f_j's (and of π). We shall denote it $\mathrm{Bl}_{\mathscr{I}}(\mathfrak{X})$.

Consider the natural map $\mathrm{Bl}_{\mathscr{I}}(\mathfrak{X}) \to \mathfrak{X}$. This is generally far from an isomorphism (e.g., the target is affine but the source almost never is). However, the associated generic fiber is an isomorphism! This follows from the next exercise.

EXERCISE 3.3.12. With notation as at the end of the previous example, let $A = k \otimes_R \mathscr{A}$. Show that the f_i's as elements of A have no common zero, so the rational domains

$$A\langle f_1/f_i, \ldots, f_n/f_i \rangle$$

make sense for $1 \leq i \leq n$. Show that the generic fibers of the charts of the formal admissible blow-up correspond to these rational domains, and that these rational domains cover $\mathfrak{X}_k = \mathrm{Sp}(A)$. Deduce that the map $\mathrm{Bl}_{\mathscr{I}}(\mathfrak{X}) \to \mathfrak{X}$ from the formal admissible blow-up induces an isomorphism on generic-fiber rigid spaces.

We conclude this lecture by briefly stating some of the main results of Bosch, Lütkebohmert, and Raynaud concerning formal models of rigid spaces. Every quasi-compact and quasi-separated rigid space over k does admit a formal model, any two formal models are dominated by a common admissible formal blow-up, and every map between quasi-compact and quasi-separated rigid spaces arises from a map between suitable formal models. More precisely, the category of quasi-compact and quasi-separated rigid spaces over k is equivalent to the localization of the category of quasi-compact admissible formal R-schemes with respect to the admissible formal blow-up morphisms. In addition, any quasi-compact admissible open subset arises from a Zariski-open subset in a suitable formal model (and something similar on the level of finite covers by quasi-compact admissible opens) and every finite collection of coherent sheaves on such a rigid space arises from a collection of formal coherent sheaf models on a suitable formal model. Also, it is often possible to transfer properties of rigid-analytic maps (such as flatness and fiber dimension) to suitable choices of formal models. This theory is powerful, because it settles in the affirmative many vexing questions of "algebraic" nature that appear to be beyond the reach of the methods of rigid geometry. The basic principle is that using suitable formal models can reduce hard questions in rigid geometry over k to known standard results in algebraic geometry over $\widetilde{k}$ or over quotient rings $R/(\pi^n)$ (by killing powers of π or all of $\mathfrak{m}$ in the structure sheaves of formal models). We refer to [**BL1**] and [**BL2**] for a systematic development of the first steps in this extremely useful theory.

EXAMPLE 3.3.13. In Berthelot's theory of rigid cohomology, an important construction is the *specialization morphism* $\mathrm{sp} : \mathfrak{X}_k \to \mathfrak{X}$ for an admissible formal R-scheme $\mathfrak{X}$. In general this is a morphism of topoi, but we just explain its definition as a map of underlying sets. Let $\{\mathrm{Spf}(\mathscr{A}_i)\}$ be a finite open affine covering of $\mathfrak{X}$, so for $A_i = k \otimes_R \mathscr{A}_i$ the affinoids $\mathrm{Sp}(A_i)$ are an open affinoid cover of $\mathfrak{X}_k$. For $x \in \mathfrak{X}_k$ we have $x \in \mathrm{Sp}(A_i)$ for some i. The associated map $A_i \to k(x)$ carries $\mathscr{A}_i$ into the valuation ring $R(x)$ of $k(x)$, and so induces a map $\mathscr{A}_i/(\pi) \to \widetilde{k}(x)$ to the residue field of the finite extension $k(x)/k$. This is a closed point $\mathrm{sp}(x) \in \mathrm{Spec}(\mathscr{A}_i/(\pi)) \subseteq \mathfrak{X}$, and it is independent of the choices made. Note that sp is "anti-continuous" in the sense that the preimage of a closed set is admissible open; this gives rise to the so-called tubes in Berthelot's theory.

EXAMPLE 3.3.14. Formal models are useful beyond the quasi-compact case. For example, in the theory of non-archimedean uniformization of Shimura curves it is convenient to work with a certain formal model for the Drinfeld half-plane over a local field.

4. Berkovich spaces I

4.1. A topological construction. Let X be a quasi-compact and quasi-separated rigid space. In the previous lecture we saw that Raynaud's theory provides formal models $\mathfrak{X}$ of X, where $\mathfrak{X}$ is an admissible formal R-scheme. It was also seen within Raynaud's theory that the collection of all such models for a fixed X naturally forms an inverse system (with formal admissible blow-ups providing a cofinal system of transition maps). In particular, although the underlying topological space $|\mathfrak{X}|$ of any particular formal model is not intrinsic to X, the inverse limit topological space

$$X^{\mathrm{ad}} = \varprojlim |\mathfrak{X}|$$

formed over the collection of all formal models of X is intrinsic to X (and is an example of a so-called Zariski–Riemann space). Hochster's work on formal models shows that this is a quasi-compact topological space (not obvious, since quasi-compactness need not be preserved under the formation of inverse limits in the non-Hausdorff case), and it is also quasi-separated (i.e., the overlap of any two quasi-compact opens is quasi-compact). More interesting is the following:

EXERCISE 4.1.1. Using Raynaud's result that maps of rigid spaces extend to maps between a suitable pair of formal models (and that such an extension is unique if it exists, when the pair of formal models is fixed), prove that X^{ad} is naturally functorial in X.

By a procedure that we will not explain here (the theory of rig-points), one can construct a natural (and functorial) injective map $X \to X^{\mathrm{ad}}$ whose image is dense. The topological space X^{ad} is the underlying space of the *adic space* associated to X in the sense of Huber [**H**]. A technical merit of the adic space associated to the quasi-compact and quasi-separated rigid space X is that its G-topology is set-theoretically quasi-compact (in the sense that any set-theoretic covering of X^{ad} by opens for the G-topology has a finite subcovering) and its associated category of sheaves of sets is equivalent to the category of sheaves of sets on X with respect to the Tate topology. There is no "admissibility" condition required on coverings as there is on X. Thus, for cohomological questions on X we can work on X^{ad}, where it is possible to use pointwise arguments. (In fact, the set of points of X^{ad} is the set of points of the topos defined by the Tate topology on X.) Unfortunately, in contrast with X, X^{ad} exhibits very un-Hausdorff features.

An intermediate space between X and X^{ad} is the topological space X^{an} underlying the *Berkovich space* associated to X. This is naturally a compact Hausdorff space with remarkable topological properties (e.g., its connected components are path-connected!), and its coherent sheaf theory is equivalent to that of X. Moreover, it has so many points that traditional stalk arguments can be carried out on X^{an} (in contrast with X). The final two lectures are devoted to explaining some basic aspects of Berkovich's theory; the main references are [**Ber1**, Ch. 1-2] and early parts of [**Ber2**]. In this lecture we will focus on definitions and examples in the affinoid setting, and in the final lecture we discuss the global theory and some simple striking applications (by no means the most impressive ones).

4.2. Modified Tate algebras and affinoid algebras. An amazing feature of Berkovich's theory is that it permits working with discs whose radius may be any positive real number at all, not just an element of $\sqrt{|k^\times|}$. Another curious feature (which does have some applications) is that it is permissible for the base field to have a *trivial* absolute value. For these reasons, at the beginning of the theory it is convenient to introduce the following innocuous-looking generalization of the concept of a Tate algebra over a non-archimedean field.

DEFINITION 4.2.1. Let k be a field complete with respect to a non-archimedean absolute value (perhaps the trivial one!). For $r_1, \ldots, r_n > 0$, define the k-algebra $T_{\underline{r}}(k) = T_{r_1, \ldots, r_n}(k)$ to be

$$k\langle r_1^{-1} X_1, \ldots, r_n^{-1} X_n \rangle = \left\{ \sum a_J X^J \in k[\![X_1, \ldots, X_n]\!] \mid |a_J| r^J \to 0 \right\}.$$

This algebra is also denoted $k\langle \underline{r}^{-1} \underline{X} \rangle$ if n and the r_i's are understood from context.

EXAMPLE 4.2.2. If k is endowed with the trivial absolute value and $r_i < 1$ for all i then $T_{\underline{r}}(k) = k[\![X_1, \ldots, X_n]\!]$. If $r_i \geq 1$ for all i then $T_{\underline{r}}(k)$ is the polynomial ring in n variables over k. For other possibilities with the r_i's, it is some mixture of the two (formal power series in some variables, polynomial in others).

EXERCISE 4.2.3. Prove that $k\langle r_1^{-1} X_1, \ldots, r_n^{-1} X_n \rangle$ is a k-Banach algebra via the norm

$$\left\| \sum a_J X^J \right\|_{\underline{r}} = \max_J |a_J| r^J,$$

and prove that this is a multiplicative norm. (Hint: Induct on n, so it may be convenient to prove a more general result for relative generalized Tate algebras over k-Banach algebras with a multiplicative norm.) In the special case that k has the trivial absolute value and $r_i < 1$ for all i, prove that when this k-algebra is identified with $k[\![X_1, \ldots, X_n]\!]$ in the evident manner then this power series ring acquires (from the norm) the maximal-adic topology.

EXAMPLE 4.2.4. Choose $r_1, \ldots, r_n > 0$ and let K/k be an analytic extension field such that $r_1, \ldots, r_n \in |K^\times|$. (We will see soon that such a K always exists.) Using the appropriate notion of completed tensor product over k for k-Banach spaces (and k-Banach algebras), there is a continuous K-algebra isomorphism $K \widehat{\otimes}_k T_{\underline{r}}(k) \simeq T_n(K)$ via $1 \widehat{\otimes} X_i \mapsto c_i X_i$ with $c_i \in K$ satisfying $|c_i| = r_i$. In this sense we think of $T_{\underline{r}}(k)$ as the coordinate ring of a closed polydisc over k with polyradius $\underline{r}$. If $r_1, \ldots, r_n \in \sqrt{|k^\times|}$ with all $r_i \leq 1$ and $|k^\times| \neq \{1\}$ then $T_{\underline{r}}(k)$ has a bounded k-algebra isomorphism to the coordinate ring of the Weierstrass domain $\{|t_1| \leq r_1, \ldots, |t_n| \leq r_n\}$ in $\mathbf{B}_k^n$.

EXAMPLE 4.2.5. Let k be a field equipped with a non-archimedean (perhaps trivial) absolute value $|\cdot|_0$. Let $|\cdot|$ be a multiplicative norm on k satisfying $|\cdot| \leq C |\cdot|_0$ for some $C > 0$. We now show that the only such norm is $|\cdot|_0$. For $a \in k$ we have $|a^n| \leq C |a^n|_0$ for all $n \geq 1$, so extracting nth roots yields $|a| \leq C^{1/n} |a|_0$. Taking $n \to \infty$ gives $|a| \leq |a|_0$. If $a \neq 0$ then we apply this to $1/a$ to get $|1/a| \leq |1/a|_0$, or equivalently $|a| \geq |a|_0$. Thus, $|\cdot| = |\cdot|_0$.

EXERCISE 4.2.6. Choose $r > 0$ with $r \notin \sqrt{|k^\times|}$, and consider

$$k\langle r^{-1} X, r X^{-1} \rangle := k\langle r^{-1} X, r Y \rangle / (XY - 1).$$

(1) Show that every element has a unique expression as a convergent sum $\sum_{n \in \mathbf{Z}} a_n X^n$ with $|a_n| r^n \to 0$ as $n \to \pm\infty$. Explain why $|a_n| r^n \neq |a_m| r^m$ whenever $n \neq m$ and $a_n, a_m \in k^\times$.

(2) Prove that the residue norm $\|\cdot\|$ induced by the multiplicative norm $\|\cdot\|_{r,1/r}$ from Exercise 4.2.3 assigns to $\sum_{n\in\mathbf{Z}} a_n X^n$ the norm $\max_n |a_n| r^n$, and that this is a multiplicative norm. Conclude that $k\langle r^{-1}X, rX^{-1}\rangle$ is a domain.

(3) Let $|\cdot| : k\langle r^{-1}X, rX^{-1}\rangle \to \mathbf{R}$ be a seminorm (norm except with perhaps nonzero kernel) that is multiplicative and also bounded in the sense that $|\cdot| \le C\|\cdot\|$ for some $C > 0$. Applying this to nth-powers and extracting nth roots, deduce that $|\cdot| \le \|\cdot\|$. Conclude that $|X| = r$ and that $|\cdot|$ on k is the given absolute value, and then that $|\cdot| = \|\cdot\|$. In particular, $k\langle r^{-1}X, rX^{-1}\rangle$ has exactly one bounded multiplicative seminorm.

The point of the preceding exercise is that (by combining it with Theorem 4.3.4(1) and the closedness of ideals) it implies that the ring $k\langle r^{-1}X, rX^{-1}\rangle$ cannot have any nonzero ideals, so it is a *field*. Hence, via $\|\cdot\|$ it is an analytic extension of k such that r is in its value group (in fact, r is the norm of the element X). By repeating this process, for any finite set of positive real numbers there is an analytic extension of k whose value group contains the finitely many such given real numbers.

At the possible risk of confusion, we now change some terminology from classical rigid geometry:

DEFINITION 4.2.7. A *k-affinoid algebra* is a k-Banach algebra $\mathscr{A}$ for which there is a continuous surjection
$$k\langle r_1^{-1}X_1, \ldots, r_n^{-1}X_n\rangle \to \mathscr{A}$$
for some $r_1, \ldots, r_n > 0$ such that the residue norm induced on $\mathscr{A}$ is equivalent to the given norm on $\mathscr{A}$. If we can take all $r_i = 1$ then $\mathscr{A}$ is called a *strict k-affinoid algebra* (these are the classical k-affinoid algebras of rigid geometry when k has a nontrivial absolute value). A *morphism* between k-affinoid algebras is a k-algebra map that is continuous.

EXERCISE 4.2.8. Let $\mathscr{A}$ be a k-Banach algebra. Define relative generalized Tate algebras
$$\mathscr{A}\langle r_1^{-1}X_1, \ldots, r_n^{-1}X_n\rangle$$
and formulate and prove a universal property (similar to the case of relative Tate algebras in rigid geometry).

The theory of completed tensor products can be extended to all k-Banach algebras. In particular, if $\mathscr{A}$ is k-affinoid in the above sense and K/k is an analytic extension field then $K\widehat{\otimes}_k\mathscr{A}$ is K-affinoid in the above sense and $\mathscr{A} \to K\widehat{\otimes}_k\mathscr{A}$ is a faithfully flat map that is an isometry onto its (closed) image. By using K with sufficiently large value group, this enables one to prove basic properties of k-affinoid algebras in the new sense (such as being noetherian, all ideals being closed, etc.) by reducing to the known strict case. For example, this technique enables one to show that if $\mathscr{C} \rightrightarrows \mathscr{A}, \mathscr{B}$ are morphisms of k-affinoid algebras then $\mathscr{A}\widehat{\otimes}_{\mathscr{C}}\mathscr{B}$ is again k-affinoid. In particular, if $\mathscr{A}$ is k-affinoid then any relative generalized Tate algebra over $\mathscr{A}$ is also k-affinoid. Also, if the absolute value on k is nontrivial then the norm-equivalence condition in the preceding definition of a k-affinoid algebra is automatically satisfied, due to the non-archimedean version of the Banach Open Mapping Theorem (Exercise 1.2.5). In general, the absence of the Banach Open Mapping Theorem in the case of trivial absolute value is the reason why the case of the trivial absolute value on the base field k often requires separate consideration.

One basic way in which these new kinds of k-affinoid algebras are a bit harder to work with than the strict (i.e., classical) type is seen in dimension theory. For example, in Exercise 4.2.6 we gave an example of a k-affinoid algebra that is a field but should correspond to the coordinate ring of the circle of radius r: after a sufficiently large ground field extension (to put r into the value group of the ground field) it becomes 1-dimensional, but the given affinoid algebra over k is 0-dimensional. Thus, to define dimension theory in the k-analytic setting, one has to first show (e.g., using Noether normalization) that dimension theory for strict affinoid algebras is well-behaved with respect to extension of the base field, and then make definitions in the general case by using an auxiliary field extension to pass to the strict case (where Krull dimension provides an adequate theory). It is a nontrivial matter to set up a geometrically satisfactory dimension theory for k-affinoid algebras that enjoys properties one expects out of analogy with the strict case and algebraic geometry, but such a theory has been developed (in some respects, quite recently, by Ducros). The main point is that ring-theoretic dimension theory alone is not an adequate viewpoint over the base field in Berkovich's theory.

4.3. Spectrum of Banach algebras. We have introduced the wider class of k-Banach algebras that will be used in the local theory of Berkovich spaces. It is now time to introduce the underlying topological spaces to be used, replacing the role of $M(A)$ in rigid geometry. To motivate what we are about to do, let us return to the setting of rigid geometry and consider a strict $\mathbf{Q}_p$-affinoid algebra A as well as its associated strict $\mathbf{C}_p$-affinoid fiber $A_{\mathbf{C}_p} = \mathbf{C}_p \widehat{\otimes}_{\mathbf{Q}_p} A$. There is a continuous map $A \to A_{\mathbf{C}_p}$ via $a \mapsto 1 \widehat{\otimes} a$, but there is no corresponding natural map of sets $M(A_{\mathbf{C}_p}) \to M(A)$ because the "transcendental points" with respect to $\mathbf{Q}_p$ in $M(A_{\mathbf{C}_p})$ have nowhere to go. We now reformulate the problem in a way that leads us to the way out of this conundrum.

For each point $x \in M(A_{\mathbf{C}_p})$, the corresponding maximal ideal $\mathfrak{m}_x \subseteq A_{\mathbf{C}_p}$ contracts to the prime ideal $\mathfrak{p}_x \subseteq A$ that is the kernel of the composite map

$$| \cdot |_x : A \to A_{\mathbf{C}_p} \xrightarrow{x} \mathbf{C}_p \xrightarrow{|\cdot|} \mathbf{R}.$$

This composite map is a seminorm on A (i.e., it satisfies all of the requirements of a norm except for possibly having nonzero kernel), it is bounded (in the sense that $| \cdot |_x$ is bounded by a constant multiple of a fixed $\mathbf{Q}_p$-Banach algebra norm on A), and it is *multiplicative*. More specifically, this corresponds to an absolute value on $\mathrm{Frac}(A/\mathfrak{p}_x)$ whose restriction to $A/\mathfrak{p}_x$ is bounded above by a constant multiple of the residue norm on $A/\mathfrak{p}_x$. Points of $M(A)$ gives rise to such structures as a special case, with the additional condition that $\mathrm{Frac}(A/\mathfrak{p}_x)$ is finite over $\mathbf{Q}_p$ (and equipped with an embedding into $\mathbf{C}_p$).

EXERCISE 4.3.1. Let A be a k-Banach algebra, and let $| \cdot | : A \to \mathbf{R}$ be a bounded multiplicative seminorm. Prove that its kernel $\mathfrak{p}$ is a prime ideal of A and that $| \cdot |$ arises from an absolute value on $\mathrm{Frac}(A/\mathfrak{p})$ that restricts to the given absolute value on k. If we let K denote the completion of $\mathrm{Frac}(A/\mathfrak{p})$ with respect to this absolute value, then show that $| \cdot |$ arises by pullback of the absolute value on K via a bounded k-algebra map from A to the analytic extension field K/k in which A generates a dense subfield. (By "bounded" k-algebra map $A \to K$ we mean boundedness in the sense of k-Banach space maps.)

Show that this procedure sets up a bijection between the set of bounded multiplicative seminorms on A and the set of bounded k-algebra maps $A \to K$ to analytic extension fields K/k in which A generates a dense subfield (and these maps are taken up to composition with isometric isomorphism in K).

It is very interesting that the consideration of bounded multiplicative seminorms makes sense even on complete normed rings that do not contain a field, such as $\mathbf{Z}$ (with the usual absolute value as its norm, or with the trivial norm). More specifically, let $(\mathscr{A}, \| \cdot \|)$ be a commutative Banach algebra (a commutative ring endowed with a submultiplicative norm with respect to which it is complete as a metric space); this could be an abstract ring endowed with the trivial norm. It makes sense to consider multiplicative seminorms $| \cdot | : \mathscr{A} \to \mathbf{R}$ that are bounded in the sense that $|f| \leq C\|f\|$ for all $f \in \mathscr{A}$ for some $C > 0$. Applying this to f^n and using multiplicativity, taking $n \to \infty$ after extracting nth roots gives the consequence $|f| \leq \|f\|$. As was seen in Exercise 4.3.1, if $\mathscr{A}$ is a k-Banach algebra then all such $| \cdot |$'s are automatically compatible with the absolute value on k. Also, if the given norm on $\mathscr{A}$ is non-archimedean then (as is easily checked by pullback to $\mathbf{Z}$) any bounded multiplicative seminorm on $\mathscr{A}$ is necessarily non-archimedean.

DEFINITION 4.3.2. The *spectrum* of a Banach algebra $(\mathscr{A}, \| \cdot \|)$ is the set $\mathscr{M}(\mathscr{A})$ of bounded multiplicative seminorms on $\mathscr{A}$. This set is given the topology for which a base of opens around a point $| \cdot |_0 \in \mathscr{M}(\mathscr{A})$ is provided by the sets

$$U_{f_1,\ldots,f_n,\varepsilon_1,\ldots,\varepsilon_n} = \{| \cdot | \in \mathscr{M}(\mathscr{A}) \,|\, ||f_j|_0 - |f_j|| < \varepsilon_j \text{ for all } j\}.$$

Equivalently, this is the weakest topology with respect to which the functions $\mathscr{M}(\mathscr{A}) \to \mathbf{R}$ defined by $| \cdot | \mapsto |f|$ for each $f \in \mathscr{A}$ are all continuous.

REMARK 4.3.3. Observe that the definition of the spectrum as a topological space is unaffected by replacing the Banach norm on $\mathscr{A}$ with an equivalent one (i.e., one that is bounded above and below by a constant positive multiple of the initial one). Hence, if k has nontrivial absolute value and $\mathscr{A}$ is a strict k-affinoid algebra then we can use any k-Banach algebra norm on $\mathscr{A}$ to unambiguously define the topological space $\mathscr{M}(\mathscr{A})$.

The preceding exercise for k-Banach algebras carries over to (commutative) Banach algebras in general, so as a set we can identify $\mathscr{M}(\mathscr{A})$ with the set of ring maps $\mathscr{A} \to K$ (up to isometric isomorphism in K) that are bounded as maps of complete normed additive groups and have target K that is a complete valued field in which $\mathscr{A}$ generates a dense subfield. This is reminiscent of how the points of an affine scheme can be interpreted as the set of "all" field-valued points of the coordinate ring; we now work with non-archimedean fields (with possibly trivial absolute value)!

The spectrum of a (commutative) Banach algebra enjoys some fundamental topological properties that we now summarize.

THEOREM 4.3.4. *Let $\mathscr{A}$ be a commutative Banach algebra.*

(1) $\mathscr{M}(\mathscr{A}) = \emptyset$ *if and only if $\mathscr{A} = 0$.*

(2) $\mathscr{M}(\mathscr{A})$ *is compact and Hausdorff.*

(3) *If $\phi : \mathscr{A} \to \mathscr{A}'$ is a bounded map between commutative Banach algebras then the pullback map $\mathscr{M}(\mathscr{A}') \to \mathscr{M}(\mathscr{A})$ defined by $| \cdot |' \mapsto | \cdot |' \circ \phi$ is continuous.*

The proof of the first two parts is an application of Zorn's Lemma, and the third part is a matter of unwinding definitions. As an interesting application of the theorem, let k be a non-archimedean field (nontrivial absolute value) and let K/k and K'/k be two analytic extension fields with at least one of them of countable type over k (in the sense of having a dense k-linear subspace of countable dimension). The Banach algebra $K \widehat{\otimes}_k K'$ is nonzero by general results on countable-type k-Banach spaces [**BGR**]. Thus, its spectrum is non-empty by Theorem 4.3.4, so we obtain a third non-archimedean extension of k into which K and K' isometrically embed. This underlies the fact that if $\mathscr{A}$ is a (perhaps non-strict) k-affinoid algebra (with k having possible trivial valuation) then for any non-archimedean extension field K/k such that $K \widehat{\otimes}_k \mathscr{A}$ is strict, the dimension of this strict K-affinoid algebra is independent of K/k.

EXAMPLE 4.3.5. A very interesting example of functoriality of the spectrum (as a topological space) is to let $\mathscr{A}$ be a k-affinoid algebra and K/k a non-archimedean extension field (with compatible absolute value). The natural map $\mathscr{A} \to \mathscr{A}_K = K \widehat{\otimes}_k \mathscr{A}$ from a k-affinoid algebra to a K-affinoid algebra induces a continuous (in fact, surjective) map $\mathscr{M}(\mathscr{A}_K) \to \mathscr{M}(\mathscr{A})$. So already we can do something that is not generally possible in rigid geometry, which is to have maps from a geometric object "over K" to one "over k". This will be improved to a morphism of analytic spaces once we endow these spectra with more structure beyond their topologies.

For each $x \in \mathscr{M}(\mathscr{A})$, let the prime ideal $\mathfrak{p}_x \subseteq \mathscr{A}$ be the kernel of this multiplicative seminorm, and let $\mathscr{H}(x)$ denote the completion of the field $\mathrm{Frac}(\mathscr{A}/\mathfrak{p}_x)$ with respect to the absolute value induced by x. For each $f \in \mathscr{A}$ we write $f(x)$ to denote the image of f under the natural map $\mathscr{A} \to \mathscr{H}(x)$ and $|f(x)|$ to denote the absolute value of $f(x)$ (using the canonical absolute value on $\mathscr{H}(x)$).

REMARK 4.3.6. There are nearly always points $x \in \mathscr{M}(\mathscr{A})$ with $\mathscr{H}(x)$ of infinite degree over k, even if k is algebraically closed. For example, for any k the space $\mathscr{M}(k\langle T \rangle)$ contains the point ξ corresponding to the Gauss norm, and $\mathscr{H}(\xi)$ is the completion of $\mathrm{Frac}(k\langle T \rangle)$ with respect to the absolute value induced by the Gauss norm on $k\langle T \rangle$. (We can replace $k\langle T \rangle$ with $k[T]$ in this latter description, for denseness reasons.) This abundance of points with large residue field (in contrast with the case of complex-analytic spaces) can be considered to be a consequence of the lack of a non-archimedean Gelfand–Mazur theorem. Even if k is discretely valued, the space $\mathscr{M}(k\langle T \rangle)$ has many points whose completed residue field is not discretely valued (see Exercise 4.3.8 below), and so it will follow that to pass between problems in rigid geometry and Berkovich's theory even over a discretely-valued field we will inexorably be led to consider rigid spaces over extension fields whose valuation ring is not a discrete valuation ring!

EXERCISE 4.3.7. Let $\mathscr{A}$ be a commutative Banach algebra. For $r > 0$ and $f, g \in \mathscr{A}$, prove that the subset $\{x \in \mathscr{M}(\mathscr{A}) \mid |f(x)| \le r|g(x)|\}$ is closed in $\mathscr{M}(\mathscr{A})$. Also prove that $f \in \mathscr{A}$ is a unit if and only if $f(x) \neq 0$ in $\mathscr{H}(x)$ for all $x \in \mathscr{M}(\mathscr{A})$, and that f is nilpotent in $\mathscr{A}$ if and only if $f(x) = 0$ in $\mathscr{H}(x)$ for all $x \in \mathscr{M}(\mathscr{A})$. (Hint: it is a consequence of Zorn's Lemma that the natural map $\mathscr{M}(\mathscr{A}) \to \mathrm{Spec}(\mathscr{A})$ assigning to each bounded multiplicative seminorm its prime ideal kernel is a surjective map.)

EXERCISE 4.3.8. Let k be a non-archimedean field (nontrivial absolute value), and let $\mathscr{A} = k\langle T \rangle$.

(1) Fix $0 \leq r \leq 1$ and define $\| \cdot \|_r$ on $\mathscr{A}$ by $\| \sum a_n T^n \|_r = \max_n |a_n| r^n$. Prove that this is a bounded multiplicative seminorm on $\mathscr{A}$. Also show that it deserves to be called the sup-norm on the closed disc of radius r (with center at 0) in the sense that $\|f\|_r$ is the supremum of the numbers $|f|_x$ as $| \cdot |_x$ ranges through all bounded k-algebra maps $\mathscr{A} \to K$ to non-archimedean fields such that the image of T in K has absolute value at most r. It is useful to think of $\| \cdot \|_r$ as the "generic point" of the closed disc of radius r centered at the origin, though it is not a generic point in the usual topological sense used for schemes.

(2) Prove that the function $[0,1] \to \mathscr{M}(\mathscr{A})$ defined by $r \mapsto \| \cdot \|_r$ is *continuous*, and so provides a path connecting the origin to the Gauss norm!

(3) Prove that the map $M(\mathscr{A}) \to \mathscr{M}(\mathscr{A})$ defined by assigning to each $x \in M(\mathscr{A})$ the seminorm induced by $\mathscr{A} \to k(x)$ is injective with dense image, and induces on $M(\mathscr{A})$ the canonical topology as the subspace topology. By using the trick of "recentering the disk", prove that any two k-rational points in $M(\mathscr{A})$ are linked by a path in $\mathscr{M}(\mathscr{A})$! (In fact, $\mathscr{M}(\mathscr{A})$ is a path-connected space, but this requires some more work to generalize the trick of linking via the Gauss norm.)

The final part of the preceding exercise holds rather more generally (but requires more effort to prove):

THEOREM 4.3.9. *Assume that k has nontrivial absolute value and let A be a strict k-affinoid algebra. The natural map of sets $M(A) \to \mathscr{M}(A)$ is injective with dense image and induces the canonical topology as the subspace topology. Moreover, if A has no nontrivial idempotents then $\mathscr{M}(A)$ is path-connected.*

4.4. Affinoid subdomains. In order to work with the spectrum of a k-affinoid algebra $\mathscr{A}$, we need to define the analogues of affinoid subdomains (including Weierstrass, Laurent, and rational domains). As in the case of rigid geometry, we will use a set-theoretic mapping property. One enormous difference is that affinoid subdomains in $\mathscr{M}(\mathscr{A})$ are not open sets in general: they are closed subsets.

DEFINITION 4.4.1. Let $\mathscr{A}$ be a k-affinoid algebra. A subset $U \subseteq \mathscr{M}(\mathscr{A})$ is an *affinoid subdomain* (more precisely, a *k-affinoid subdomain*) if there exists a bounded map $i : \mathscr{A} \to \mathscr{A}'$ of k-affinoid algebras such that two conditions hold: $\mathscr{M}(i)$ has image contained in U, and for any bounded map $\phi : \mathscr{A} \to \mathscr{B}$ of k-affinoid algebras with $\mathscr{M}(\mathscr{B}) \to \mathscr{M}(\mathscr{A})$ having image contained in U, there is a unique factorization of ϕ through i via a bounded map $\mathscr{A}' \to \mathscr{B}$ of $\mathscr{A}$-algebras.

As in rigid geometry, this definition uniquely determines $\mathscr{A}'$ up to a unique bounded $\mathscr{A}$-algebra isomorphism, so we can write $\mathscr{A}_U$ to denote this $\mathscr{A}$-algebra (with norm determined up to equivalence); it is called the *coordinate ring* of U. By chasing appropriate field-valued points and using the stronger universal property given in Example 4.4.3, one shows that $\mathscr{M}(\mathscr{A}_U) \to U$ is a homeomorphism and that for any analytic extension field K/k the affinoid K-algebra $K \widehat{\otimes}_k \mathscr{A}_U$ has the analogous universal property for the preimage U_K of U under the surjective map $\mathscr{M}(K \widehat{\otimes}_k \mathscr{A}) \to \mathscr{M}(\mathscr{A})$. In particular, the concept of affinoid domain is well-behaved with respect to extension of the base field. Somewhat less evident (but unsurprising) is that $\mathscr{A}_U$ is $\mathscr{A}$-flat. By using completed tensor products, it may be proved exactly as in rigid geometry that an overlap of finitely many affinoid subdomains is again an affinoid subdomain.

EXERCISE 4.4.2. We now work out the basic examples of Weierstrass, Laurent, and rational domains in the setting of Berkovich spaces. One new feature is that we can use arbitrary positive real scaling factors on the non-strict inequalities used to define such domains. (In the rigid-analytic case we saw that it only made sense to impose scaling factors in $\sqrt{|k^\times|}$, and that the use of such factors provided no extra generality than not using them.)

(1) Choose $f_1, \ldots, f_n, g_1, \ldots, g_m$ in a k-affinoid algebra $\mathscr{A}$, and positive real numbers $r_1, \ldots, r_n, s_1, \ldots, s_m$. Prove that the subset

$$\{x \in \mathscr{M}(\mathscr{A}) \mid |f_i(x)| \leq r_i, |g_j(x)| \geq s_j \text{ for all } i, j\}$$

is an affinoid subdomain with coordinate ring

$$\mathscr{A}\{\underline{r}^{-1}\underline{X}, \underline{s}\underline{Y}\}/(X_1 - f_1, \ldots, X_n - f_n, g_1 Y_1 - 1, \ldots, g_m Y_m - 1)$$

(usually denoted $\mathscr{A}\langle \underline{r}^{-1}\underline{f}, \underline{s}\underline{g}^{-1}\rangle$).

(2) Let $f_1, \ldots, f_n, g \in \mathscr{A}$ be elements which generate the unit ideal (equivalently, they have no common zero on $\mathscr{M}(\mathscr{A})$). Choose $r_1, \ldots, r_n > 0$. Prove that the subset

$$\{x \in \mathscr{M}(\mathscr{A}) \mid |f_1(x)| \leq r_1|g(x)|, \ldots, |f_n(x)| \leq r_n|g(x)|\}$$

is an affinoid subdomain with coordinate ring

$$\mathscr{A}\{r_1^{-1}X_1, \ldots, r_n^{-1}X_n\}/(gX_1 - f_1, \ldots, gX_n - f_n)$$

(usually denoted $\mathscr{A}\langle \underline{r}^{-1}\underline{f}/g\rangle$).

These affinoid subdomains are respectively called *Laurent domains* and *rational domains*. A Laurent domain is called *Weierstrass* if there are no g_j's. Prove that every point of $\mathscr{M}(\mathscr{A})$ has a base of closed neighborhoods consisting of Laurent domains.

Recall that in rigid geometry (i.e., when working with strict k-affinoid algebras, with k having nontrivial absolute value), the natural map $A \to A\langle a_1, \ldots, a_n\rangle$ to the coordinate ring of a Weierstrass domain has dense image. The same obviously holds in our new setting (for $\mathscr{A} \to \mathscr{A}\langle \underline{r}^{-1}\underline{a}\rangle$), but Berkovich proved more: if the natural map $\mathscr{A} \to \mathscr{A}_U$ from a k-affinoid algebra to the coordinate ring of an affinoid subdomain has dense image then U is *necessarily* a Weierstrass domain!

EXAMPLE 4.4.3. Let $\mathscr{A}$ be a k-affinoid algebra and $U \subseteq \mathscr{M}(\mathscr{A})$ be an affinoid subdomain with associated coordinate ring $\mathscr{A}_U$. Choose $x \in U = \mathscr{M}(\mathscr{A}_U) \subseteq \mathscr{M}(\mathscr{A})$. Let us show that the completed residue fields at x for $\mathscr{A}$ and $\mathscr{A}_U$ coincide (via the natural map). In other words, if $\mathscr{H}(x)$ denotes the completion of the fraction field of $\mathscr{A}/\ker(|\cdot|_x)$ then we claim that the map $\mathscr{A} \to \mathscr{H}(x)$ factors (uniquely) through $\mathscr{A} \to \mathscr{A}_U$ (and so we then get the desired equality of completed residue fields, due to denseness and absolute-value compatibility reasons). To prove this, it is enough to show that affinoid subdomains satisfy their universal mapping property with respect to maps from $\mathscr{A}$ to arbitrary k-Banach algebras, not merely to k-affinoid algebras. For Laurent and rational domains this is clear. In the general case, the important theorem of Gerritzen–Grauert from rigid geometry (Theorem 2.2.5) can be proved to remain true in the setting of the spaces $\mathscr{M}(\mathscr{A})$, and from this it may be deduced that affinoid subdomains do indeed satisfy the desired stronger universal mapping property.

The theorems of Tate and Kiehl from rigid geometry carry over to our new setting so as to provide a theory of coherent sheaves on affinoids, though these are not sheaves in the ordinary sense since they are evaluated on closed subsets such as affinoid domains. By using appropriate extension of the base field to reduce to the strict cases that were considered by Tate and Kiehl, the key basic result (in a slightly weaker form) is as follows:

THEOREM 4.4.4. *Let $\mathscr{A}$ be a k-affinoid algebra and M a finite $\mathscr{A}$-module. Let $\mathfrak{U}$ be a finite collection of affinoid subdomains of $X = \mathscr{M}(\mathscr{A})$ that cover X. Define $M_U := \mathscr{A}_U \otimes_{\mathscr{A}} M$ for any affinoid subdomain U of X. The Čech complex $C^{\bullet}(\mathfrak{U}, M)$ built on the modules $M_{U_{i_1} \cap \cdots \cap U_{i_n}}$ for $U_{i_1}, \ldots, U_{i_n} \in \mathfrak{U}$ is an exact sequence.*

4.5. Relative interior and boundary. We conclude our discussion of the preliminary parts of the affinoid Berkovich theory (prior to the introduction of sheaves, to be discussed in the final lecture) by explaining some new concepts with no good analogue in rigid geometry. It may be appealing to think that in $\mathscr{M}(k\langle T \rangle)$, the points x for which $|T|_x = 1$ should be the "boundary points" whereas the points x for which $|T|_x < 1$ should be the "interior points". Such a distinction is not meaningful because the automorphism group of a Tate algebra acts transitively on its set of k-rational points by recentering the polydisc. Nonetheless, Berkovich's theory does have an intrinsic theory of relative boundary and relative interior for morphisms which roughly exhibits properties of boundaries and interiors in topology. These notions require some experience to become familiar, and before giving the relevant definitions and examples in the affinoid case we consider a simple example.

EXERCISE 4.5.1. Let $\xi \in \mathscr{M}(k\langle T \rangle)$ be the point corresponding to the Gauss norm, and let D be the subset of points $x \in \mathscr{M}(k\langle T \rangle)$ for which $|T|_x = 1$.

(1) Prove that the interior of D in $\mathscr{M}(k\langle T \rangle)$ is $D - \{\xi\}$. In particular, the compact subset $D \subseteq \mathscr{M}(k\langle T \rangle)$ becomes an open subset upon deletion of the single point $\xi \in D$. (Hint: For each $|\cdot|_x$ on $k\langle T \rangle$, its restriction to $R[T]$ has an ideal of topological nilpotents that contains $\mathfrak{m}[T]$ and so corresponds to an ideal in $R[T]/\mathfrak{m}[T] = \widetilde{k}[T]$. Show that this ideal is prime and that the resulting map $\mathscr{M}(k\langle T \rangle) \to \mathrm{Spec}(\widetilde{k}[T]) = \mathbf{A}^1_{\widetilde{k}}$ has ξ as the only point over the generic point of the affine line over $\widetilde{k}$.)

(2) Prove that $\mathscr{M}(k\langle T \rangle) - \{\xi\}$ is topologically a disjoint union of open nonempty preimages of the closed points of $\mathbf{A}^1_{\widetilde{k}}$. (Thus, the connectedness properties of $\mathscr{M}(k\langle T \rangle)$ depend crucially on the presence of the point ξ!)

DEFINITION 4.5.2. Let $\phi : \mathscr{A} \to \mathscr{A}'$ be a bounded map of k-affinoid algebras. The *relative interior* of $\mathscr{M}(\mathscr{A}')$ over $\mathscr{M}(\mathscr{A})$, which we shall denote as $\mathrm{Int}(\mathscr{M}(\mathscr{A}')/\mathscr{M}(\mathscr{A}))$, is

$$\{x' \in \mathscr{M}(\mathscr{A}') \mid \mathscr{A} \to \mathscr{H}(x') \text{ is inner with respect to } \mathscr{A}\},$$

where "inner" means that there is a bounded surjection of $\mathscr{A}$-algebras $\mathscr{A}\{r^{-1}\underline{X}\} \twoheadrightarrow \mathscr{A}'$ (inducing a residue norm on $\mathscr{A}'$ that is equivalent to the given one) such that $|X_j|_{x'} < r_j$ for all j. (Geometrically, this says that a point $x' \in \mathscr{M}(\mathscr{A}')$ lies in the relative interior precisely when there is a closed immersion of $\mathscr{M}(\mathscr{A}')$ into the product of $\mathscr{M}(\mathscr{A})$ and a closed polydisc of polyradius $\underline{r}$ such that x' avoids the boundary of the polydisc.)

The complement of $\mathrm{Int}(\mathscr{M}(\mathscr{A}')/\mathscr{M}(\mathscr{A}))$ in $\mathscr{M}(\mathscr{A}')$ is the *relative boundary* of $\mathscr{M}(\mathscr{A}')$ over $\mathscr{M}(\mathscr{A})$; it is denoted $\partial(\mathscr{M}(\mathscr{A}')/\mathscr{M}(\mathscr{A}))$.

EXERCISE 4.5.3. Prove that $\mathrm{Int}(\mathcal{M}(\mathscr{A}')/\mathcal{M}(\mathscr{A}))$ is open in $\mathcal{M}(\mathscr{A}')$.

EXAMPLE 4.5.4. We give several (non-obvious) examples to illustrate the interest in the concepts of relative interior and relative boundary.

(1) If $U \subseteq \mathcal{M}(\mathscr{A})$ is an affinoid subdomain with associated coordinate ring $\mathscr{A}_U$ then $\mathrm{Int}(\mathcal{M}(\mathscr{A}_U)/\mathcal{M}(\mathscr{A}))$ is the topological interior of U with respect to $\mathcal{M}(\mathscr{A})$.

(2) The "naive" version of the Maximum Modulus Principle is trivially true on the compact Hausdorff space $\mathcal{M}(\mathscr{A})$: for any $f \in \mathscr{A}$ the continuous $\mathbf{R}$-valued function $x \mapsto |f(x)|$ on $\mathcal{M}(\mathscr{A})$ attains a maximal value. Less evident is that the maximum must be attained on the relative boundary $\partial(\mathcal{M}(\mathscr{A})/\mathcal{M}(k))$ when $\mathscr{A}$ has no factor rings that are k-finite (which ensures that the relative boundary meets every connected component). In general the maximum is attained on a certain distinguished finite subset (called the *Shilov boundary*) that is independent of f (and lies in the relative boundary when $\mathscr{A}$ has no factor rings that are k-finite).

(3) If $\mathscr{A} \to \mathscr{A}'$ is a morphism of k-affinoid algebras then the relative boundary $\partial(\mathcal{M}(\mathscr{A}')/\mathcal{M}(\mathscr{A}))$ is empty if and only if $\mathscr{A} \to \mathscr{A}'$ is finite and (as is automatic when k has nontrivial absolute value) has closed image whose subspace and residue norms are equivalent.

(4) Let $\mathscr{B}$ be a k-affinoid algebra, and $\mathscr{B} \to \mathscr{A}$ and $\mathscr{B} \to \mathscr{A}'$ be two morphisms of k-affinoid algebras. Let $\mathscr{A} \to \mathscr{A}'$ be a bounded $\mathscr{B}$-algebra map. We say that this latter map is *inner with respect to* $\mathscr{B}$ if there is a bounded surjection of $\mathscr{A}$-algebras

$$\mathscr{B}\{\underline{r}^{-1}\underline{X}\} \twoheadrightarrow \mathscr{A}$$

inducing the quotient topology and such that the map

$$\mathcal{M}(\mathscr{A}') \to \mathcal{M}(\mathscr{A}) \subseteq \mathcal{M}(\mathscr{B}\{\underline{r}^{-1}\underline{X}\})$$

lands inside of the locus of points $|\cdot|$ satisfying $|X_j| < r_j$ for all j. One can show that this property (which motivates the terminology "inner") is equivalent to the condition that $\mathcal{M}(\mathscr{A}') \to \mathcal{M}(\mathscr{A})$ lands inside of $\mathrm{Int}(\mathcal{M}(\mathscr{A})/\mathcal{M}(\mathscr{B}))$.

The theory of the relative boundary (especially when it is empty) will underlie the theory of properness in the final lecture, where properness will be defined essentially as the condition "topological properness and the absence of a relative boundary".

5. Berkovich spaces II

In this final lecture we discuss sheaf theory on $\mathcal{M}(\mathscr{A})$ and then globalize to define k-*analytic spaces* (also called *Berkovich spaces*). We also discuss the relationship between k-analytic spaces and rigid-analytic spaces (when k has nontrivial absolute value), and we briefly address the theory of étale maps. We also mention some elegant concrete applications of Berkovich's theory.

5.1. Globalization revisited. To make $\mathcal{M}(\mathscr{A})$ into an object that globalizes, we have to define its "structure sheaf" in a manner that makes sense to evaluate on affinoid subdomains, which are generally closed sets that are not open (with respect to the compact Hausdorff topology on $\mathcal{M}(\mathscr{A})$).

Consider a quasi-compact and separated rigid space X. This can be described by gluing together finitely many affinoids $\mathrm{Sp}(A_i)$ along affinoid open overlaps $\mathrm{Sp}(A_{ij})$. If we try to carry out the same gluing topologically with the spaces $\mathscr{M}(A_i)$ and their affinoid subdomains $\mathscr{M}(A_{ij})$, the resulting topological space X will be compact and Hausdorff, but there may be points $x \in X$ around which no $\mathscr{M}(A_i)$ is a neighborhood in X. However, if $V_1 = \mathscr{M}(A_{i_1}), \dots, V_n = \mathscr{M}(A_{i_n})$ are the $\mathscr{M}(A_i)$'s that contain x then the V_j's are compact Hausdorff subsets of X such that $x \in \cap V_j$ and $\cup V_j$ is a neighborhood of x in X. In general, a Berkovich space will be a kind of "structured" locally Hausdorff space X such that each $x \in X$ has a neighborhood of the form $V_1 \cup \cdots \cup V_n$ with $x \in \cap V_j$, each V_j compact and Hausdorff, and the V_j's endowed with "compatible k-affinoid structure" (with respect to the overlaps $V_j \cap V_{j'}$). To axiomatize the properties of such collections of V_j's, we shall use the following notion.

DEFINITION 5.1.1. A *quasi-net* on a locally Hausdorff topological space X is a collection τ of compact Hausdorff subsets $V \subseteq X$ such that each $x \in X$ has a neighborhood of the form $\cup V_j$ for finitely many $V_j \in \tau$ with $x \in \cap V_j$.

EXAMPLE 5.1.2. The (local) finiteness aspect of quasi-nets is the key reason why they can provide a workable substitute for usual coverings by open sets. A suggestive topological example to keep in mind (though not logically relevant to k-analytic geometry) is the closed unit disk X in $\mathbf{C}$ decomposed into a finite union of non-overlapping closed sectors V_j. The most interesting points x from the viewpoint of being a quasi-net are the points on edges where two sectors meet, and especially the center of the disk.

To define the gluing process for imposing a k-analytic structure on a topological gluing of $\mathscr{M}(A)$'s along affinoid (or more general kinds of) overlaps, we cannot work within the framework of ringed spaces because the building blocks of the theory will be compact subsets rather than open subsets. The following definition introduces the replacement for the structure of a ringed space.

DEFINITION 5.1.3. Let X be a locally Hausdorff topological space. A *k-affinoid atlas* on X is the data consisting of a quasi-net τ on X such that
  (1) for all $U, U' \in \tau$, $\{V \in \tau \mid V \subseteq U \cap U'\}$ is a quasi-net on $U \cap U'$ (think of the collection of all affinoid subdomains in a rigid space as a motivating analogy),
  (2) to each $V \in \tau$ there is assigned a k-affinoid algebra $\mathscr{A}_V$ and a homeomorphism $V \simeq \mathscr{M}(\mathscr{A}_V)$ such that if $V' \in \tau$ and $V' \subseteq V$ then V' is a k-affinoid subdomain of $\mathscr{M}(\mathscr{A}_V)$ with coordinate ring $\mathscr{A}_{V'}$ (as a Banach $\mathscr{A}_V$-algebra).
The triple $(X, \mathscr{A}, \tau)$ is a *k-analytic space*. If all $\mathscr{A}_V$ are strictly k-analytic then this triple is called a *strictly k-analytic space*.

EXAMPLE 5.1.4. Let $\mathscr{A}$ be a k-affinoid algebra. Letting $X = \mathscr{M}(\mathscr{A})$ and letting τ be the collection of k-affinoid subdomains (with $V \mapsto \mathscr{A}_V$ the usual assignment of coordinate rings to such subsets), we get a k-analytic space. If we instead take $\tau = \{X\}$ with $\mathscr{A}_X = \mathscr{A}$ then we get "another" k-analytic space. Clearly we want to consider these two k-analytic spaces to be naturally isomorphic (and to call these kinds of k-analytic spaces the *k-affinoid spaces*). If $\mathscr{A}$ is a strict k-affinoid algebra then we can take τ to be the collection of k-affinoid subdomains with strict coordinate ring, and this also should give rise to a k-analytic space that is naturally isomorphic to the other two. Making this idea precise requires defining the concept of *morphism* of k-analytic spaces.

Because we are not working with sheaves of rings on the underlying topological space X, we cannot easily pass to the sheaf language so as to hide the atlas as is possible in differential geometry. Thus, the definition of morphism (and in particular, isomorphism) between k-analytic spaces is rather subtle (in comparison with the definition of a morphism of ringed spaces). Rather than give the definition of a morphism of k-analytic spaces, we shall describe how one constructs morphisms in practice. Suppose that $(X, \mathscr{A}, \tau)$ and $(X', \mathscr{A}', \tau')$ are two k-analytic spaces and $\phi : X' \to X$ is a continuous map such that for all $V' \in \tau'$ and $V \in \tau$ with $V' \subseteq \phi^{-1}(V)$ there is given a compatible k-Banach algebra map $\mathscr{A}_V \to \mathscr{A}_{V'}$ that is transitive in the pair (V, V'). This data defines what is called a *strong morphism* $(X', \mathscr{A}', \tau') \to (X, \mathscr{A}, \tau)$, and more general kinds of morphisms can be defined by a gluing process at the level of affinoids in the quasi-net and their affinoid subdomains. The key point is that to construct the most general morphisms that one wants to have, it is necessary to define a calculus of fractions to formally invert certain kinds of strong morphisms, much as one does (with quasi-isomorphisms of complexes) in the theory of derived categories. We omit the details, and instead present the key example of the kind of strong morphism that has to be "inverted", corresponding to the formation of a maximal atlas in differential geometry:

EXAMPLE 5.1.5. Let $(X, \mathscr{A}, \tau)$ be a k-analytic space. Let $\overline{\tau}$ be the set of $V \subseteq X$ such that V is a k-affinoid subdomain in $\mathscr{M}(\mathscr{A}_{V'}) = V'$ for some $V' \in \tau$ containing V. Up to unique isomorphism there is a unique k-affinoid atlas structure $(X, \overline{\mathscr{A}}, \overline{\tau})$ extending $(X, \mathscr{A}, \tau)$. Let $\widehat{\tau}$ be the collection of compact Hausdorff subsets $W \subseteq X$ such that W is covered by a finite collection of $W_i \in \overline{\tau}$ such that

(1) $W_i \cap W_j \in \overline{\tau}$ for all i and j,
(2) the natural map $\overline{\mathscr{A}}_{W_i} \widehat{\otimes}_k \overline{\mathscr{A}}_{W_j} \to \overline{\mathscr{A}}_{W_i \cap W_j}$ is a surjection inducing a residue norm equivalent to the given norm on the target (this norm condition is automatically satisfied if k has nontrivial absolute value),
(3) the k-Banach subalgebra $\widehat{\mathscr{A}}_{\{W_i\}} \subseteq \prod \overline{\mathscr{A}}_{W_i}$ consisting of elements with the same image under both (continuous) maps $\prod_i \overline{\mathscr{A}}_{W_i} \rightrightarrows \prod_{i,j} \overline{\mathscr{A}}_{W_i \cap W_j}$ is k-affinoid, and the canonical map of sets $W \to \mathscr{M}(\widehat{\mathscr{A}}_{\{W_i\}})$ is a homeomorphism identifying each $W_j \subseteq W$ with a k-affinoid subdomain having coordinate ring $\overline{\mathscr{A}}_{W_j}$ as a Banach $\widehat{\mathscr{A}}_{\{W_i\}}$-algebra.

It can be shown that $\widehat{\mathscr{A}}_{\{W_i\}}$ only depends on W (and so is independent of the choice of $\{W_i\}$) up to norm equivalence; in particular, if $W \in \overline{\tau}$ then this k-Banach algebra is naturally isomorphic to the originally given $\overline{\mathscr{A}}_W$. Thus, we may and do write $\widehat{\mathscr{A}}_W$ to denote this k-affinoid algebra for general W as above. The assignment $\widehat{\mathscr{A}} : W \mapsto \widehat{\mathscr{A}}_W$ for $W \in \widehat{\tau}$ is a k-affinoid atlas, and so $(X, \widehat{\mathscr{A}}, \widehat{\tau})$ is a k-analytic space. The canonical strong morphism $(X, \mathscr{A}, \tau) \to (X, \widehat{\mathscr{A}}, \widehat{\tau})$ is an *isomorphism* of k-analytic spaces (with respect to the correct definition of morphism, which essentially creates inverses for maps of this type). The sense in which this procedure is like the formation of a maximal atlas in differential geometry is that $\widehat{\widehat{\tau}} = \widehat{\tau}$ and

$\widehat{\widehat{\mathscr{A}}} = \widehat{\mathscr{A}}$. For this reason, $(\widehat{\tau}, \widehat{\mathscr{A}})$ is called the *maximal k-affinoid atlas* on the given k-analytic space.

In the definition of k-analytic spaces as a category (including morphisms), one can consider the category obtained by using only strictly k-analytic affinoids in the data of τ, $\overline{\tau}$, and $\widehat{\tau}$. This is the (sub)category of *strictly k-analytic spaces*, and it is a natural category of interest when promoting a rigid-analytic space to a k-analytic space in a functorial manner. Likewise, if K/k is an extension of complete fields endowed with compatible absolute values then we can define the notion of a *morphism* from a K-analytic space to a k-analytic space by using maps of k-Banach algebras $\mathscr{A} \to \mathscr{B}$ from a k-affinoid algebra to a K-affinoid algebra (over $k \to K$) and the associated maps of topological spaces $\mathscr{M}(\mathscr{B}) \to \mathscr{M}(\mathscr{A})$. In this way, we can consider as a *single category* all K-analytic spaces for all K/k with a fixed k, and morphisms induce continuous maps on underlying topological spaces. There is nothing like this in rigid-analytic geometry for infinite-degree extensions K/k.

It is *not* obvious that a pair of strictly k-analytic spaces cannot admit a morphism as k-analytic spaces that is not already a morphism in the subcategory of strictly k-analytic spaces, since going outside of the strictly k-analytic category permits the possibility of gluing morphisms between non-strict affinoids that "cover" the spaces. It is a recent (hard) theorem of Temkin that the category of strictly k-analytic spaces is in fact a full subcategory of the category of k-analytic spaces (i.e., one gets no new morphisms between strictly k-analytic spaces by viewing them as merely k-analytic spaces). Thus, we do not need to distinguish these two kinds of morphisms when working with strictly k-analytic spaces.

An interesting class of k-analytic spaces is those for which every point admits an affinoid neighborhood. This will include the k-analytic spaces associated to arbitrary algebraic k-schemes, but it does not include the k-analytic spaces that one obtains from arbitrary quasi-compact and (quasi-)separated rigid spaces (by a process to be described later). Thus, this class of k-analytic spaces (which were the only ones considered in [**Ber1**]) are rather special and not sufficient for the purposes of using Berkovich's theory to study questions in rigid-analytic geometry. Nonetheless, these special spaces do play an important role in the theory (e.g., in the definition of properness, to be given later), and so they deserve a name:

DEFINITION 5.1.6. A k-analytic space $(X, \mathscr{A}, \tau)$ is *good* if every $x \in X$ has a neighborhood V with $V \in \widehat{\tau}$ (i.e., V is an affinoid neighborhood of x in the sense of the "maximal" atlas).

EXAMPLE 5.1.7. In Berkovich's theory there is a natural analytification functor from algebraic k-schemes to k-analytic spaces (characterized by a universal mapping property similar to the case of rigid-analytic geometry, to be discussed later), and the k-analytic spaces obtained from algebraic k-schemes are always good spaces. In contrast, we will later also discuss a natural functor from (certain) rigid-analytic spaces to k-analytic spaces, and the output of this construction can fail to be good.

For a quasi-compact and quasi-separated rigid space X, it is a difficult matter to describe in rigid-analytic terms when the associated k-analytic space X^{an} (in the sense of Example 5.2.3) is a good space; see Example 5.2.4 for a simple example for which X^{an} is not good. In proofs of "local" theorems for k-analytic spaces, it is typical to first settle the affinoid case, then the good case, and finally (by appropriate gluing arguments with quasi-nets) the general case. In [**Ber2**] the more general (i.e., possibly not good) kind of k-analytic spaces as defined above was introduced.

In order to explain how to associate a k-analytic space to an algebraic k-scheme or to a (reasonable) rigid-analytic space, and more generally to work with k-analytic spaces, we need to define an analogue of the admissible open subsets within a rigid space. These distinguished subsets of a k-analytic space are rarely open, and so we need to use a terminology other than "admissible open":

DEFINITION 5.1.8. Let $X = (X, \mathscr{A}, \tau)$ be a k-analytic space. A k-*analytic domain* in X is a subset $Y \subseteq X$ such that for all $y \in Y$ there exists $V_1, \ldots, V_n \in \widehat{\tau}$ with $y \in \cap V_j$ and $\cup V_j$ a neighborhood of y in Y (in particular, $V_j \subseteq Y$ for all j).

EXAMPLE 5.1.9. If Y is a k-analytic domain in $X = (X, \mathscr{A}, \tau)$ then the assignment $Y' \mapsto \widehat{\mathscr{A}_{Y'}}$ for Y' in the quasi-net $\widehat{\tau} \cap Y$ on Y naturally gives Y a structure of k-analytic space with respect to which a morphism $X' \to X$ of k-analytic spaces factors through $Y \subseteq X$ set-theoretically if and only if it does so as k-analytic spaces (in which case such a factorization is unique). This is the analogue of endowing an admissible open subset of a rigid space with a natural structure of rigid space. Any open subset of X is a k-analytic domain, but many interesting k-analytic domains (such as the affinoid ones) are generally not open.

5.2. Fiber products, analytification, and the G-topology. By using k-analytic domains, one can glue k-analytic spaces and morphisms between them in a manner similar to what is done with ringed spaces. Here are some basic and important examples of this gluing process (which we will not formally define, because it is a bit lengthy to explain due to the role of quasi-nets for spaces that are not good).

EXAMPLE 5.2.1. Let us explain fiber products and extension of the base field. First we consider the affinoid case. For a pair of morphisms $\mathscr{A} \rightrightarrows \mathscr{A}', \mathscr{A}''$ of k-affinoid algebras we get a k-affinoid algebra $\mathscr{A}' \widehat{\otimes}_{\mathscr{A}} \mathscr{A}''$, and for an analytic extension field K/k (i.e., one with compatible absolute value with respect to which it is complete) we get a K-affinoid algebra $\mathscr{A}_K = K \widehat{\otimes}_k \mathscr{A}$. The natural morphisms of k-analytic spaces $\mathscr{M}(\mathscr{A}' \widehat{\otimes}_{\mathscr{A}} \mathscr{A}'') \rightrightarrows \mathscr{M}(\mathscr{A}'), \mathscr{M}(\mathscr{A}'')$ have the same composition to $\mathscr{M}(\mathscr{A})$, and this makes $\mathscr{M}(\mathscr{A}' \widehat{\otimes}_{\mathscr{A}} \mathscr{A}'')$ into a fiber product $\mathscr{M}(\mathscr{A}') \times_{\mathscr{M}(\mathscr{A})} \mathscr{M}(\mathscr{A}'')$ in the category of k-analytic spaces. (The verification of this rests on the ability to uniquely determine and define morphisms from a k-analytic space by specification of the morphism on a suitable covering by affinoid k-analytic domains.) Likewise, the natural morphism $\mathscr{M}(\mathscr{A}_K) \to \mathscr{M}(\mathscr{A})$ from a K-affinoid space to a k-affinoid space serves as a fiber product $\mathscr{M}(\mathscr{A}) \times_{\mathscr{M}(k)} \mathscr{M}(K)$ in the category of analytic spaces over arbitrary analytic extension fields K'/K. This latter extension of the ground field functor is naturally transitive with respect to further extension of the base field.

These constructions in the affinoid case can be uniquely globalized to define fiber products of arbitrary k-analytic spaces and extension of the ground field functors for any K/k. Moreover, the formation of fiber products of k-analytic spaces is naturally compatible with any extension of the ground field. We omit any discussion of the details, except to make several remarks. First, the gluing process in these globalizations is a bit more involved than the analogous kind of gluing used in the case of schemes (though it follows a similar strategy), and fiber products are well-behaved with respect to k-analytic domains: for a pair of maps $X', X'' \rightrightarrows X$ of k-analytic spaces and a compatible collection of k-analytic domains $Y \subseteq X$, $Y' \subseteq X'$, and $Y'' \subseteq X''$, the natural map $Y' \times_Y Y'' \to X' \times_X X''$ is a k-analytic domain. The other aspect we wish to emphasize is that the topological space $|X' \times_X X''|$ has more subtle behavior than that of fiber products of complex-analytic spaces. As for schemes, the natural map $|X' \times_X X''| \to |X'| \times_{|X|} |X''|$ is generally *not* bijective (even if k is algebraically closed), essentially because of the nature of completed tensor products of analytic extension fields of k, but this map is a proper surjection of topological spaces (ultimately because of the k-affinoid case); recall that a map of topological spaces is *proper* if it is separated (i.e., has diagonal that is a closed embedding) and universally closed. Moreover, a point in $X' \times_X X''$ can fail to have a base of neighborhoods of the type $Y' \times_Y Y''$ for k-analytic domains $Y \subseteq X$, $Y' \subseteq X'$, $Y'' \subseteq X''$ (even if $X = Y = \mathscr{M}(k)$).

The notion of a *closed immersion* between k-analytic spaces can be defined in a manner that is similar to the rigid-analytic case (though the definition is a little more involved for spaces that are not good). In particular, closed immersions of k-analytic spaces are closed embeddings on topological spaces. Applying this concept of closed immersion to diagonal maps, we can define the notion of *separatedness* for a morphism $f : X' \to X$ of k-analytic spaces. Beware that, unlike for schemes, a diagonal map with closed image need not be a closed immersion. (For example, there are compact Hausdorff k-analytic spaces X that are not separated over $\mathscr{M}(k)$. Any quasi-compact and quasi-separated rigid space that is not separated gives rise to such an X, by the procedure in Example 5.2.3 below. The distinction between the Hausdorff condition and separatedness is due to the distinction between topological fiber products and k-analytic fiber products, as well as to the fact that diagonal maps for k-analytic spaces can lack a kind of "immersion" property that they always have in the categories of schemes and complex-analytic spaces.)

EXERCISE 5.2.2. Let $f : X' \to X$ be a map of k-analytic spaces, and let $\Delta_f : X' \to X' \times_X X'$ be the diagonal. Prove that the induced map $|f| : |X'| \to |X|$ on underlying topological spaces is separated (in the sense that its relative diagonal is a closed embedding) if and only if the preimage under $|f|$ of a Hausdorff subset of $|X|$ is Hausdorff. (Hint: k-analytic spaces are locally Hausdorff, and the natural map $|X' \times_X X'| \to |X'| \times_{|X|} |X'|$ is a proper surjection of topological spaces.) Also prove that if f is separated then $|f| : |X'| \to |X|$ is separated. In particular, if X is separated over $\mathscr{M}(k)$ then $|X|$ is Hausdorff. The converse is *false*.

EXAMPLE 5.2.3. The relations between k-analytic spaces, algebraic k-schemes, and rigid-analytic spaces over k go as follows. First assume that k has a nontrivial absolute value. For any quasi-compact and quasi-separated rigid space X over k, we can describe X in terms of gluing of finitely many k-affinoids along quasi-compact admissible opens. By choosing a finite (necessarily admissible) covering of each of these quasi-compact overlaps by affinoid opens, we have a gluing description of X in terms of finitely many k-affinoid spaces and affinoid subdomains.

This can be carried over to the k-analytic category, and it yields a compact Hausdorff strictly k-analytic space X^{an} that is called the *analytification* of X (in the sense of k-analytic spaces). The underlying set of X is the set of points $x \in X^{\mathrm{an}}$ such that $[\mathscr{H}(x) : k] < \infty$ (ultimately because of the affinoid case). This construction is independent of the choice of the covering, and it can be made into a functor from the category of quasi-compact and quasi-separated rigid spaces to the category of compact Hausdorff strictly k-analytic spaces. Spaces of this latter type may fail to be good! (See Example 5.2.4.) This is why the good spaces (the only ones considered in [**Ber1**]) are inadequate for a satisfactory general theory.

The functor $X \rightsquigarrow X^{\mathrm{an}}$ from quasi-compact and quasi-separated rigid spaces to compact Hausdorff strictly k-analytic spaces is an equivalence of categories, and it is compatible with the formation of fiber products and extension of the ground field. A map $f : X' \to X$ between quasi-compact and quasi-separated rigid spaces is a closed immersion if and only if f^{an} is a closed immersion, and so applying this to diagonal maps gives that f is separated if and only if f^{an} is separated. Thus, if X is a quasi-compact and quasi-separated rigid space that is not separated (such as a gluing of a closed unit disk to itself along a proper subdisk) then X^{an} is a strictly k-analytic space that is compact and Hausdorff but not separated (over $\mathscr{M}(k)$).

This procedure can be carried out without quasi-compactness of the spaces, provided that we impose a local finiteness condition on both sides. We thereby get an equivalence of categories between the category of quasi-separated rigid spaces admitting a locally finite (for the Tate topology) admissible covering by affinoid opens and the category of paracompact Hausdorff strictly k-analytic spaces. There is also a natural analytification functor from algebraic k-schemes to strictly k-analytic spaces (compatible with fiber products and extension of the base field), and this is compatible with the analytification functors from algebraic k-schemes to rigid spaces over k and from (suitable) rigid spaces over k to strictly k-analytic spaces.

EXAMPLE 5.2.4. Assume that k has nontrivial absolute value and consider the admissible open locus

$$\{(t,t') \in \mathbf{B}_k^2 \,|\, |t| = 1\} \cup \{(t,t') \in \mathbf{B}_k^2 \,|\, |t'| = 1\} \subseteq \mathbf{B}_k^2$$

that is a union of two affinoid subdomains and is the complement of the open unit polydisc. The associated k-analytic space $\mathscr{X}$ is a k-analytic domain in $\mathscr{M}(k\langle t, t'\rangle)$ that is not good. More specifically, the Gauss point ξ of $\mathscr{M}(k\langle t, t'\rangle)$ lies in $\mathscr{X}$ and does not admit a k-affinoid neighborhood in $\mathscr{X}$. One way to justify this is to use Temkin's theory of reductions of germs of k-analytic spaces. To briefly explain how this goes, we first note that the specialization morphism in Example 3.3.13 can be generalized to the context of k-analytic spaces, in which case the map of sets sp : $\mathfrak{X}_k^{\mathrm{an}} \to \mathfrak{X}$ for an admissible formal R-scheme $\mathfrak{X}$ will typically have image containing non-closed points in $\mathfrak{X}$. In general, if $x \in \mathfrak{X}_k^{\mathrm{an}}$ is a point with associated specialization $\widetilde{x} \in \mathfrak{X}$ then Temkin's theory shows that x admits a k-affinoid neighborhood in $\mathfrak{X}_k^{\mathrm{an}}$ if and only if the Zariski closure of $\widetilde{x}$ in $\mathfrak{X}_{\mathrm{red}}$ has normalization S that is proper over an affine $\widetilde{k}$-scheme of finite type (or equivalently, for which $\Gamma(S, \mathscr{O}_S)$ is finitely generated over $\widetilde{k}$ and $S \to \mathrm{Spec}(\Gamma(S, \mathscr{O}_S))$ is proper). In the above example we take $\mathfrak{X}$ to be the complement of the origin in $\mathrm{Spf}(R\{t, t'\})$, so $\mathfrak{X}_{\mathrm{red}}$ is the complement U of the origin in $\mathbf{A}_{\widetilde{k}}^2$. For $x = \xi$ we have that $\widetilde{x} \in \mathfrak{X}_{\mathrm{red}} = U$ is the generic point, so ξ has no k-affinoid neighborhood in $\mathscr{X}$ because $\Gamma(U, \mathscr{O}_U) = \widetilde{k}[t, t']$ and the inclusion $U \to \mathrm{Spec}(\widetilde{k}[t, t'])$ is not proper.

EXAMPLE 5.2.5. Assume that k has nontrivial absolute value and let $X = \mathscr{M}(k\langle T\rangle)$ and $U = X - \{\xi\}$, where ξ is the Gauss norm. These are both paracompact Hausdorff strictly k-analytic spaces, and under the fully faithful functor from such

k-analytic spaces into the category of rigid-analytic spaces over k we get that the associated rigid spaces X_0 and U_0 are as follows: $X_0 = \mathbf{B}_k^1$ and U_0 is a disjoint union of "twisted" open unit discs labeled by the closed points of $\mathbf{A}_{\overline{k}}^1$. (If k is algebraically closed then U_0 is a disjoint union of ordinary open unit discs.) Thus, although $U \to X$ is an open immersion, the corresponding map $U_0 \to X_0$ is a bijective local isomorphism that is not an isomorphism: this rigid-analytic map identifies U_0 with the disjoint union of the "open residue discs" for $\mathbf{B}_k^1$ (which are a non-admissible cover of $\mathbf{B}_k^1$). This gives a rigid-analytic interpretation for the operation of removing the non-classical point ξ from the space $\mathscr{M}(k\langle T\rangle)$.

EXERCISE 5.2.6. Let X be a quasi-compact and quasi-separated rigid space. What rigid-analytic conditions on X are equivalent to the property that the associated k-analytic space X^{an} is good? This is a difficult exercise.

REMARK 5.2.7. A remarkable feature of k-analytic spaces is that they have very nice topological properties. For example, a k-analytic space is locally path-connected (so it is connected if and only it is path-connected!), and the topological dimension (in any of the usual senses of topological dimension theory) is bounded above by the analytic dimension (defined in terms of coordinate rings of affinoid subdomains). This relation with topological dimension theory is extremely useful in the study of étale cohomology for analytic spaces, where there is an interesting spectral sequence that relates étale cohomology of an étale sheaf to topological cohomology of the underlying topological sheaf and Galois cohomology for the stalks at the the residue fields of points of the space. Coupled with bounds on the cohomological dimension of such fields, one gets vanishing results in étale cohomology via vanishing results in topological sheaf theory (using topological dimension theory).

On k-analytic spaces $X = (X, \mathscr{A}, \tau)$ we can use k-analytic subdomains to define a G-topology that is analogous to the Tate topology on rigid-analytic spaces. The objects of the G-topology on X are the k-analytic subdomains $Y \subseteq X$, and a covering $\{Y_i\}$ of such a Y by k-analytic subdomains of X is a set-theoretic covering with the "quasi-net" property that each $y \in Y$ has a neighborhood in Y of the form $Y_{i_1} \cup \cdots \cup Y_{i_n}$ with $y \in \cap_j Y_{i_j}$. This encodes the local finiteness property of admissible coverings in rigid-analytic geometry. We write X_G to denote X endowed with the G-topology. There is a *unique* way to define a sheaf $\mathscr{O}_{X_G} : Y \mapsto \mathscr{A}_Y$ on the G-topology of X such that for all $V \in \widehat{\tau}$ it recovers the coordinate ring $\mathscr{A}_V$ of the k-affinoid subdomain V. This is what underlies the procedure by which every k-analytic subdomain $Y \subseteq X$ naturally acquires a structure of k-analytic space such that it satisfies the set-theoretic mapping property that a morphism $X' \to X$ factoring through Y set-theoretically uniquely does so as k-analytic spaces. When we restrict the sheaf $\mathscr{O}_{X_G}$ on the G-topology to the collection of open subsets of X then we get a sheaf of k-algebras $\mathscr{O}_X$ on the topological space X. The pair $(X, \mathscr{O}_X)$ is a locally ringed space that is functorial in the k-analytic space X.

If X is not a good space then it is difficult to work with the stalks $\mathscr{O}_{X,x}$ when x lacks an affinoid neighborhood. This is the reason why flatness is a difficult concept to work with for general morphisms between k-analytic spaces. (In fact, there are examples of flat maps $\mathscr{A} \to \mathscr{A}'$ between non-strict k-affinoid algebras such that $K \widehat{\otimes}_k \mathscr{A} \to K \widehat{\otimes}_k \mathscr{A}'$ is not flat for some infinite-degree extension K/k! This is related to the existence of coordinate rings of "curves" such as $k\langle r^{-1}X, rX^{-1}\rangle$

whose associated topological space consists of a single point but which become 1-dimensional after a sufficiently large extension of the base field.) Often one studies "local" problems by first treating the good case (where $\mathscr{O}_{X,x}$ and its residue field can be computed using an affinoid neighborhood and so have some nice properties), and then bootstrapping to the general case by gluing arguments. For example, building up from the affinoid case (with Kiehl's theory of coherent sheaves in the strict case) permits one to construct a satisfactory theory of *coherent sheaves* on X_G and on the locally ringed space $(X, \mathscr{O}_X)$. An important example of a coherent sheaf is $\Omega^1_{X_G/Y_G}$ on X_G for any map of k-analytic spaces $f : X \to Y$ (and $\Omega^1_{X/Y}$ on $(X, \mathscr{O}_X)$ if X and Y are good). This sheaf of relative 1-forms is defined in terms of the coherent ideal sheaf of the relative diagonal in the separated (e.g., affinoid) case, much like in the case of schemes, and it is defined in general by a gluing procedure.

If X is a good k-analytic space then the categories of coherent sheaves on X_G and $(X, \mathscr{O}_X)$ are naturally equivalent. In general, the theory of coherence underlies the theory of *finite morphisms* (using coherent sheaves of algebras), and the coherent sheaf $\Omega^1_{X_G/Y_G}$ underlies the theories of smooth and étale morphisms of k-analytic spaces. The definition of a *finite morphism* in general (for spaces that are not necessarily good) illustrates how to work pointwise and to bootstrap from the affinoid case via quasi-nets, so let us give this definition.

DEFINITION 5.2.8. A map $f : X' \to X$ of k-analytic spaces is *finite* if, for all $x \in X$, there exist k-affinoid subdomains $V_1, \ldots, V_n \subseteq X$ such that $\cup V_j$ is a neighborhood of x in X, $x \in \cap V_j$, and each $V_j' = f^{-1}(V_j) \subseteq X'$ is a k-affinoid subdomain in X' with $\mathscr{A}_{V_j} \to \mathscr{A}_{V_j'}$ a finite map of Banach $\mathscr{A}_{V_j}$-algebras such that $\mathscr{A}_{V_j'}$ has its norm equivalent to the natural one induced by the finite $\mathscr{A}_{V_j}$-module structure (this final condition being automatic when k has nontrivial absolute value).

REMARK 5.2.9. The requirement on the pairs (V_j, V_j') in the preceding definition can be shown to be satisfied by the pair $(V, f^{-1}(V))$ for any k-affinoid subdomain $V \subseteq X$ when $f : X' \to X$ is a finite morphism.

Let us now illustrate the power of k-analytic spaces to give an intuitive proof of a fact in rigid geometry that is difficult to prove within the framework of rigid geometry. Assume that k has a nontrivial absolute value, and let A be a strict k-affinoid algebra. Let $X = \mathrm{Sp}(A)$ and $Z = \mathrm{Sp}(A/I)$ for an ideal $I \subseteq A$. Let $U \subseteq X$ be an admissible open subset such that $Z \subseteq U$. Finally, let $\{f_1, \ldots, f_n\}$ be a set of generators of I. I claim that for some $\varepsilon > 0$ (with $\varepsilon \in \sqrt{|k^\times|}$), the "tube" $\{|f_1| \leq \varepsilon, \ldots, |f_n| \leq \varepsilon\}$ around Z is contained in U. This can be proved by the methods of rigid geometry, but the only proof along such lines which I know is long and complicated. A short proof was given by Kisin via Raynaud's formal models, but let us now give an appealing short proof via Berkovich spaces. Without loss of generality we can assume that the open subset U is quasi-compact, so it is a finite union of rational domains in X by the Gerritzen–Grauert theorem. Thus, U^{an} is a k-analytic domain in X^{an}. The space X^{an} is compact Hausdorff, so the decreasing collection of neighborhoods $N_\varepsilon = \{|f_j| \leq \varepsilon\}$ of the closed set $Z^{\mathrm{an}} = \{f_j = 0\} = \cap_\varepsilon N_\varepsilon$ must be cofinal. Thus, it suffices to show that U^{an} is a neighborhood of Z^{an} in X^{an}. Since U^{an} is a (possibly non-affinoid) k-analytic domain in X^{an}, by the global theory of the relative interior (which we have only discussed for morphisms between affinoid spaces, but can be developed for morphisms between any pair of k-analytic spaces) it follows that the topological interior $\mathrm{int}_{X^{\mathrm{an}}}(U^{\mathrm{an}})$ of U^{an} in X^{an} is equal to the relative interior $\mathrm{Int}(U^{\mathrm{an}}/X^{\mathrm{an}})$. But the closed immersion $Z^{\mathrm{an}} \to X^{\mathrm{an}}$ is a finite map, so $Z^{\mathrm{an}} = \mathrm{Int}(Z^{\mathrm{an}}/X^{\mathrm{an}})$. There is a general transitivity property of relative interior with respect to a composition of morphisms, and applying this in the case of the composition $Z^{\mathrm{an}} \to U^{\mathrm{an}} \to X^{\mathrm{an}}$ yields

$$
\begin{aligned}
Z^{\mathrm{an}} = \mathrm{Int}(Z^{\mathrm{an}}/X^{\mathrm{an}}) &= \mathrm{Int}(Z^{\mathrm{an}}/U^{\mathrm{an}}) \cap \mathrm{Int}(U^{\mathrm{an}}/X^{\mathrm{an}}) \\
&\subseteq \mathrm{Int}(U^{\mathrm{an}}/X^{\mathrm{an}}) \\
&= \mathrm{int}_{X^{\mathrm{an}}}(U^{\mathrm{an}}),
\end{aligned}
$$

as desired.

5.3. Proper and étale morphisms. Another interesting application of k-analytic spaces within rigid geometry is in the theory of properness. Recall that although proper morphisms between rigid spaces do enjoy properties as in algebraic and complex-analytic geometry (such as coherence of higher direct images, and the theory of cohomology and base change), such maps are not as easy to work with; for example, it is not at all obvious that properness is preserved under composition. However, this latter problem was solved by Temkin by comparison with the definition of properness for k-analytic spaces. The key definition is as follows:

DEFINITION 5.3.1. A map $f : X' \to X$ of k-analytic spaces is *proper* if the map $|f| : |X'| \to |X|$ on underlying topological spaces is proper and for all morphisms $Y = \mathscr{M}(\mathscr{A}) \to X$ from k-affinoid spaces the pullback $Y' = X' \times_X Y$ is a good space and the relative boundary $\partial(Y'/Y)$ is empty. (Since we have not defined the relative boundary in a non-affinoid setting, let us at least give a concrete translation of the condition that $\partial(Y'/Y)$ is empty: for any $y' \in Y'$ and k-affinoid domain $U' \subseteq Y'$ that is a neighborhood of y' in the good space Y', we have that y' lies in the relative interior $\mathrm{Int}(U'/Y)$ as defined in our discussion of the affinoid theory.)

Roughly speaking, a proper map is one that is topologically proper and boundaryless in a relative sense; this is akin to the idea of a compact manifold. In the setting of proper k-schemes and coherent sheaves on them, Serre's methods can be carried over to prove versions of the GAGA theorems for analytification in the sense of k-analytic spaces. Also, a morphism of k-analytic spaces is finite if and only if it is proper and has finite fibers (exactly as for schemes and complex-analytic spaces).

Elementary properties of the relative interior in the affinoid case ensure that properness in the k-analytic category is stable under composition (and under base change and change of the ground field). Thus, the following theorem of Temkin elegantly disposes of the general problem of stability of properness under composition in rigid-analytic geometry (which was settled earlier by Lütkebohmert in the discretely-valued case).

THEOREM 5.3.2. *If k has a nontrivial absolute value then a map $f : X' \to X$ between quasi-compact and separated rigid spaces is proper in the sense of rigid-analytic spaces if and only if f^{an} is proper in the sense of k-analytic spaces.*

We conclude by briefly discussing some aspects of the theory of étale maps in the k-analytic category; this is exhaustively developed in [**Ber2**].

DEFINITION 5.3.3. A map $f : X' \to X$ of k-analytic spaces is (locally) *quasi-finite* if, for all $x' \in X'$, there exist open subsets $U' \subseteq X'$ around x' and $U \subseteq X$ around $f(x')$ such that $U' \subseteq f^{-1}(U)$ and $f : U' \to U$ is finite. If moreover for every affinoid domain $V \subseteq U$ the affinoid pullback $V' = V \times_U U'$ has coordinate ring that is flat over the coordinate ring of V then f is called *flat quasi-finite*.

This definition of (local) quasi-finiteness is inspired by the local structure theorem for maps with discrete fibers between complex-analytic spaces, and it is very different from the definition of quasi-finiteness in algebraic geometry; the Zariski topology is too coarse to permit using a definition as above in the algebraic theory. As a partial "justification" for the above definition of quasi-finiteness in the k-analytic setting, we remark that if $\phi : \mathscr{X}' \to \mathscr{X}$ is a map between k-schemes of finite type then it is quasi-finite (in the sense of algebraic geometry) if and only if $\phi^{\mathrm{an}} : \mathscr{X}'^{\mathrm{an}} \to \mathscr{X}^{\mathrm{an}}$ is quasi-finite in the sense of the preceding definition. Likewise, one has that ϕ is flat quasi-finite if and only if ϕ^{an} is flat quasi-finite. If $f : X' \to X$ is a flat quasi-finite map then for each $x \in X$ the fiber $f^{-1}(x)$ as an $\mathscr{H}(x)$-analytic space has the form $f^{-1}(x) = \coprod_i \mathscr{M}(\mathscr{B}_i)$ for $\mathscr{H}(x)$-finite algebras $\mathscr{B}_i$.

Flatness for a morphism of k-analytic spaces is generally a subtle matter, but for quasi-finite maps it is not too hard to develop the theory of flatness via the definition introduced above because such maps locally (on the source and target) are finite maps, and so one can work algebraically with finite maps between coordinate rings. For a general map of k-analytic spaces the condition of flatness in the sense of ringed spaces is not a good notion.

DEFINITION 5.3.4. A map $f : X' \to X$ between k-analytic spaces is *étale* if it is flat quasi-finite and for each $x \in X$ the fiber $f^{-1}(x)$ as a $\mathscr{H}(x)$-analytic space is a disjoint union of analytic spaces of the form $\mathscr{M}(\mathscr{K}_i)$ where $\mathscr{K}_i$ is a finite separable extension field of $\mathscr{H}(x)$.

Étale maps are always open on underlying topological spaces, and an *étale cover* is defined as in algebraic geometry: a collection of étale maps whose (open) images cover the target space. Also, the property of being étale is preserved under base change and change of the base field, and any X-map between étale k-analytic spaces over X is necessarily étale. By using étale covers one can define an *étale topology* on the category of étale objects over any k-analytic space. It can be shown that all representable functors on the category of k-analytic spaces are sheaves for the étale topology (proved in [**Ber2**] under a mild restriction).

Prior to the work of Huber and Berkovich there was some work by deJong and van der Put [**dJvP**] on étale cohomology for rigid spaces, but the approaches to étale cohomology via adic and k-analytic spaces have developed into theories with a vast range of applicability, roughly paralleling the development for schemes but with proofs that require many new ideas and exhibit an interesting interplay between algebraic and analytic points of view. A notable feature of the theory via k-analytic spaces, ultimately resting on the nice topological structure of k-analytic spaces (especially the role of paracompactness in the theory), is that étale cohomology with proper supports in the k-analytic theory can be defined exactly as in ordinary topological sheaf theory (as opposed to the case of schemes, where one has to bring in compactifications, which may not exist in analytic settings): it is the derived functor of the functor of sections with proper support! There are also comparison isomorphisms with étale cohomology in algebraic geometry via the analytification functor, analogous to Artin's comparison results for étale and topological cohomology over $\mathbf{C}$ but resting on proofs that lead to simplifications even in the classical case over $\mathbf{C}$.

Bibliography

[Ber1]   V. Berkovich, *Spectral theory and analytic geometry over non-Archimedean fields*, Mathematical Surveys and Monographs, vol. 33, American Mathematical Society, 1990.

[Ber2]   V. Berkovich, *Étale cohomology for non-Archimedean analytic spaces*, Publ. Math. IHES, **78** (1993), pp. 7–161.

[B]   S. Bosch, *Lectures on formal and rigid geometry*, SFB **478** (preprint; June, 2005), http://www.math1.uni-muenster.de/sfb/about/publ/heft378.ps.

[BGR]   S. Bosch, U. Günzter, R. Remmert, *Non-Archimedean analysis*, Springer-Verlag, 1984.

[BL1]   S. Bosch, W. Lütkebohmert, *Formal and rigid geometry* I, Math. Annalen, **295** (1993), pp. 291–317.

[BL2]   S. Bosch, W. Lütkebohmert, *Formal and rigid geometry* II, Math. Annalen, **296** (1993), pp. 403–429.

[EGA]   J. Dieudonné, A. Grothendieck, *Éléments de géométrie algébrique*, Publ. Math. IHES, **4, 8, 11, 17, 20, 24, 28, 32**, (1960-7).

[D]   A. Ducros, *Espaces analytiques p-adiques au sens de Berkovich*, Sém. Bourbaki, Exp. 958, 2005-06.

[FvP]   J. Fresnel, M. van der Put, *Rigid-analytic geometry and its applications*, Progress in Mathematics **218**, Birkhäuser, Boston (2004).

[FK]   K. Fujiwara, F. Kato, *Rigid geometry and applications*, in Moduli spaces and arithmetic geometry, Adv. Stud. Pure Math., **45**, Math. Soc. Japan, Tokyo, 2006, pp. 327–386.

[H]   R. Huber, *Étale cohomology of rigid analytic varieties and adic spaces*, Aspects of Mathematics **E30**, Friedrich Vieweg & Sohn, Braunschweig (1996).

[dJvP]   A.J. de Jong, M. van der Put, *Étale cohomology of rigid analytic spaces*, Doc. Math., **1** (1996), pp. 1–56.

[Liu]   Q. Liu, *Un contre-exemple au "critére cohomologique d'affinoidicité"*, C.R. Acad. Sci. Paris Sér. I Math., **307** (1988), no. 2, pp. 83–86.

[S]   J-P. Serre, *Lie groups and Lie algebras* (5th ed.), Lecture Notes in Mathematics **1500**, Springer–Verlag, Berlin (2006).

[T]   M. Temkin, *On local properties of non-Archimedean analytic spaces II*, Israeli Journal of Math., **140** (2004), pp. 1–27.

CHAPTER 2

The p-adic upper half plane

Samit Dasgupta and Jeremy Teitelbaum

Introduction

The p-adic upper half plane $\mathfrak{X}$ is a rigid analytic variety over a p-adic field K, on which the group $\mathrm{GL}_2(K)$ acts, that Mumford introduced (as a formal scheme) as part of his efforts to generalize Tate's p-adic uniformization of elliptic curves to curves of higher genus. The $\mathbf{C}_p$–valued points of $\mathfrak{X}$ are just $\mathbf{P}^1(\mathbf{C}_p) - \mathbf{P}^1(K)$, with $\mathrm{GL}_2(K)$ acting by linear fractional transformations. Mumford showed that the appropriate generalization of Tate's elliptic curves – the "totally split" curves of higher genus – could be constructed as the quotient of the space $\mathfrak{X}$ by appropriate discrete groups $\Gamma \in \mathrm{GL}_2(K)$. Mumford's work acquired even greater significance for number theorists when Cerednik and Drinfeld showed that an important class of modular curves – the Shimura curves – could be constructed by p-adic uniformization by choosing for the discrete group Γ an appropriate arithmetic subgroup coming from a definite quaternion algebra over $\mathbf{Q}$. More recently, the p-adic upper half plane has figured prominently in recent developments in arithmetic geometry. In Section 1 of these notes, we will construct the space $\mathfrak{X}$ as a rigid variety and describe some of its most fundamental geometric properties, and in subsequent sections we will explore some of this more recent work.

Our focus in Section 2 will be the analytic theory of $\mathfrak{X}$, and in particular the relationship between spaces of functions on the p-adic upper half plane and distributions on $\mathbf{P}^1(K)$, which is the "boundary" of $\mathfrak{X}$. One main result will be the construction of the Poisson integral for $\mathfrak{X}$; in a manner analogous to the classical Poisson transform, this integral allows one to recover rigid analytic functions on $\mathfrak{X}$ from appropriate boundary distributions by integrating against a kernel function.

In Sections 3 and 4, we establish connections between number theory and the geometry of the p-adic upper half plane, with particular emphasis on the relationship between the p-adic upper half plane and $\mathcal{L}$-invariants. If E is an elliptic curve over $\mathbf{Q}$ with split multiplicative reduction at p and analytic Mordell-Weil rank zero, then [**25**] conjectured and [**19**] proved that the p-adic L-function of the modular form f corresponding to E vanishes to order 1 and the special value of $L_p'(1)$ differs from the classical special value by the number

$$\mathcal{L}(f) = \frac{\log_p(q_E)}{\mathrm{ord}_p(q_E)}$$

The first author was partially supported by NSF grant DMS-0653023. The second author was supported by NSF grant DMS-0245410. This work was partially supported by NSF grant DMS-0602287. The authors thank Henri Darmon and Christophe Breuil for helpful comments on the manuscript.

where q_E is the Tate period of the curve E at p. The paper [**25**] made a weak conjecture (the exceptional zero conjecture) about the relationship between the special values of the p-adic and classical L-functions of higher weight modular forms, and in an attempt to make that conjecture more precise, different mathematicians introduced a whole collection of more general $\mathcal{L}$-invariants associated to such forms. Many of these $\mathcal{L}$-invariants – all of which are now known to be equal – are related in some way to the p-adic upper half plane, and after a general discussion of $\mathcal{L}$-invariants we focus in particular on three such: one defined by the second author of these notes; one defined by Orton; and one defined by Breuil. Much of Section 3 is devoted to Orton's proof of the exceptional zero conjecture using her invariant, while Section 4 discusses Breuil's invariant and its relationship with the cohomology of modular curves and a possible p-adic Langlands correspondence.

1. Geometry of the p-adic upper half plane

1.1. Basic notations. We let K denote a finite extension of the p-adic numbers $\mathbf{Q}_p$, and we let G be the group $\mathrm{GL}_2(K)$. If o_K denotes the ring of integers in K, then we write G_o for the maximal compact subgroup $\mathrm{GL}_2(o_K)$ in G. We let π be a uniformizing parameter for o_K and write $|\cdot|$ for the normalized p-adic absolute value on K extending the p-adic absolute value on $\mathbf{Q}_p$. We will also use the additive valuation

$$\omega : K \to \mathbf{Z}$$

normalized so that $\omega(\pi) = 1$.

Let V be a fixed two dimensional vector space over K, viewed as a space of row vectors, on which G acts on the left by the formula

$$g([x,y]) = [x,y] \begin{pmatrix} a & b \\ c & d \end{pmatrix}^{-1}.$$

When we refer to $\mathbf{P}^1$ we mean specifically $\mathbf{P}(V)$ with its G-action. We let Ξ_0 and Ξ_1 be the dual elements in V^* to the standard basis vectors $[1,0]$ and $[0,1]$ in V; they are "homogeneous coordinates" on $\mathbf{P}^1$. A linear form in Ξ_0 and Ξ_1 is called unimodular if at least one of its two coefficients is a unit in o_K and the other coefficient lies in o_K.

The coordinate function

$$z = \frac{\Xi_0}{\Xi_1}$$

is acted on by a matrix $g = \begin{pmatrix} a & b \\ c & d \end{pmatrix} \in G$ through the formula

$$
\begin{aligned}
g_*(z)([x,y]) &= z(g^{-1}([x,y])) \\
&= z([x,y]g) \\
&= z([ax+cy, bx+dy]) \\
&= \frac{az+c}{bz+d}.
\end{aligned}
$$

1.2. The p-adic upper half plane. The central object of interest in this series of lectures is the p-adic upper half plane $\mathcal{X}$, a rigid analytic space whose L-points are given by the rule

$$\mathcal{X}(L) = \mathbf{P}^1(L) \backslash \mathbf{P}^1(K)$$

for complete extension fields L of K.

1.2.1. *An admissible covering.* To construct $\mathcal{X}$, we need to describe an admissible covering that defines its rigid structure. We will describe an increasing sequence of affinoid subdomains $\mathcal{X}_n^-$ in $\mathbf{P}^1$, for integer $n \geq 1$, and some related admissible domains $\mathcal{X}_n$, so that $\mathcal{X}$ is the union of the $\mathcal{X}_n^-$ and the $\mathcal{X}_n$. Essentially, $\mathcal{X}_n^-$ is constructed by deleting from $\mathbf{P}^1$ smaller and smaller balls around the rational points.

We will describe this in a coordinate-free way that may not be the simplest approach in dimension 1 but is easier to generalize to higher dimension (see [**35**]).

Given $x \in \mathbf{P}^1(\mathbf{C}_p)$, we may choose homogeneous coordinates $x = [x_0, x_1]$ for x that are unimodular, meaning that both coordinates are integral, but at least one is not divisible by π. For a real number $r > 0$, let

$$B(x, r) = \{y \in \mathbf{P}^1(\mathbf{C}_p) : \omega(y_0 x_1 - y_1 x_0) \geq r\},$$

where we always take a unimodular representative $[y_0, y_1]$ of y. Also define

$$B^-(x, r) = \{y \in \mathbf{P}^1(\mathbf{C}_p) : \omega(y_0 x_1 - y_1 x_0) > r\}.$$

LEMMA 1.2.1. *Let x and x' be two elements of $\mathbf{P}^1(K)$, and let n be a positive integer. Then $B(x, n) \cap B(x', n) \neq \emptyset$ if and only if $[x_0, x_1] \equiv \lambda[x_0', x_1'] \pmod{\pi^n}$ for some unit $\lambda \in o_K^*$.*

PROOF. Suppose $y \in B(x, n) \cap B(x', n)$. Then we have the equations:

$$\begin{aligned} \omega(x_1 y_0 - y_1 x_0) &\geq & n, \\ \omega(x_1' y_0 - y_1 x_0') &\geq & n. \end{aligned}$$

Suppose for convenience that y_0 is a unit. Then we can conclude that

$$\omega(x_0' x_1 - x_1' x_0) \geq n.$$

This means that the vectors $[x_0, x_1]$ and $[x_0', x_1']$ are linearly dependent modulo π^n, which is the claim. Conversely, if the vectors are linearly dependent mod π^n, we may construct a y in the intersection of the two sets by choosing a unimodular representative of the kernel of the appropriate matrix made out of x and x'. $\square$

DEFINITION 1.2.2. For each integer $n > 0$, let $\mathcal{P}_n$ be a set of representatives for the points of $\mathbf{P}^1(K)$ modulo π^n. Let $\mathcal{X}_n$ be the set

$$\mathcal{X}_n := \mathbf{P}^1(\mathbf{C}_p) \backslash \bigcup_{x \in \mathcal{P}_n} B(x, n).$$

Let $\mathcal{X}_n^- \subset \mathcal{X}_n$ be the set

$$\mathcal{X}_n^- := \mathbf{P}^1(\mathbf{C}_p) \backslash \bigcup_{x \in \mathcal{P}_n} B^-(x, n-1).$$

Let

$$\mathcal{X} = \bigcup_n \mathcal{X}_n = \bigcup_n \mathcal{X}_n^-.$$

We can make the sets $\mathcal{X}_n$ and $\mathcal{X}_n^-$ more explicit. Fix an integer $n \geq 1$. Then we can choose representatives for $\mathcal{P}_n$ as follows:

$[a_i, 1]$, where $\{a_i\}_{i=0}^{q^n - 1}$ is a set of representatives in o_K for $o_K / \pi^n o_K$;

$[1, b_i]$, where $\{b_i\}_{i=0}^{q^{n-1} - 1}$ is a set of representatives in πo_K for $\pi o_K / \pi^n o_K$.

Then it follows from the definitions that $\mathfrak{X}_n$ is the set of points $x \in \mathbf{P}^1$ defined by the inequalities

$$\omega(z(x) - a_i) < n \qquad i = 0, \ldots, q^n - 1,$$
$$\omega\left(\frac{1}{z(x)} - b_i\right) < n \qquad i = 0, \ldots, q^{n-1} - 1.$$

It will be useful later to have slightly different inequalities for the covering domains. In particular, it is easy to check that if $\omega(b) < n$, then

$$\omega(1/z - b) < n \Leftrightarrow \omega(z - 1/b) < n - 2\omega(b).$$

Consequently we can (choosing $b_0 = 0$) rewrite the system of inequalities defining $\mathfrak{X}_n$ as

$$
\begin{array}{lll}
& \omega(z - a_i) < n & \text{for } i = 0, \ldots, q^n - 1, \\
(1.2.3) & \omega(z - 1/b_i) < n - 2\omega(b_i) & \text{for } i = 1, \ldots, q^{n-1} - 1, \\
& \omega(z) > -n, &
\end{array}
$$

and, for $\mathfrak{X}_n^-$,

$$
\begin{array}{lll}
& \omega(z - a_i) \leq n - 1 & \text{for } i = 0, \ldots, q^n - 1, \\
(1.2.4) & \omega(z - 1/b_i) \leq n - 1 - 2\omega(b_i) & \text{for } i = 1, \ldots, q^{n-1} - 1, \\
& \omega(z) \geq 1 - n. &
\end{array}
$$

PROPOSITION 1.2.5. $\mathfrak{X}$ *is an admissible open subdomain of* $\mathbf{P}^1$ *and the coverings of* $\mathfrak{X}$ *by the families* $\{\mathfrak{X}_n\}_{n=1}^\infty$ *and* $\{\mathfrak{X}_n^-\}_{n=1}^\infty$ *are admissible coverings. In the latter case, the covering is by open affinoid domains.*

PROOF. See the discussion following Lemma 3 in [**35**]. $\qquad\square$

1.2.2. *The ring* $\mathcal{O}_\mathfrak{X}$ *of entire functions on* $\mathfrak{X}$. The ring of entire functions on $\mathcal{O}_\mathfrak{X}$ is the projective limit of the affinoid algebras $\mathcal{O}(\mathfrak{X}_n^-)$ as $n \to \infty$:

$$\mathcal{O}_\mathfrak{X} := \varprojlim_n \mathcal{O}(\mathfrak{X}_n^-).$$

Many important function-theoretic properties of $\mathfrak{X}$ flow from two key facts:

(1) $\mathfrak{X}$ is a (smooth, one-dimensional) rigid analytic Stein space;
(2) the restriction maps between the affinoid algebras $\mathcal{O}(\mathfrak{X}_n^-)$ are compact maps. (Recall that a continuous linear map $f : A \to B$ between Banach spaces is called compact if the image of the unit ball in A has compact closure in B).

The compactness property (2) of the transition maps is a fairly general phenomenon. At its core is the following special case. Consider the affinoid ball of points z with $\omega(z) \geq -1$, with its associated affinoid algebra $K\langle \pi T\rangle$. Consider also the restriction map to the sub-affinoid of points z with $\omega(z) \geq 0$, and its affinoid algebra $K\langle T\rangle$. Ignoring the ring structure, we see that the image in $K\langle T\rangle$ of the unit ball in $K\langle \pi T\rangle$ is the subspace of power series $\sum a_n T^n$, with $a_n \in o_K$, and whose coefficients satisfy

$$\omega(a_n) \geq n \text{ for all } n \geq 0.$$

One can verify that the norm topology on $K\langle T\rangle$ identifies this subset with the space of sequences $(\pi^n a_n)$ with $a_n \in o_K$, equipped with its product topology. As a product of compact sets this space is clearly compact by Tychonoff's theorem.

(See [**34**], the example following Remark 12.8). For a proof of compactness in our more general situation, see [**38**], Proposition 4.

A Fréchet space is a locally convex topological vector space that is complete and metrizable. The topology on a Fréchet space can be given by a countable family of semi-norms. Such spaces arise naturally as projective limits of Banach spaces – see [**34**, Chapter I, Section 8] for a general discussion.

In functional analysis, one can equip the space of continuous linear forms on a locally convex topological vector space with many different topologies. One of the most important of these is the "strong topology," which is the topology of uniform convergence on bounded subsets. If V is a topological vector space, we let V' be the vector space of continuous linear forms and V'_b be this space equipped with the strong topology (the "b" is for "bounded"). Recall that a topological vector space is reflexive if V is isomorphic to $(V'_b)'_b$ by the natural map from V to V'' given by evaluation. All of these topics are thoroughly treated in [**34**].

The principal consequence of this compactness is the following result describing $\mathcal{O}_{\mathcal{X}}$ as a topological vector space.

PROPOSITION 1.2.6. $\mathcal{O}_{\mathcal{X}}$ *is a reflexive Fréchet space. The topology comes from the family of norms on the Banach algebras* $\mathcal{O}(\mathcal{X}_n^-)$.

For a proof, see [**34**], Proposition 16.5.

We next briefly recall the definition of a Stein space, following Kiehl ([**22**]):

DEFINITION 1.2.7. *A rigid space* X *is called a (quasi)-Stein space if there is an increasing sequence* $U_1 \subset U_2 \subset \cdots$ *of open affinoid subdomains of* X *forming an admissible covering such that the transition maps* $\mathcal{O}(U_i) \to \mathcal{O}(U_{i-1})$ *have dense image.*

One can see that this density property holds for the transition maps $\mathcal{O}(\mathcal{X}_n^-) \to \mathcal{O}(\mathcal{X}_{n-1}^-)$ for $n \geq 2$ by considering the set of algebraic rational functions on $\mathbf{P}^1$ whose polar divisors are supported on the K-rational points. This set of rational functions forms a dense subring in each $\mathcal{O}(\mathcal{X}_n^-)$.

With regard to coherent sheaves, a Stein space behaves somewhat like an affine variety does in algebraic geometry. In particular, we have the following theorem (see [**22**]):

PROPOSITION 1.2.8. *Let* $\mathcal{M}$ *be a coherent sheaf on* $\mathcal{X}$. *Then* $H^i(\mathcal{X}, \mathcal{M}) = 0$ *for* $i > 0$ *(Theorem B), and, if* $M = H^0(\mathcal{X}, \mathcal{M})$ *then the map* $\mathcal{O}_{\mathcal{X}} \otimes M \to \mathcal{M}(\mathcal{X}_i^-)$ *has dense image for any* $i \geq 1$ *(Theorem A).*

One can do even better. To give a coherent sheaf $\mathcal{M}$ on $\mathcal{X}$ is the same as giving, for each i, a finitely generated module M_i for $\mathcal{O}(\mathcal{X}_i^-)$. The global sections of this sheaf are an $\mathcal{O}_{\mathcal{X}}$-module

$$M = \operatorname{proj\,lim} M_i.$$

As is explained in [**16**] (see also [**42**]), one can recover the M_i as $\mathcal{O}(\mathcal{X}_i^-) \otimes M$. One can also characterize those M arising as global sections of a coherent sheaf by requiring that these tensor products be finitely generated.

1.3. The Reduction Map.

1.3.1. *The Bruhat-Tits Tree.* Our next task will be to introduce the Bruhat-Tits tree, which functions as a beautiful combinatorial approximation of $\mathfrak{X}$. We work always with the fixed two dimensional vector space V^* over K. By a lattice L in V^* we mean a free rank-two o_K module in V^*. We say two lattices L_1 and L_2 are equivalent if there is a scalar $a \in K$ so that $L_1 = aL_2$.

DEFINITION 1.3.1. Let X be the graph whose vertices are equivalence classes $[L]$ of lattices $L \subset V^*$, where two vertices x and y are joined by an edge if $x = [L_1]$ and $y = [L_2]$ with

$$\pi L_1 \subsetneq L_2 \subsetneq L_1.$$

PROPOSITION 1.3.2. *The graph X is a homogeneous tree of degree $q + 1$.*

PROOF. The degree assertion follows from the fact that the edges leaving a given vertex $[L_1]$ correspond to the distinct lattices L_2 satisfying the adjacency relation

$$\pi L_1 \subsetneq L_2 \subsetneq L_1$$

and these in turn are in bijection with the one-dimensional $o_K/\pi o_K$-subspaces in the two-dimensional $o_K/\pi o_K$-vector space $L_1/\pi L_1$. There are $q+1$ such subspaces, so there are $q+1$ adjacent vertices.

Suppose that X is not a tree. A cycle in X would be minimally represented by a chain of lattices

$$L' \subsetneq L_d \subsetneq L_{d-1} \subsetneq \cdots L_1 \subsetneq L$$

where $L' = \pi^r L$ for some positive integer r and where none of the intermediate lattices are equivalent. Because L/L' is not a cyclic o_K-module, there is a smallest i such that L/L_i is cyclic but L/L_{i+1} is not. It follows that L_{i-1}/L_{i+1} is a non-cyclic, length 2 o_K-module, so $L_{i+1} = \pi L_{i-1}$. This contradicts the minimality of the representation of the cycle, and so we conclude that X has no cycles.   □

As constructed so far, the tree X is a combinatorial object. If we view each edge of X (with its bounding vertices) as a copy of the unit interval, we obtain a topological space called the geometric realization of X. Since it is this geometric realization that we are principally interested in, we will just go ahead and use the letter X to refer to it. A point on the edge in X joining the vertices $[L]$ and $[L']$ is determined by its barycentric coordinates: for $t \in [0, 1]$, we write $x = (1-t)[L]+t[L']$ to indicate the point "at distance t from the vertex $[L]$ in the direction of $[L']$."

The group G acts on the lattices in V^* and on the tree X. The stabilizer of a lattice class $[L]$ is the subgroup of G generated by the center of G and by the compact open subgroup $\mathrm{GL}(L) \subset G$. If L is the lattice spanned by Ξ_0 and Ξ_1, then $\mathrm{GL}(L)$ is just $G_o = \mathrm{GL}_2(o_K) \subset G = \mathrm{GL}_2(K)$.

From now on, we let L_0 and L_1 be the lattices $\langle \Xi_0, \Xi_1 \rangle$ and $\langle \Xi_0, \pi\Xi_1 \rangle$ respectively. We will also write v_0 for the vertex of X corresponding to $[L_0]$, and e_0 for the edge of X running from v_0 to the vertex corresponding to L_1.

1.3.2. *Norms.* The tree X parameterizes norms on the two dimensional vector space V^* in a natural way. This description actually pre-dates Bruhat-Tits (see [**18**]).

DEFINITION 1.3.3. A norm on V^* is a function $\gamma : V^* \to \mathbf{R} \cup \{\infty\}$ such that

- $\gamma(x) = \infty$ if and only if $x = 0$,
- $\gamma(ax) = \omega(a) + \gamma(x)$ for $a \in K$,
- $\gamma(x + y) \geq \inf\{\gamma(x), \gamma(y)\}$.

Two norms γ_1 and γ_2 on V^* are considered equivalent if $\gamma_1 - \gamma_2 = C$ for some constant $C \in \mathbf{R}$. Given a point $x \in X$, we may associate an equivalence class of norms on V^*. There are two cases to consider:

Case 1: x is a vertex. In this case, choose a lattice L representing x and let

$$\gamma(w) = -\inf\{n \in \mathbf{Z} : \pi^n w \in L\}.$$

Alternatively, choose a basis ℓ_0, ℓ_1 for L and define

$$\gamma(a\ell_0 + b\ell_1) = \inf\{\omega(a), \omega(b)\}.$$

Case 2: $x = (1 - t)[L] + t[L']$. In this case, choose a basis ℓ_0, ℓ_1 for L such that L' is spanned by $\ell_0, \pi\ell_1$. Define

$$\gamma(a\ell_0 + b\ell_1) = \inf\{\omega(a), \omega(b) - t\}.$$

Notice that, in Case 2, the construction is consistent with Case 1 when $t = 0$ or $t = 1$. It's not hard to check, too, that Case 2 is compatible with different choices of lattices in the equivalence classes.

PROPOSITION 1.3.4. *This construction establishes a bijection between the set of equivalence classes of norms on V^* and the points of the space X.*

PROOF. We will construct an inverse map to the construction given above. Let γ be any norm on V^*. By translating γ in its equivalence class, we may assume that there is some $x \in V^*$ with $\gamma(x) = 0$. Let L' be the unit ball for γ:

$$L' = \{x \in V^* : \gamma(x) \geq 0\}.$$

Choose a (finite) set of representatives R in L for the projective space $\mathbf{P}(L'/\pi L')$. The norm γ is determined by its values on elements of R, all of which lie in $[0, 1)$. To see this, write any $w \in V^*$ as $w = u\pi^m r + \pi^{m+1}w'$ with $u \in o_K^*$ and $w' \in L'$. Then $\gamma(w) = m + \gamma(r)$. Now, if $\gamma(r) = 0$ for all $r \in R$, then γ is the norm associated to the lattice L as in Case 1 above. One can check further that, in this case, all norms equivalent to γ have unit balls equivalent to L', so the association of L' to γ makes sense. On the other hand, if there exists a point r with $\gamma(r) > 0$, then that r is unique. Indeed, if there were two such elements r and r', then these elements span L', from which it follows that $\gamma(x) > 0$ for all $x \in L'$, contrary to hypothesis. Now set $L = L' + r/\pi$. The norm γ, in this case, corresponds to the norm coming from Case 2 with the given lattice classes $[L]$ and $[L']$ and $t = 1 - \gamma(r)$. In this case, one checks that for norms equivalent to γ, the unit balls are either equivalent to L or to L'; and that if one follows the recipe given here to construct a point in X, one gets the same point regardless of which of these two possibilities holds for the chosen representative norm. $\qquad\square$

1.3.3. *The group action.* The group action of G on X translates to the action on norms through the rule $(g \cdot \gamma)(x) = \gamma(g^{-1}x)$.

LEMMA 1.3.5. *Some properties of the group action on X are:*

(1) *The group G permutes the vertices and edges of X transitively.*

(2) *The stabilizer of the lattice L_0 spanned by the standard coordinates Ξ_0 and Ξ_1, and the corresponding norm, is the subgroup $K^* \mathrm{GL}_2(o_K)$ in $\mathrm{GL}_2(K)$.*

(3) *If an element of G fixes the two endpoints of an edge, then it fixes the edge pointwise; the stabilizer of the edge in X corresponding to the lattice pair $\pi L_0 \subset L_1 \subset L_0$, where $L_1 = \langle \Xi_0, \pi \Xi_1 \rangle$ is, mod the center of G, the "Iwahori subgroup"*

$$B = \{g \in G_o : g \equiv \begin{pmatrix} a & b \\ 0 & d \end{pmatrix} \bmod \pi\}.$$

(4) *B is of index two in its normalizer; this normalizer is generated by B and any element of G that interchanges the two boundary vertices of the basic edge in (3). One such element is*

$$n = \begin{pmatrix} 0 & 1 \\ \pi & 0 \end{pmatrix}.$$

1.3.4. *Ends.* Let

$$([\Lambda_0], [\Lambda_1], \dots)$$

be an infinite, non-backtracking sequences of adjacent vertices, which we can think of as an infinite ray in the tree heading off to ∞. Two such sequences are equivalent if they differ by a finite initial sequence of vertices, i.e.,

$$([\Lambda_0], [\Lambda_1], \dots) \sim ([\Lambda_0'], [\Lambda_1'], \dots)$$

if $[\Lambda_n] = [\Lambda_{n+m}']$ for some fixed $m \in \mathbf{Z}$, and all n large enough. An equivalence class of such sequences is called an "end" of the tree. The set of ends of X is denoted $\mathrm{Ends}(X)$ and represents the set of points "at infinity" for the tree X.

To an oriented edge e running from $[\Lambda_0]$ to $[\Lambda_1]$, we associate the subset

$$U(e) = \{x \in \mathrm{Ends}(X) : x = ([\Lambda_0], [\Lambda_1], \dots)\}.$$

The collection of sets $U(e)$, as e runs through the oriented edges of X, form the basis for a topology on $\mathrm{Ends}(X)$.

Given an end $x = ([\Lambda_0], [\Lambda_1], \dots)$, we can construct a representing sequence of lattices

$$\Lambda_0 \supsetneq \Lambda_1 \supsetneq \Lambda_2 \supsetneq \cdots$$

with the property that Λ_i/Λ_{i+1} is isomorphic to $o_K/\pi o_K$. Since the sequence has no backtracking, the argument that we used in Proposition 1.3.2 tells us that Λ_0/Λ_i is a cyclic o_K-module of length i for each $i \geq 1$, and the same is true for $\Lambda_i/\pi^i \Lambda_0$. As a result we may choose $\ell_i \in \Lambda_0 \backslash \pi \Lambda_0$ such that

$$\Lambda_i = o_K \ell_i + \pi^i \Lambda_0.$$

Similarly,

$$\Lambda_{i+1} = o_K \ell_{i+1} + \pi^{i+1} \Lambda_0.$$

Because $\Lambda_{i+1} \subsetneq \Lambda_i$ and both ℓ_i and ℓ_{i+1} belong to $\Lambda_0 \backslash \pi \Lambda_0$, we must have

$$\ell_{i+1} \equiv a\ell_i \pmod{\pi^i \Lambda_0},$$

for some $a \in o_K^*$. We conclude that we may choose the ℓ_i to form a coherent sequence converging to a nonzero element ℓ of the intersection $\cap \Lambda_i$, and that this intersection is one-dimensional. The kernel of ℓ is a point of $\mathbf{P}^1$, denoted $N(x)$.

LEMMA 1.3.6. *The map $N : \mathrm{Ends}(X) \to \mathbf{P}^1$ is a G-equivariant homeomorphism.*

PROOF. Let $L_0 = o_K \Xi_0 + o_K \Xi_1$ as above. Given a point $[x, y]$ in $\mathbf{P}^1$, written with unimodular coordinates, let $\ell = -y\Xi_0 + x\Xi_1 \in L_0$. The end

$$(L_0, o_K \ell + \pi L_0, o_K \ell + \pi^2 L_0, \ldots)$$

maps, under N, to the point $[x, y]$. This shows the map is surjective. Conversely, we showed above that, if ℓ is a generator for the intersection of the sequence of lattices Λ_i representing an end

$$([\Lambda_0], [\Lambda_1], [\Lambda_2], \ldots).$$

with $\Lambda_0 = L_0$, then we must have

$$\Lambda_i = o_K \ell + \pi^i L_0,$$

and so the map N is bijective.

To complete the proof, observe that the image under N of the open set $U(e_0)$ determined by the edge e_0 is the set of points (unimodular as always) $[x, y]$ such that $ax + by = 0$ for some $a, b \in o_K$ with $a\Xi_0 + b\Xi_1 \equiv \Xi_0 \pmod{\pi o_K}$. This is precisely the open set $\{[x, 1] : x \in \pi o_K\} \subset \mathbf{P}^1$. The G-equivariance of the map can be checked from the definitions, and using G-equivariance, one may conclude that N is open and continuous; since it is bijective we conclude that it is a homeomorphism. $\quad\square$

1.3.5. *Group action on the ends.* The group G acts transitively on the ends. Furthermore:

(1) The stabilizer of an end is a Borel subgroup in G. In particular, the stabilizer of the end

$$([L_0], [\Xi_0 + \pi L_0], [\Xi_0 + \pi^2 L_0], \ldots)$$

is the subgroup

$$P = \{g = \begin{pmatrix} a & b \\ 0 & d \end{pmatrix} : g \in G\}.$$

(2) By (1), we may identify the ends (or $\mathbf{P}^1$) with G/P. To be completely explicit, the point $[x, y] \in \mathbf{P}^1$ corresponds to the coset

$$\begin{pmatrix} y & * \\ -x & * \end{pmatrix} P \in G/P.$$

In this identification, the open set corresponding to the edge e_0 is $BP/P \subset G/P$.

1.3.6. *The reduction map.* Given a point $x \in \mathcal{X}(\mathbf{C}_p)$ represented by homogenous coordinates $[a, b]$, we obtain a norm γ_x (defined up to equivalence) on V^* by setting

$$\gamma_x(\ell) = \omega(\ell(a, b))$$

for ℓ a linear form in V^*. The fact that this is indeed a norm is just a restatement of the fact that the coordinates a and b are linearly independent over K, which holds because the point x belongs to

$$\mathcal{X}(\mathbf{C}_p) = \mathbf{P}^1(\mathbf{C}_p) \backslash \mathbf{P}^1(K).$$

The map $x \mapsto [\gamma_x]$ from $\mathcal{X}(\mathbf{C}_p)$ to X is called the reduction map:

$$r : \mathcal{X} \to X.$$

LEMMA 1.3.7. *The reduction map is G-equivariant, so $g(\gamma_x)(\ell) = \gamma_{gx}(\ell)$. Let $[L_0]$ be, as usual, the lattice spanned over o_K by Ξ_0 and Ξ_1, and L_1 the sublattice spanned by Ξ_0 and $\pi\Xi_1$. Then the inverse image under the reduction map of the vertex $[L_0]$ in $\mathfrak{X}$ is the affinoid subdomain*

$$r^{-1}([L_0]) = \{[x, 1] : x \in \mathbf{C}_p \text{ and } \omega(x - t) = 0 \text{ for all } t \in o_K\}.$$

The inverse image of the open edge e_0 running from $[L_0]$ to $[L_1]$ is the admissible annulus

$$r^{-1}(e) = \{[x, 1] : x \in \mathbf{C}_p \text{ and } 1 > \omega(x) > 0\}.$$

PROOF. The G-equivariance is a simple calculation. Let us therefore analyze the fibers of the reduction map. Consider first $r^{-1}(L_0)$. By definition, this is the set of points $[x, y]$ in unimodular coordinates such that

$$(1.3.8) \qquad\qquad \omega(ax + by) = \inf\{\omega(a), \omega(b)\}$$

for all a and b in o_K. This equation is a fancy way to write the requirement that, writing $z = x/y$, we must have $\omega(az + b) = 0$ as a and b run through o_K. This is precisely the condition defining our affinoid in the lemma. For the inverse image of the edge, the condition we seek is

$$\omega(ax + by) = \inf\{\omega(a) + t, \omega(b)\}$$

for some real $1 > t > 0$. We may conclude that y is a unit (which might as well be 1) and that $\omega(x) = t$. Conversely, if $y = 1$ and $\omega(x) = t$, then we obtain the desired norm. Letting t vary between 0 and 1 gives us the full result. $\qquad\square$

The affinoids $\mathfrak{X}_n^-$ that we constructed to form an admissible covering of $\mathfrak{X}$ are the inverse images under reduction of the subtrees of X made up of vertices and edges at distance at most $n - 1$ from the fixed central vertex v_0.

It's also worth observing that points $[x, y] \in \mathbf{P}^1(K)$ give rise to seminorms on V^*, and that the kernel of such a seminorm corresponds to an end of X. One can extend the reduction map from $\mathfrak{X}$ to all of $\mathbf{P}^1$, with the boundary points mapping to the ends – all in a G-equivariant way. For this approach to the higher dimensional building, see [**47**].

1.3.7. *The holomorphic discrete series.* We conclude this lecture by introducing certain spaces of functions on $\mathfrak{X}$ that are closely related to modular forms. Let k be an even integer. Define $\mathcal{O}(k)$ to be the ring of entire functions on $\mathfrak{X}$, equipped with the G-action:

$$g_* f = \frac{\det(g)^{k/2}}{(bz + d)^k} f\left(\frac{az + c}{bz + d}\right).$$

These spaces are called "the holomorphic discrete series" for $\mathrm{GL}_2(K)$. For their general construction, see [**32**].

In the important special case $k = 2$, we have a G-isomorphism

$$
\begin{array}{ccc}
\mathcal{O}(2) & \to & \Omega^1_{\mathfrak{X}} \\
f & \mapsto & f\,dz.
\end{array}
$$

where we write $\Omega^1_{\mathfrak{X}}$ for the global sections of the sheaf of one-forms on $\mathfrak{X}$. More generally, for $k > 0$ and even we can identify $\mathcal{O}(k)$ with $(\Omega^1_{\mathfrak{X}})^{\otimes k/2}$.

LEMMA 1.3.9. *The $(k-1)$-fold derivative map gives a G-equivariant map*

$$\mathcal{O}(2-k) \quad \to \quad \mathcal{O}(k)$$

$$f \quad \mapsto \quad \left(\frac{d}{dz}\right)^{k-1} f.$$

The kernel of this map is the (finite dimensional) space of polynomials in z of degree at most $k-1$.

DEFINITION 1.3.10. We let $H_{DR}(k)$ be the cokernel of the derivative map defined above.

When $k = 2$, the Stein property of $\mathcal{X}$ implies that $H_{DR}(2) = H^1_{DR}(\mathcal{X})$.

2. Boundary distributions and integrals

2.1. Locally analytic functions and distributions. The space of rigid analytic functions on $\mathcal{X}$ is isomorphic, via an integral transform, to a space of distributions on the boundary $\mathbf{P}^1$ of $\mathcal{X}$. This result, due originally to Morita, is a kind of p-adic analogue of the Poisson kernel from classical complex analysis. In order to introduce the space of distributions that concerns us, we need a brief digression on locally analytic functions.

2.1.1. *Locally analytic functions.* We first define locally analytic functions and manifolds. Given $\mathbf{a} = (a_1, \ldots, a_n) \in K^n$ and $\mathbf{r} \in \mathbf{R}^n$, let $B(\mathbf{a}, \mathbf{r})$ be the closed polydisc:

$$B(\mathbf{a}, \mathbf{r}) = \{(x_i)_{i=1}^n \in K^n : \omega(x_i - a_i) \geq r_i \text{ for } i = 1, \ldots, n\}.$$

A K-analytic function on such a disc (with values in a complete field L containing K) is given by a convergent power series

$$f(\mathbf{x}) = \sum_I c_I (\mathbf{x} - \mathbf{a})^I.$$

Here the sum is over n-tuples $I = (i_1, \ldots, i_n)$ with $i_j \geq 0$, the coefficients $c_I \in L$, and

$$(\mathbf{x} - \mathbf{a})^I = \prod_{j=1}^n (x_j - a_j)^{i_j}.$$

The convergence condition is

$$\omega(c_I) + \sum_{j=1}^n r_j i_j \to \infty \text{ as } |I| \to \infty$$

where $|I| = \sum_{j=0}^n i_j$. Let us call the space of such analytic functions $\mathcal{A}_L(B(\mathbf{a}, \mathbf{r}))$. It is a Banach space with respect to the norm

$$\omega(f) = \inf_I \left\{ \omega(c_I) + \sum_{j=1}^n r_j i_j \right\}.$$

More generally, a K-analytic map between such discs is a map given by a collection of power series of this form.

DEFINITION 2.1.1. Let M be a paracompact topological space.

(1) A K-analytic chart (M_i, ϕ_i) for M is an open set M_i together with a homeomorphism

$$\phi_i : M_i \to B_i = B(0, \mathbf{r}) \subset K^d$$

for some radius $\mathbf{r}$.

(2) Two charts are compatible if the map

$$\phi_i \circ \phi_j^{-1} : B_j \to B_i$$

is given by an analytic function.

(3) A collection of compatible charts is called an atlas for M.

(4) M, together with a maximal atlas, is called a locally K-analytic manifold.

The atlas on M allows us to identify the analytic functions on the set M_i with those on the ball B_i via ϕ_i. We will write $\mathcal{A}(M_i, \phi_i)$ for this space of power series. One can show that any covering of a K-analytic manifold can be refined to a pairwise disjoint covering.

We will be interested in the following spaces viewed as K-analytic manifolds:

(1) The group $G = \mathrm{GL}_2(K)$ and the various subgroups G_o, B, and P introduced so far;

(2) The projective space $\mathbf{P}^1$.

Note the difference between $\mathbf{P}^1$ viewed as a rigid analytic space and as a K-analytic manifold!

DEFINITION 2.1.2. Let M be a K-analytic manifold. The locally analytic functions on M (with values in a field L) are defined as follows. To each covering of M by disjoint charts (M_i, ϕ_i) we associate the space of functions

$$C^{an}(\{M_i, \phi_i\}) = \prod_i \mathcal{A}_L(M_i, \phi_i)$$

with its product topology. We define

$$C^{an}(M, L) = \varinjlim C^{an}(\{M_i, \phi_i\})$$

where the limit is over finer and finer coverings. We equip this space with the direct limit topology.

Two other important function spaces associated with a K-analytic manifold are the locally constant (or *smooth*) functions $C^\infty(M, K)$ and the continuous functions $C(M, K)$. We have

$$C^\infty(M, L) \subset C^{an}(M, L) \subset C(M, L).$$

The locally constant functions are closed in $C^{an}(M, L)$, and the analytic functions are dense in $C(M, L)$.

When M is compact, the coverings used to construct $C^{an}(M, L)$ are finite, and the spaces

$$\prod \mathcal{A}_L(M_i, \phi_i)$$

are Banach spaces. By the same reasoning that we sketched in Section 1.2.2, the transition maps in this direct limit are compact. A topological vector space that is the direct limit of a sequence of Banach spaces with compact transition maps is called a vector space *of compact type*.

With these preliminary remarks, the following results hold for spaces of compact type and their duals:

(1) Compact type spaces are reflexive and complete.
(2) Closed subspaces of compact type spaces are of compact type.
(3) The quotient V/U of a vector space of compact type by a closed subspace is of compact type.
(4) If $V = \varinjlim V_i$, with the V_i Banach spaces and the transition maps compact, then the strong dual V_b' is a Fréchet space and satisfies $V_b' = \varprojlim (V_i)_b'$.

See [**34**] Section 16 for points (1) and (4). For point (2), see [**23**] Theorem 7' and 8.

PROPOSITION 2.1.3. *Suppose that M is compact and L is locally compact. Then $C^{an}(M, L)$ is a vector space of compact type, hence reflexive and complete.*

The space $D(M, L)$ of analytic distributions on M is, by definition, the strong dual $C^{an}(M, L)_b'$ of the analytic functions on M.

2.1.2. *Locally analytic principal series representations.* Of particular interest to us in these lectures are the following spaces of locally analytic functions. For each even integer k, let χ_k be the character of the Borel subgroup P defined by the formula

$$\chi_k \left(\begin{pmatrix} a & b \\ 0 & d \end{pmatrix} \right) = (a/d)^{(k/2)}.$$

The "locally analytic induction" $\mathrm{ind}_P^G(\chi_k)$ is the space

$$\mathrm{ind}_P^G(\chi_k) = \{ f \in C^{an}(G, K) : f(gp) = \chi_k^{-1}(p)f(g) \text{ for } p \in P \text{ and } g \in G \}.$$

This space carries a G-action on the left by the rule $g(f)(h) = f(g^{-1}h)$. It is an example of a locally analytic G-representation; for more on such representations, see [**39**] and [**42**].

We will interpret the representation $\mathrm{ind}_P^G(\chi_k)$ as the space of "locally meromorphic functions on $\mathbf{P}^1$ having poles only at infinity of order at most $-k$." To be a bit clearer about what we mean, recall that, in Section 1.3.5, we pointed out that $\mathbf{P}^1 \xrightarrow{\sim} G/P$ via the identification

$$[a, b] \mapsto \begin{pmatrix} b & * \\ -a & * \end{pmatrix} P.$$

At the cost of singling out the point $[1, 0]$ as the point at infinity, we can pullback a function $f \in \mathrm{ind}_P^G(\chi_k)$ to K via the map

$$x \mapsto u(x) = \begin{pmatrix} 1 & 0 \\ -x & 1 \end{pmatrix}.$$

This function is locally analytic on K. However, it enjoys the stronger properties that there is an integer N such that

- f is locally analytic on the set of $x \in K$ with $\omega(x) \geq N$, and
- we have a convergent expansion

$$f(z) = \sum_{i \geq k} c_i z^{-i}$$

on the set where $\omega(z) < N$.

The group action on $\operatorname{ind}_P^G(\chi_k)$ becomes the action

$$g(f) = \frac{(ad - bc)^{k/2}}{(bz + d)^k} f\left(\frac{az + c}{bz + d}\right),$$

which clearly preserves the space of functions satisfying the conditions above. This is the space of functions we say are locally analytic, except have a pole of order at most $-k$ at infinity.

DEFINITION 2.1.4. For $k \leq 0$, Let $C^{an}(K, k)$ be the space of "locally analytic functions on K with at most a pole of order $-k$ at infinity", as defined above, with its associated group action. The topology on $C^{an}(K, k)$ can be defined as follows. Choose an integer N and a finite set $S_N = \{a_j\}$ of elements in K, so that one can cover K by the finite collection of balls

$$D(a_j, N) := \{x \in K : \omega(x - a_j) \geq N\},$$

together with

$$D(\infty, N) := \{x \in K : \omega(x) \leq N\}.$$

The analytic functions $\mathcal{A}(D(a_j, N))$ on $D(a_j, N)$ are given by convergent power series

$$f(x) = \sum_{i=0}^{\infty} c_i(x - a_j)^i$$

where $\omega(c_i) + iN \to \infty$ as $i \to \infty$. $\mathcal{A}(D(a_j, N))$ is a Banach space for the norm

$$\omega(f) = \inf\{\omega(c_i) + iN\}.$$

The "analytic functions with poles at infinity" $\mathcal{A}(D(\infty, N))$ on $D(\infty, N)$ are given by

$$f(x) = \sum_{i=k}^{\infty} c_i x^{-i}$$

where $\omega(c_i) + iN \to \infty$ as $i \to \infty$, with the norm

$$\omega(f) = \inf\{\omega(c_i) + iN\}.$$

The topology on $C^{an}(K, k)$ is the direct limit topology:

$$C^{an}(K, k) := \varinjlim_{N \to \infty} \prod_{a \in S_N \cup \{\infty\}} \mathcal{A}(a, N).$$

If $k \leq 0$ and even, and we look at $\operatorname{ind}_P^G(\chi_k)$ as $C^{an}(K, k)$, then we can identify two G-invariant subspaces inside it. The first one is the finite dimensional space

$$P_{-k} := \text{the space of polynomials of degree at most } -k.$$

The second is the space of functions $f(z)$ that are "locally polynomial functions" on K of degree at most $-k$, meaning that, for some covering of K by $D(a, N)$ (including $D(\infty, N)$), the restriction of f to each disc is a polynomial of degree at most $-k$. We let $C^{la}(K, k)$ denote this space. (The la stands for "locally algebraic.")

For the sake of concreteness it is also worth noticing that, even though elements of $C^{an}(K, k)$ are not strictly speaking functions on $\mathbf{P}^1$ (because they have poles at ∞), every element of the quotient space $C^{an}(K, k)/P_{-k}$ can be represented by a unique, truly locally analytic function on $\mathbf{P}^1$ that vanishes at ∞ — just subtract off the polar part at infinity using P_{-k}.

The methods of the paper [**37**] prove that the following sequence is exact, and that the representation $C^{an}(K, k)$ has length 3:

$$(2.1.5) \qquad 0 \to C^{la}(K, k)/P_{-k} \to C^{an}(K, k)/P_{-k} \xrightarrow{(\frac{d}{dx})^{1-k}} C^{an}(K, 2-k) \to 0.$$

The locally algebraic part of the representation $C^{an}(K, k)$ (for $k \leq 0$) can be decomposed as a tensor product of the smooth representation $C^{\infty}(\mathbf{P}^1, K)/K$ and the finite dimensional representation P_{-k}.

2.2. The Integral Transform and Morita Duality. The locally analytic representations $\mathrm{ind}_P^G(\chi_k)$ discussed in the previous section are closely related to the topological vector spaces $\mathcal{O}(k)$ that come from functions on the p-adic upper half plane. One way to formulate this relationship, which we call Morita duality (see [**26**]), is by an integral transform.

Suppose that $\lambda : \mathcal{O}(k) \to K$ is a continuous linear functional. We may construct from λ a function $I_k(\lambda)$ on K via the formula

$$I(\lambda)(x) = \lambda \left(\frac{1}{z - x} \right).$$

Our goal in this section is to prove the following theorem.

THEOREM 2.2.1. *For $k \geq 2$ and even, the map I_k yields a topological isomorphism*

$$\mathcal{O}(k)'_b \to C^{an}(K, 2-k)/P_{k-2}.$$

To prove the theorem, we will proceed in stages. First, we check the G-action. Substituting in the various definitions, we obtain:

$$g(I_k(g^{-1}(\lambda)))(x) = \frac{(bx + d)^{k-2}}{(ad - bc)^{k/2-1}} \lambda \left(\frac{(ad - bc)^{k/2-1}}{(bz + d)^k} \frac{(bz + d)(bx + d)}{z - x} \right).$$

Now using the fact that

$$(bz + d)(bx + d) = (bz + d)^2 - (z - x)b(bz + d)$$

one obtains

$$g(I_k(g^{-1}(\lambda)))(x) = \lambda \left(\left(\frac{bx + d}{bz + d} \right)^{k-2} \left(\frac{1}{z - x} + c(z) \right) \right)$$

where $c(z)$ is independent of x. Finally,

$$\left(\frac{bx + d}{bz + d} \right)^{k-2} = 1 + (z - x)H,$$

where H is a polynomial in x of degree $k - 2$, with coefficients rational functions in z. Thus

$$g(I_k(g^{-1}(\lambda)))(x) \equiv \lambda(x) \pmod{P_{k-2}},$$

proving G-equivariance (formally).

Next, we prove that the function $I_k(\lambda)$ belongs to $C^{an}(K, 2 - k)$. Functional analysis tells us that any continuous linear form λ on $\mathcal{O}(k)$ is induced by a continuous linear form $\lambda_n : \mathcal{O}(\mathcal{X}_n^-) \to K$ for some integer n. More precisely, we have the following relationship between these spaces and their strong duals (see [**34**], Proposition 16.5; the subscript b refers to the strong topology):

$$(2.2.2) \qquad (\mathcal{O}_\mathcal{X})'_b = (\varprojlim \mathcal{O}(\mathcal{X}_n^-))'_b \xrightarrow{\sim} \varinjlim (\mathcal{O}(\mathcal{X}_n^-))'_b.$$

Let us choose representatives $\{a_i\}$ for $o_K/\pi^n o_K$ and $\{b_j\}$ for $\pi o_K/\pi^n o_K$ as in Section 1.2.1, with $b_0 = 0$. The balls $D(a_i, n)$, $D(1/b_j, n - 2\omega(b_j))$, and $D(\infty, n)$ form a covering of K as in the discussion following Definition 2.1.4. Suppose that $x \in D(a_i, n)$. Then

$$\frac{1}{z - x} = \frac{1}{(z - a_i) - (x - a_i)} = \sum_{j=0}^{\infty} \frac{(x - a_i)^j}{(z - a_i)^{j+1}},$$

the geometric series converging when $z \in X_n^-$. The continuity of λ on $\mathcal{O}(X_n^-)$ means that

$$\omega(\lambda(z - a_i)^{-j-1}) \geq C - (n - 1)(j + 1)$$

for some constant C. For $x \in D(a_i, n)$, we have

$$\lambda\left(\frac{1}{z - x}\right) = \sum_{j=0}^{\infty} \lambda\left(\frac{1}{(z - a_i)^{j+1}}\right)(x - a_i)^j,$$

and the series on the right converges because

$$C - (n - 1)(j + 1) + nj = C + j + 1 - n \to \infty \text{ as } j \to \infty.$$

This exhibits $I_k(\lambda)$ as an analytic function on $D(a_i, n)$. On $D(\infty, n)$, we see that

$$\lambda\left(\frac{1}{z - x}\right) = \lambda\left(\frac{1}{x(z/x - 1)}\right) = -\sum_{j=0}^{\infty} \lambda(z^j) x^{-j-1}$$

and the inequalities bounding λ, defining $D(\infty, n)$ and $\mathcal{O}(X_n^-)$ guarantee convergence as in the case considered earlier.

Notice that this calculation for $I_k(\lambda)$ on $D(\infty, n)$ actually proves more — the function $I_k(\lambda)(x)$ vanishes at the point ∞. This will enable us to settle the next step in our proof, which is to show that $I_k(\lambda)$ is injective. Because $I_k(\lambda)$ vanishes at infinity, it belongs to P_{k-2} only if it is identically zero. From the computations above, we see that $I_k(\lambda) = 0$ if and only if λ vanishes on the functions z^j and $1/(z - a)^j$ for all non-negative integers j and all $a \in K$. This in turn implies that λ vanishes on the rational functions in z having poles at rational points of $\mathbf{P}^1(K)$. Since these rational functions are dense in each $\mathcal{O}(X_n^-)$, and λ is continuous, it follows that λ must be identically zero.

In light of the functional-analytic fact (2.2.2), the continuity of I_k is implicit in the calculation above. More precisely, we showed that I_k, restricted to those λ which factor through $\mathcal{O}(X_n^-)$, is a bounded linear functional from this Banach space to the Banach space of locally analytic functions that are analytic for the specific covering we used in the calculation. This implies that I_k is continuous.

We must show that I_k is surjective, and that it is a topological isomorphism. In fact, the second claim follows from the first by the open mapping theorem (see [34], Proposition 8.8). To prove surjectivity, we will construct linear forms λ of a particular form. This will require something of a digression.

2.2.1. *Residues.* Given an edge e of the tree X, we know that the fiber $r^{-1}(e)$ of the reduction map at e is an admissible open set that is an annulus. Let us look for the moment at the particular admissible open set

$$U = r^{-1}(e_0) = \{[a, 1] : 1 > \omega(a) > 0\} \subset X$$

described in Lemma 1.3.7. This space is a union of affinoid subdomains

$$U_n = \{[a,1] : 1 - (1/n) \geq \omega(a) \geq (1/n)\}$$

as $n \to \infty$, and the space of rigid functions on U is the Fréchet space arising as the projective limit of the corresponding affinoid algebras:

$$\mathcal{O}(U) = \varprojlim \mathcal{O}(U_n).$$

In concrete terms, $\mathcal{O}(U)$ consists of power series

$$f(z) = \sum_{z \in \mathbf{Z}} c_j z^j$$

that converge on each U_n. This condition amounts to the requirement that

$$\omega(c_j) + j/n \to \infty$$

for all n and $j \to \infty$, and

$$\omega(c_j) + j(1 - 1/n) \to \infty$$

as for all n as $j \to -\infty$. The family ρ_n of seminorms defining the topology are

$$\rho_n(f) = \inf\{\omega(c_j) + j/n, \omega(c_j) + j(1 - 1/n)\}.$$

PROPOSITION 2.2.3. *The (rigid) DeRham cohomology of the annulus U is one dimensional and spanned by dz/z.*

PROOF. The only obstacle to formal integration of a rigid function on U to obtain another rigid function is dz/z. □

Let Res be the isomorphism $\mathrm{Res} : H^1_{DR}(U) \to K$ such that $\mathrm{Res}(dz/z) = 1$.

DEFINITION 2.2.4. Given an oriented edge e of X and a rigid analytic one-form $f(z)dz$ in $\Omega^1(\mathcal{X})$, we define $\mathrm{Res}_e(fdz)$ to be $\mathrm{Res}(g^{-1}(fdz)|U)$ where $g \in G$ is any element such that $ge_0 = e$.

This definition makes sense because the Iwahori group B acts trivially on $H^1_{DR}(U)$ and preserves the orientation of e_0. The normalizer of B, which reverses the edge e_0, sends z to π/z. It follows that $\mathrm{Res}_{e'}(fdz) = -\mathrm{Res}_e(fdz)$ if e and e' are opposite to one another.

2.2.2. *Surjectivity of the integral transform.* We can now use the residue map to prove that our integral transform I_k is surjective. Because of the G-equivariance of I_k, it suffices to prove that any analytic function on $o_K \subset K$ is in the image of I_k. Let

$$f(x) = \sum_{j=0}^{\infty} b_j x^j,$$

where $\omega(b_j) \to \infty$ as $j \to \infty$, be the desired target function. We will find a linear form λ such that $I_k(\lambda) = f$. To do this, let f_a be the restriction of f to the disk $D(a,1) = a + \pi o_K$, where a runs through a set of representatives for $o_K/\pi o_K$. The function f_0 is given by the same power series as our original function f. The other functions f_a can be written as power series

$$f_a(z) = \sum_{j=0}^{\infty} b_j^a (x - a)^j$$

where the coefficients b_j^a satisfy $\omega(b_j^a) \to \infty$ as $j \to \infty$. Note that these representative power series on the disks $D(a, 1)$ are overconvergent — they converge on the bigger disk $D(0, 0)$. Each f_a is a translate, under the group action, of an overconvergent power series on $D(0, 1)$ like f_0. Thus, to prove surjectivity, it suffices to prove that f_0 is in the image of I_k. For this, we use the following lemma.

LEMMA 2.2.5. *Let* $\{b_j\}_{j=0}^\infty$ *be a sequence of elements of* K *with the property that, for some integer* $n > 0$, $\omega(b_j) + j/n \to \infty$. *Then*

$$f \mapsto \sum_{j=0}^\infty b_j \operatorname{Res}_e(z^j f)$$

is a continuous linear form on $\mathcal{O}(U)$.

PROOF. This follows from a computation with the semi-norms defining the topology on $\mathcal{O}(U)$. $\qquad\square$

The lemma applies to the particular coefficients of our analytic function f. We compute

$$I_k(\lambda)(x) = \sum_{j=0}^\infty b_j \operatorname{Res}\left(\frac{z^j}{z - x}\right).$$

Now we distinguish two cases. When $x \in D(0, 1)$ and $z \in U$, we have $\omega(x) > \omega(z)$ and we see from the geometric series that

$$\frac{z^j}{z - x} = \sum_{\ell=0}^\infty z^{j-1-\ell} x^\ell.$$

As a result, we have

$$I_k(\lambda)(x) = \sum_{j=0}^\infty b_j x^j = f(x).$$

On the other hand, when $x \notin D(0, 1)$, but $z \in U$, we have $\omega(x) < \omega(z)$ and thus we obtain the expansion

$$\frac{z^j}{z - x} = \sum_{\ell=0}^\infty z^{\ell+j} x^{-\ell-1}$$

and all residues of this function vanish. Therefore $I_k(\lambda)$ is supported on $D(0, 1)$, where it agrees with $f(x)$. This proves surjectivity, and completes the proof of Theorem 2.2.1.

2.2.3. *The Poisson Kernel.* The Poisson Kernel J_k is the transpose of the map I_k:

$$J_k : (C^{an}(K, 2 - k)/P_{k-2})'_b \to \mathcal{O}(k).$$

PROPOSITION 2.2.6. *Let* μ *be a continuous linear form on* $C^{an}(K, 2 - k)$ *vanishing on* P_{k-2}. *The transpose* J_k *is given by*

$$J_k(\mu)(z) = \int_{\mathbf{P}^1(K)} \frac{1}{z - x} \, d\mu.$$

PROOF. Much of this calculation reproduces what we did in the proof of the main theorem. For example, the G-equivariance follows by essentially the same argument that we used earlier. We need to prove that $J_k(\mu)$ is rigid analytic on $\mathcal{X}$, and that $\lambda(J_k(\mu)) = \mu(I_k(\lambda))$. The second of these properties is formal once we know the analyticity, so we will focus on that. Choose a large integer N and representatives a_i for $o_K/\pi^N o_K$ and b_j for $\pi o_K/\pi^N o_K$ (with $b_0 = 0$) so that the balls $D(a_i, N)$, $D(1/b_j, N - 2\omega(b_j))$ for $j \neq 0$, and $D(\infty, N)$ cover K, as in (1). Then

$$J_k(\mu)(z) = \sum \int_D \frac{1}{z - x} d\mu$$

where the sum is over the discs in the covering. For a typical such disc $D(a, N)$, we have

$$\frac{1}{z - x} = \frac{1}{(z - a) - (x - a)} = \sum_{\ell=0}^{\infty} \frac{(x - a)^\ell}{(z - a)^{\ell+1}}$$

converging when $\omega(x - a) > \omega(z - a)$, in particular when $z \in \mathcal{X}_N^-$. The continuity of the distribution μ means that

$$\omega\left(\int_{D(a,N)} (x - a)^\ell d\mu\right) \geq C + N\ell$$

for some constant C. Applying this to the sum, we obtain

$$\int_{D(a,n)} \frac{1}{z - x} d\mu = \sum_{\ell=0}^{\infty} \frac{\int_{D(a,n)} (x - a)^\ell d\mu}{(z - a)^{\ell+1}}.$$

Since $\omega(z - a) \leq N - 1$ on $\mathcal{X}_N^-$, we see that this series gives a rigid function on $\mathcal{X}_N^-$. Assembling the different discs shows that $J_k(\mu)$ is in fact rigid analytic on $\mathcal{X}_N^-$. $\square$

COROLLARY 2.2.7. *Let $f \in \mathcal{O}(k)$ be a rigid function, and choose $N > 0$. Let a_i, b_j be chosen as in the proof of the theorem (or as in equation (1)). Then f restricted to $\mathcal{X}_N^-$ has a "partial fraction expansion"*

$$f(z) = \sum_{j=0}^{\infty} c_j^\infty z^j + \sum_{i=0}^{q^N-1} \sum_{\ell=1}^{\infty} \frac{c_\ell^i}{(z - a_i)^\ell} + \sum_{i=1}^{q^{N-1}-1} \sum_{\ell=1}^{\infty} \frac{d_\ell^i}{(z - 1/b_i)^\ell}.$$

2.2.4. *Morita Duality.* We have shown that there is a duality pairing (first established by Morita ([**26**]) for $k \geq 2$:

$$\mathcal{O}(k) \times C^{an}(K, 2 - k)/P_{k-2} \to K$$

given by

$$\langle F(z), f(x)\rangle = I_k^{-1}(f)(F) = J_k^{-1}(F)(f).$$

We can refine our understanding of this duality by looking more closely at the Jordan-Holder factors of the locally analytic representation on the right. Let us look at the restriction of the pairing to the subspace of locally polynomial functions:

$$\mathcal{O}(k) \to (C^{la}(K, 2 - k)/P_{2-k})'_b.$$

Functional analysis tells us that this map is surjective. Furthermore, if we refer back to the definition of locally analytic functions on $\mathbf{P}^1(K)$, we see that the topology induced on the locally polynomial subspace $C^{la}(K, k)$ comes from viewing $C^{la}(K, k)$ as the direct limit of its finite dimensional subspaces. (This is because the set of locally polynomial functions relative to a fixed covering of $\mathbf{P}^1(K)$ is finite dimensional, and these finite dimensional subspaces are cofinal with all such subspaces.) The continuous dual, with respect to this topology, is just the full linear dual. The subspace $(C^{la}(K,k)/P_{k-2})'$ consists of all linear functionals on $C^{la}(K,k)$ that vanish on P_{k-2}. To make this map more explicit, let us extract the following piece of information from the proof of the main theorem.

LEMMA 2.2.8. *Let $P(z)$ be a polynomial of degree at most $k-2$, and let $\lambda(f) = \mathrm{Res}_e(P(z)f(z)dz)$. Then $I_k(\lambda)$ is the function in $C^{an}(K, 2-k)/P_{k-2}$ equal to $P(x)$ on $D(0,1) = \pi o_K$ and zero elsewhere.*

Now we can make the dual of the locally polynomial functions completely explicit. To give a linear form λ on $C^{la}(K, 2-k)/P_{k-2}$, it suffices to know the values $\lambda(P(x)|U(e))$ for all polynomials $P(x)$ of degree at most $k-2$ and all open sets $U(e)$ corresponding to edges e of X. We can collect this information in a function

$$c_\lambda : \mathrm{Edges}(X) \to \mathrm{Hom}(P_{k-2}, K)$$

defined by $c_\lambda(e) = (P(x) \mapsto \lambda(P(x)|U(e)))$.

To make sure that the linear form λ vanishes on P_{k-2}, we need two properties:

(1) $c_\lambda(e') = -c_\lambda(e)$, when e' is the edge obtained by reversing e.
(2) c_λ is harmonic, meaning

$$\sum_{e \mapsto v} c_\lambda(e) = 0$$

where the sum is over the edges leaving a given vertex.

DEFINITION 2.2.9. Let M be an abelian group. Then a function $c : \mathrm{Edges}(X) \to M$ is called an (M-valued) harmonic cocycle if it satisfies the two conditions given above.

In our special case, given $F(z) \in \mathcal{O}(k)$ (with $k \geq 2$ as usual), we define a function

$$\begin{aligned} c_F : \mathrm{Edges}(X) &\to \mathrm{Hom}(P_{k-2}, K) \\ c_F(e)(P(x)) &= \langle F, P(x)|U(e)\rangle. \end{aligned}$$

The function c_F is determined by the residue map – indeed, suppose that $e' = ge$, where e is the original basic edge used to define Res_e. Then

$$\begin{aligned} c_F(e')(x^j) &= \langle F, x^j|U(e')\rangle = \langle F, g((g^{-1}(x^j)|U(e))\rangle \\ &= \langle g^{-1}(F), g^{-1}(x^j)|U(e)\rangle. \end{aligned}$$

Substituting in the definitions of the group actions, and remembering that the space of polynomials P_{k-2} is a subset of the space of locally algebraic functions $C^{la}(K, 2-k)$ we see that

$$(2.2.10) \qquad\qquad c_F(e') = \mathrm{Res}_e(g_*^{-1}(z^j F(z)dz))$$

where the group action is the usual action on differentials — ignoring k. For this reason, we call the map $F \mapsto c_F$ the residue map.

DEFINITION 2.2.11. Let $C_{har}(k)$ be the space of harmonic functions on the edges of the tree X with values in $\mathrm{Hom}(P_{k-2}, K)$.

Referring back to Equation 2.1.5, we have the following commutative diagram for $k \geq 2$:

$$
\begin{array}{ccc}
(C^{an}(K, 2-k)/P_{k-2})'_b & \longrightarrow & (C^{la}(K, 2-k)/P_{k-2})'_b \longrightarrow 0 \\
\downarrow & & \downarrow \\
\mathcal{O}_\mathcal{X}(k) & \xrightarrow{\quad \text{Res} \quad} & C_{har}(k) \longrightarrow 0
\end{array}
$$

The remaining question for understanding $\mathcal{O}_\mathcal{X}(k)$, for $k \geq 2$ is to understand the kernel of the residue map. This is answered by the following result

THEOREM 2.2.12. *The kernel of the residue map is the image of $\mathcal{O}_\mathcal{X}(2-k)$ in $\mathcal{O}_\mathcal{X}(k)$ (see Lemma 1.3.9). In particular:*

(1) *This image is closed;*
(2) $C_{har}(k) \xrightarrow{\sim} H_{DR}(k)$;
(3) $\mathcal{O}_\mathcal{X}(2-k)/P_{k-2} \xrightarrow{\sim} C^{an}(K, k)'_b$ *for $k \geq 0$ and even.*

PROOF. It is easy to see that the image of $\mathcal{O}_\mathcal{X}(2-k)$ in $\mathcal{O}_\mathcal{X}(k)$ belongs to the kernel of the residue map. Indeed, if

$$
f = \sum_{i=-\infty}^{\infty} c_i z^i
$$

then

$$
\frac{d^j f}{dz^j} = \sum_{i=-\infty}^{-j-1} b_i z^i + \sum_{i=0}^{\infty} b_i z^i
$$

for some constants b_i — from this it is clear that $\text{Res}(P(z)f(z)dz) = 0$ for all polynomials $P(z)$ of degree at most $j - 1$. Conversely, a function $g(z)$ on U is in the image of the $(1-k)^{th}$ derivative if and only if the coefficients c_i, for $i = -1, \ldots, k-1$, are all zero. Since that image is G-equivariant, we can conclude that $\text{Res}_e(P(x)F(x)dx) = 0$ for all e if F is in the image. Working with the derivatives, one can further check that the following diagram commutes:

$$
\begin{array}{ccc}
C^{an}(K, 2-k) & \longrightarrow & C^{an}(K, k) \longrightarrow 0 \\
\downarrow & & \downarrow \\
\mathcal{O}_\mathcal{X}(k)'_b & \longrightarrow & \mathcal{O}_\mathcal{X}(2-k)'_b
\end{array}
$$

where the upper arrow comes from Equation 2.1.5 and the lower one is the dual map to the derivative map in Lemma 1.3.9. Therefore, the key point is that $\mathcal{O}_\mathcal{X}(2-k)$ has closed image. We won't give all the details of the proof of this, but it follows from the "partial fractions" decomposition given in Corollary 2.2.7. The idea is to see that, if $\text{Res}(z^j f(z))$ vanishes on all edges e, then the terms of the form $1/(z-a)^j$ in the partial fractions decomposition vanish for $0 \leq j \leq k-2$. Then one can formally integrate the partial fractions decomposition $k-1$ times and obtain a rigid function that still converges on some $\mathcal{X}_n^-$. These integrals can be glued together because the obstruction to doing so lies in $H^1(\mathcal{X}, \mathcal{O})$, which is zero by the Stein property. $\square$

2.3. Bounded distributions. The bounded harmonic functions, relative to a suitably chosen norm, play a special role in the analytic theory of the p-adic upper half plane. To explore this, choose a norm on P_{k-2} (for $k \geq 2$) that is invariant by the Iwahori group B which stabilizes our standard edge e_0. (There are many such choices; for example, the sup-norm on the coefficients of the polynomials in P_{k-2} will do). We will use the same notation ω for this norm, and for the associated dual norm on $\mathrm{Hom}(P_{k-2}, K)$.

If $c \in C_{har}(k)$, we say that c is *bounded* if

$$(2.3.1) \qquad \omega(c) = \inf_{g \in G/B} \omega(g^{-1}(c(ge_0)))$$

exists. Notice that $\omega(c)$ is well-defined, because the B-invariance of the norm on $\mathrm{Hom}(P_{k-2}, K)$ means that the terms in the infimum are independent of the choice of coset representatives. We write $C_{har}^b(k)$ for the space of bounded harmonic functions — they form a Banach space with respect to the given G-invariant norm.

The bounded elements in $C_{har}(2)$ are the harmonic functions whose values are p-adically bounded.

The boundedness condition translates into an estimate for the "integrals" of locally polynomial functions in $C^{la}(K, 2-k)$. This, in turn, leads to the following version of the "Theorem of Amice-Velu-Vishik."

THEOREM 2.3.2. *Suppose that c is a bounded harmonic function in $C_{har}(k)$. Then there is a unique continuous linear form $\lambda_c : C^{an}(K, 2-k) \to K$ that vanishes on P_{k-2} and satisfies the following conditions:*

(1) $\lambda_{gc} = g(\lambda_c)$.
(2) $\lambda_c(P(x)|U(e_0)) = c(e_0)(P(x))$ *for $P(x)$ of degree at most $k-2$.*
(3) *There is a constant A such that, for all $n \geq 0$, $m \geq 0$, and $a \in o_K$, we have*

$$\omega(\lambda_c((x-a)^n | a + \pi^m o_K)) \geq A + m(j - 1 - k/2).$$

(4) *We have*

$$\lambda_c\left(\left(\sum_{m=0}^{\infty} c_m(x-a)^m\right) | a + \pi^m o_K\right) = \sum_{m=0}^{\infty} c_m \lambda((x-a)^m | a + \pi^m o_K).$$

PROOF. We only sketch the proof. A computation with the various group actions shows that conditions (1) and (2) give us a well-defined way to compute $\lambda_c(P(x)|U)$ for any compact open set U in $\mathbf{P}^1$ and any polynomial $P \in P_{k-2}$. The harmonicity of c implies that λ_c vanishes on P_{k-2}. To integrate a locally polynomial function, choose $g \in G$ carrying U to the standard open set $U(e_0)$ and compute

$$\lambda_c(P(x)|U) = \lambda_c(P(x)|g^{-1}(U(e_0))) = \lambda_{gc}(g^{-1}(P)|U(e_0)).$$

(one checks that this does not depend on the choice of g.) The boundedness property of c turns into the estimate (3), at least for $0 \leq n \leq k-2$. If e is an edge such that $U(e)$ does not contain ∞, then it must be of the form $a + \pi^m o_K$. In that case, we wish to estimate, for $0 \leq n \leq k-2$, the value of λ_c:

$$(2.3.3) \qquad \lambda_c((x-a)^n | a + \pi^m o_K)) = \lambda_c(g^{-1}([\pi^{(j-(k-2)/2)m} x]^n | o_K)$$

where

$$g = \begin{pmatrix} 1 & 0 \\ a\pi^{-m} & \pi^{-m} \end{pmatrix}.$$

Let A be $\omega(c) = \omega(gc)$. Then

$$\omega(g(\lambda_c)(x^n|o_K)) = c(e_0)(x^n) \geq A.$$

Combined with equation 2.3.3 we obtain (3).

Finally, we show how to compute $\lambda_c(f|U(e_0))$ for locally analytic f. Cover $U(e_0) = \pi o_K$ by open sets $a + \pi^m o_K$ for some large m. On each open set, let $P_{a,m}$ be the truncation of the Taylor expansion of f on the disc $a + \pi^m o_K$ obtained by discarding terms of degree greater than $k - 2$. Define

$$S_m = \sum_a \lambda_c(P_{a,m}|a + \pi^m o_K)$$

using the fact that we know how to integrate polynomials of low degree. Then the estimate (3) implies that the limit, as $m \to \infty$, of S_m exists. This gives our integral. See [25, Section 11] for one proof with details. See [6, Theorem 2.5] for another proof. $\qquad\square$

COROLLARY 2.3.4. *Let* $\mathcal{O}_\mathcal{X}(k)^b$ *be the space of rigid functions* F *such that* $\mathrm{Res}(F)$ *is a bounded harmonic function. The residue map gives an isomorphism between* $\mathcal{O}_\mathcal{X}(k)^b$ *and* $C_{har}(k)^b$; *the inverse of this map is the Poisson integral.*

PROOF. The kernel function $\frac{1}{z-x}$ is locally analytic; given a bounded harmonic function, we can apply the corresponding linear form to it. The proof that the result is rigid analytic is another argument with the geometric series that relies on the estimate (3) to obtain convergence. $\qquad\square$

The bounded functions $\mathcal{O}_\mathcal{X}(k)$ can be characterized differently. The spaces $\mathcal{O}(\mathcal{X}_n^-)$ are Banach spaces; fix one such n and let $\omega(F)$ denote, for the moment, the norm of a function F restricted to $\mathcal{O}(\mathcal{X}_n^-)$.

THEOREM 2.3.5. *The residues* $\mathrm{Res}(F)$ *are bounded, and* $F \in \mathcal{O}_\mathcal{X}(k)^b$ *if and only if* $\omega(gF) \geq C$ *for some constant* C *and all* $g \in G$.

PROOF. See [10]. $\qquad\square$

2.4. Discrete groups, modular forms, and uniformization. The p-adic upper half plane was originally introduced by Mumford as a way to construct families of algebraic curves lying at the boundary of moduli space. Mumford showed that, for appropriate discrete subgroups $\Gamma \subset G$, the quotient $\mathcal{X}/\Gamma$ has the structure of an algebraic curve.

The work of Cerednik and Drinfeld made clear the arithmetic significance of Mumford's p-adic uniformization theory. They showed that one could construct Shimura curves — modular curves parameterizing abelian surfaces with quaternionic multiplication — via p-adic methods.

We will recall a few of the features of this theory. For more of the story, see the work of Gerritzen and van der Put ([17]) or Mumford's original paper ([27]). For the arithmetic theory and uniformization of Shimura curves, see Drinfeld's original (7 page) paper ([12]) or the book by Boutot and Carayol that explains that paper in detail ([4]).

Choose a definite quaternion algebra B over $\mathbf{Q}$ with discriminant N. From the theory of such algebras, we know that N must be a squarefree integer with an odd number of prime divisors. Now choose a prime p not dividing N and fix an isomorphism $B \otimes_{\mathbf{Q}} \mathbf{Q}_p \tilde{\rightarrow} M_2(\mathbf{Q}_p)$. Finally, pick a maximal $\mathbf{Z}[1/p]$ order $A \subset B$.

The strong approximation theorem tells us that all such A are conjugate in B. The units A^* of A form a discrete subgroup Γ of G. More generally, one can choose a non-maximal $\mathbf{Z}[1/p]$-order A' in A and let Γ' be the units of A'. The groups Γ' form a family of congruence subgroups of Γ.

The main results of p-adic uniformization in this setting say that:

(1) The groups Γ' act discontinuously on the tree X. For A' small enough, this action is free, Γ' is a finitely generated free group, and X/Γ' is a finite graph.

(2) The quotient $S_N(\Gamma') = \mathcal{X}/\Gamma'$ exists as a rigid space; it can be embedded in projective space as a closed rigid subvariety, and therefore is an algebraic curve.

(3) The quotient algebraic curve $S_N(\Gamma')$ is a Shimura curve. It classifies two-dimensional, principally polarized abelian varieties with endomorphism ring equal to a maximal order in the indefinite quaternion algebra with discriminant Np and with level structure determined by $A' \subset A$.

(4) The curve $S_N(\Gamma')$ is totally split over $\mathbf{Q}_p$, meaning that it has a regular model over $\mathbf{Z}_p$ with the property that all of the components of this model are reduced rational curves and all intersection points of components are ordinary double points. The intersection graph of this configuration is exactly X/Γ'.

(5) The genus of $S_N(\Gamma')$ is the genus of the graph X/Γ'. (The genus of a graph is the number of independent cycles in the graph, or more formally the rank of the first homology group of its geometric realization.)

Of particular interest to us are the spaces $\mathcal{O}_{\mathcal{X}}(k)^{\Gamma'}$ where Γ' is a congruence group associated to a quaternion algebra. For $k \geq 2$, elements of this space are "modular forms for Γ'" – that is, functions satisfying the condition

$$f(\gamma z) = (az + c)^k \det(\gamma)^{-k/2} f(z) \qquad \text{for } \gamma \in \Gamma.$$

Such a modular form of (even) weight k corresponds to a global section of the $k/2$-fold tensor power of the canonical bundle $(\Omega^1)^{k/2}$ on $S_N(\Gamma')$.

PROPOSITION 2.4.1. *The residue map* $\mathrm{Res} : \mathcal{O}_{\mathcal{X}}(k)^{\Gamma'} \to C_{har}(k)^{\Gamma'}$ *is an isomorphism.*

Proof: The essential point is that the quotient graph X/Γ' has finitely many edges. If we choose finitely many representative edges $e_1, \ldots, e_m$ for this quotient, then the value $c(e)$ of a harmonic function on a general edge is determined by its value on one of these finitely many edges. It follows that the norm $\omega(c)$ is bounded below. In other words, any Γ'-invariant harmonic function is bounded. As a result, we can use Corollary 2.3.4 to construct a preimage for c. This proves surjectivity. When $k = 2$, the space of harmonic functions $C_{har}(2)$ is just the space of harmonic functions on the graph X/Γ', and this is g-dimensional where g is the genus of X/Γ'. On the other hand, the elements of $\mathcal{O}_{\mathcal{X}}(k)^{\Gamma'}$ give rise to holomorphic differential forms on X/Γ', and that space is also g dimensional — therefore the map is injective. When $k > 2$, the space of invariant harmonic cocycles is determined by specifying, on each edge of X/Γ', an element of P_{k-2}; while, for each vertex, one obtains $k-1$ linear relations. Thus the dimension of the space $C_{har}(k)^{\Gamma'}$ is at least $(k-1)(E-V) = (k-1)(g-1)$ where E and V are the number of vertices and edges, respectively in the quotient graph X/Γ'. The Riemann-Roch theorem implies that the space $\mathcal{O}_{\mathcal{X}}(k)^{\Gamma'}$, corresponding to the $k/2$-tensor power of the canonical bundle, has dimension $(k-1)(g-1)$. By dimension counting we see the map is surjective in each case.

2.5. Hecke operators. The quaternion algebra B has an associated Hecke algebra. Without attempting to work out the whole theory of this algebra, we will indicate the key idea. As above, we let A denote a fixed, maximal $\mathbf{Z}[1/p]$-order, and A' be a sub-$\mathbf{Z}[1/p]$-order of A.

For any unramified prime ℓ of B, the order $A_\ell = A \otimes \mathbf{Z}_\ell$ can be assumed isomorphic to the ring $M_2(\mathbf{Z}_\ell)$. For any ℓ outside of a finite set S containing the ramified primes, we have $A'_\ell = A_\ell$. From the strong approximation theorem and the adelic theory of quaternion algebras (see [**46**, III.4-5]) we see that, for $\ell \notin S$, there are exactly $\ell + 1$ inequivalent left ideals of A' of index ℓ, and these ideals are principal. Let $x_1, \ldots, x_{\ell+1}$ be generators for these ideals. A unit γ in A' permutes these ideals:
$$A' x_i \gamma = A' x_j.$$
Suppose now that $f \in \mathcal{O}_{\mathcal{X}}(k)^{\Gamma'}$. Then
$$(T(\ell)f)(z) = \sum_{i=1}^{\ell+1} x_i(f)$$
is again Γ'-invariant, because, for $\gamma \in \Gamma'$, the left multiplication by g permutes the x_i.

For ℓ outside the finite set S, the operators $T(\ell)$ generate a commutative algebra $\mathbf{T}$ called the Hecke algebra. Since the units $B^\times$ of B act on the tree X through the embedding $B^\times \hookrightarrow \mathrm{GL}_2(\mathbf{Q}_p)$, one obtains an action of $\mathbf{T}$ on the edges of X and therefore on $C_{har}(k)$. Tracing through the definitions, and using the G-equivariance of the residue map, we obtain:

PROPOSITION 2.5.1. *The residue map*
$$\mathcal{O}_{\mathcal{X}}(k)^{\Gamma'} \to C_{har}(k)^{\Gamma'}$$
is a Hecke module isomorphism.

In practical terms, this means that one can compute the Hecke module structure of the spaces of modular forms on the upper half plane by working combinatorially on the tree.

It is also worth noting that the action of the Hecke operators on the finite dimensional spaces $C_{har}(k)^{\Gamma'}$ arises in the theory of classical automorphic forms for B. The matrices representing this action are called *Brandt matrices* and there is extensive literature on them. See [**46**, Exercise II.5.8], as well as the papers by Pizer and collaborators ([**29**]).

3. $\mathcal{L}$-invariants and modular symbols

Now we change course dramatically, and begin a discussion leading to the connection between the global arithmetic of modular forms and the p-adic analysis we've discussed so far in these lectures.

We will rely implicitly on a fairly significant chunk of the theory of classical modular forms. Beyond the basic definitions of modular forms and the theory of Hecke operators, we will make extensive use of the theory of modular symbols and the connection between periods of modular forms and special values of L-functions. The literature on all of these topics is vast. For the foundations, one may consult Shimura's famous book [**44**]. The beginning of the paper [**25**] develops some of the elementary theory of modular symbols and L-functions.

The work of Mazur–Swinnerton-Dyer ([**24**], see also [**25**]) explains how to attach to an eigenform f of even weight k and level M a p-adic L-function $L_p(f, \chi, s)$ that interpolates the "algebraic parts" $L^{alg}(f, \chi, j)$, for $j = 0, \ldots, k - 2$ of the special values of the classical L-function of f and its twists by Dirichlet characters χ. The resulting L-function plays a central role in the Iwasawa theory of modular forms and in the p-adic Birch–Swinnerton-Dyer conjecture.

Let χ be a Dirichlet character with $\chi(p) = w = \pm 1$, and let f be an eigenvector for the Hecke operators. In the special case $M = Np$, where N is an integer not divisible by the prime p, and the form f is an eigenvector for the Atkin-Lehner U_p-operator with eigenvalue $a_p = w \cdot p^{(k-2)/2}$, [**25**] showed that the order of vanishing of the p-adic L-function $L_p(f, \omega^{\frac{k-2}{2}} \chi, s)$ at $s = (k-2)/2$ is one higher than that of the classical L-function $L(f, \chi, s)$ at $s = k/2$ (the two functions having different traditional normalizations on the variable s). Here ω is the Teichmuller character. The "exceptional zero conjecture" proposed in [**25**] asserted that there is an invariant $\mathcal{L}(f)$, depending only on the local Galois representation associated to f, such that

$$L_p'(f, \omega^{\frac{k-2}{2}} \chi, (k-2)/2) = \mathcal{L}(f) L(f, \chi, (k-2)/2)^{alg}.$$

When $k = 2$ and f is the modular form associated to an elliptic curve E, the assumption that p precisely divides the level M and that $a_p = \pm 1$ means that E has multiplicative reduction at p. In that case, [**25**] presented numerical evidence that

$$\mathcal{L}(f) = \frac{\log(q)}{\mathrm{ord}(q)}$$

where q is the Tate period of the elliptic curve E at p and log is the p-adic logarithm. This weight 2 form of the conjecture was proved by Greenberg–Stevens ([**19**]) using Hida theory.

In the higher weight case, in the period since [**25**], a number of different candidates for the invariant $\mathcal{L}(f)$ have been proposed. (See [**7**] for more background and references.) These include:

(1) An invariant $\mathcal{L}_T(f)$ built by taking advantage of the theory of p-adic uniformization of Shimura curves — we will discuss this in more detail later;

(2) An invariant $\mathcal{L}_C(f)$ built using Coleman's theory of p-adic integration on modular curves;

(3) An invariant $\mathcal{L}_{FM}(f)$ due to Fontaine–Mazur built using Fontaine's classification of p-adic representations;

(4) An invariant $\mathcal{L}_O(f)$ due to Darmon (in weight two) and Orton (in general) using "modular form-valued distributions" (also to be discussed later in this lecture);

(5) An invariant $\mathcal{L}_B(f)$ due to Breuil that derives from his investigations of p-adic Langlands theory (discussed in the next lecture).

All of these invariants are known to be equal:

(1) $\mathcal{L}_T = \mathcal{L}_{FM} = \mathcal{L}_C$ by Coleman-Iovita ([**5**]) and Iovita-Spiess ([**21**]).

(2) $\mathcal{L}_O = \mathcal{L}_B$ by Breuil ([**2**]).

(3) $\mathcal{L}_B = \mathcal{L}_{FM}$ by Colmez ([**8**]).

(4) $\mathcal{L}_O = \mathcal{L}_T$ by Bertolini, Darmon, and Iovita ([**1**]).

The Exceptional Zero Conjecture itself has been proved in general by Stevens (for the Coleman invariant), by Kato, Kurihara and Tsuji (for the Fontaine–Mazur invariant), by Darmon and Orton for $\mathcal{L}_O(f)$, by Emerton using Breuil's invariant, and by Bertolini–Darmon–Iovita using $\mathcal{L}_T(f)$. Stevens's and the Kato–Kurihara–Tsuji result remain unpublished, but one can consult [**9**] for information. For the other results, see [**14**], [**28**],[**11**], and [**1**].

3.1. $\mathcal{L}_T(f)$ and p-adic uniformization. As an application of the theory developed in Lectures I and II, let us describe the construction of the invariant $\mathcal{L}_T(F)$ when F is a modular form for a Shimura curve. In other words, we are in the situation of section 2.4. We begin with a definite quaternion algebra B of discriminant N and a prime p not dividing N. Let A' be an order contained in a fixed maximal $\mathbf{Z}[1/p]$-order A in B, and let Γ' be the discrete group of units in A'. Let F be a modular form of weight k (k even and $k \geq 2$) for Γ'. Assume that F is an eigenform for the Hecke algebra of the quaternion algebra B.

From our integration theory, associated to this F we have a distribution λ_F on $C^{an}(K, 2-k)/P_{k-2}$. We define two elements of $H^1(\Gamma', \operatorname{Hom}(P_{k-2}, \mathbf{C}_p))$ using this distribution. Fix any point $z \in \mathfrak{X}(\mathbf{C}_p)$ and set :

$$h^F_{\log}(\gamma, P(x)) = \lambda_F\left(P(x)\log\left(\frac{x - \gamma(z)}{x - z}\right)\right),$$

$$h^F_{\operatorname{ord}}(\gamma, P(x)) = \lambda_F\left(P(x)\operatorname{ord}\left(\frac{x - \gamma(z)}{x - z}\right)\right).$$

The functions $P(x)\log(\frac{x-\gamma(z)}{x-z})$ and $P(x)\operatorname{ord}(\frac{x-\gamma(z)}{x-z})$ both belong to $C^{an}(K, 2-k)$, since both are locally analytic and have the correct pole order at infinity. The fact that $h^F_{\log}$ and $h^F_{\log}$ are cocycles that depend on z only up to a coboundary is a straightforward calculation.

One can interpret $h^F_{\log}$ as a period of the form $F(z)dz$ on $S_N(M)$. Using the expression of F as a Poisson integral, we have (formally):

$$h^F_{\log}(\gamma, P(x)) = \int_z^{\gamma(z)} \int_{\mathbf{P}^1} \frac{1}{z - x} d\lambda_F$$

as follows from a change in the order of integration. Using the theory of Coleman integration, one can give meaning to this integral, and in fact this argument is legitimate — see [**45**].

Similarly, one can interpret h^F_{ord} as a period on the tree. Choose z so that $r(z)$ is a vertex v on the tree. Then one has the following.

LEMMA 3.1.1. *The integral defining h^F_{ord} reduces to a sum on the tree:*

$$h^F_{\operatorname{ord}}(\gamma, P(x)) = \sum_{v \mapsto \gamma(v)} c_e(P(x))$$

where the sum is over the edges e on the minimal path joining v to $\gamma(v)$.

PROOF. Let e be an oriented edge in the tree X, and let $s, t \in \mathcal{X}$ be points whose reductions are the source and terminal vertices of e, respectively. Then for $u \in \mathbf{P}^1(K)$, one readily verifies that

$$\operatorname{ord}\left(\frac{u-t}{u-s}\right) = \begin{cases} -1 & \text{if } u \in U(\bar{e}), \\ 0 & \text{otherwise,} \end{cases}$$

where $U(\bar{e})$ is the open subset of $\mathbf{P}^1(K)$ associated to the oppositely oriented edge of e (see Section 1.3.4). Choosing points $z = z_0, z_1, \ldots, z_n = \gamma(z)$ reducing to successive vertices on the path from z to $\gamma(z)$ we obtain

$$\begin{aligned} h_{\text{ord}}^F(\gamma, P(x)) &= \sum_{i=1}^n \lambda_F \left(P(x) \operatorname{ord}\left(\frac{x - z_{i+1}}{x - z_i}\right) \right) \\ &= -\sum_{i=1}^n c(\bar{e}_i)(P(x)) \\ &= \sum_{i=1}^n c(e_i)(P(x)) \end{aligned}$$

where e_i joins the reductions of z_i and z_{i+1}. $\square$

THEOREM 3.1.2. *(Schneider, de Shalit) The two maps* $h_{\text{ord}} : F \mapsto h_{\text{ord}}^F$ *and* $h_{\log} : F \mapsto h_{\log}^F$ *are homomorphisms*

$$\mathcal{O}(k)^{\Gamma'} \to H^1(\Gamma', \operatorname{Hom}(P_{k-2}, \mathbf{C}_p))$$

commuting with the natural action of the Hecke algebra $\mathbf{T} \otimes \mathbf{C}_p$ *on both sides. Furthermore,* h_{ord} *is an isomorphism.*

PROOF. See [10] and [31]. $\square$

With this theorem, we can construct the invariant $\mathcal{L}_T(F)$. Theorem 3.1.2 and the fact that the F-isotypic component of $\mathcal{O}(k)^{\Gamma'}$ is 1-dimensional yields:

DEFINITION 3.1.3. There is a unique $\mathcal{L}_T(F) \in \mathbf{C}_p$ such that

$$h_{\log}^F - \mathcal{L}_T(F) h_{\text{ord}}^F = 0$$

in $H^1(\Gamma', \operatorname{Hom}(P_{k-2}, \mathbf{C}_p))$. This is called the $\mathcal{L}_T$–invariant of the form F.

3.2. Modular symbols. To develop the additional theory of $\mathcal{L}$-invariants following Breuil and Darmon, we must undertake a digression into the theory of modular symbols, and also develop some of the ideas of Darmon's integration on $\mathcal{X} \times \mathcal{H}$, where $\mathcal{H}$ is the classical upper half plane. We follow, in part, Breuil's presentation ([2]) in this discussion.

Fix a normalized newform f of even weight $k \geq 2$ on $\Gamma_0(M)$ for some integer M. We assume that $T_\ell f = a_\ell f$ for $(\ell, M) = 1$. The eigenvalues a_ℓ generate an extension E of $\mathbf{Q}$ with ring of integers R. We will view E as a subfield of $\mathbf{C}_p$.

In working with these formulae, one caveat is necessary. It is traditional in the theory of modular forms to work with the right action (the "slash" action) on modular forms given by the formula:

$$f(z)|_g = (cz + d)^{-k} \det(g)^{k/2 - 1} f\left(\frac{az + b}{cz + d}\right).$$

Since we have consistently worked with left actions, we use the associated left action
$$g(f)(z) = f(z)|_{g^t}$$
where g^t is the transpose of g.

The following theorem of Shimura is the starting point of the theory we will describe.

THEOREM 3.2.1. *There are nonzero periods $\Omega_f^{\pm} \in \mathbf{C}$ such that, for any $P(z) \in P_{k-2}(R)$ and any rational number r, we have*
$$\left(\int_r^\infty f(z)P(z)dz \right)^{\pm} := \pi i \left(\int_r^\infty f(z)P(z)dz \pm \int_{-r}^\infty f(z)P(-z)dz \right) \in R\Omega_f^{\pm} \in \mathbf{C}.$$

Let D be the set of divisors on $\mathbf{P}^1(\mathbf{Q})$, and let D_0 be the subspace of divisors of degree zero. We associate to our form f the "modular symbol"
$$\phi_f^{\pm} \in \mathrm{Hom}(D_0, \mathrm{Hom}(P_{k-2}(E), E))$$
by defining
$$\phi_f^{\pm}([r] - [s])(P) := \frac{1}{\Omega_f^{\pm}} \left(\int_s^r f(z)P(z)dz \right)^{\pm}$$
$$= \frac{1}{\Omega_f^{\pm}} \left(\int_s^\infty f(z)P(z)dz \right)^{\pm} - \frac{1}{\Omega_f^{\pm}} \left(\int_r^\infty f(z)P(z)dz \right)^{\pm}.$$

The modular symbol enjoys the following invariance property for $g \in \Gamma_0(M)$:
$$\phi_f^{\pm}([g(r)] - [g(s)])(P) = \left(\int_{g(r)}^{g(s)} f(z)P(z)dz \right)^{\pm}$$
$$= \left(\int_r^s f(g^{-1}(z))P(g^{-1}(z))dg^{-1}(z) \right)^{\pm}$$
$$= \left(\int_r^s (-bz+a)^{k-2} f(z)P(\frac{dz-c}{-bz+a})dz \right)^{\pm}$$
$$= \phi_f^{\pm}([r] - [s])(g^{-1}(P)(z))^{\pm}$$
so that
$$\phi_f^{\pm} \in \mathrm{Hom}(D_0, \mathrm{Hom}(P_{k-2}, E))^{\Gamma_0(M)}.$$

Theorem 3.2.1 implies that for any element $[r] - [s]$ of D_0, the corresponding linear form
$$\phi_f([r] - [s]) \in \mathrm{Hom}(P_{k-2}(E), E)$$
is bounded.

3.2.1. *Modular symbols and L-values.* Both the algebraic part of the L-function associated to a modular form and its p-adic L-function may be expressed in terms of modular symbols.

DEFINITION 3.2.2. The algebraic part of the special value(s) of the classical L-function associated to f and a Dirichlet character χ of conductor c is given by the formula
$$L^{alg}(f, \chi, j) = \frac{c^{j+1} j!}{(-2\pi i)^j \tau(\overline{\chi}) \Omega_f^{w_\infty}} L(f, \overline{\chi}, j+1),$$
where $w_\infty = \chi(-1)$.

A computation using the expression for the L-function of f as the Mellin transform of the modular form f yields the following formula expressing L^{alg} in terms of modular symbols.

LEMMA 3.2.3. *We have*

$$L^{alg}(f,\chi,j) = \sum_{\nu \in (\mathbf{Z}/c\mathbf{Z})^\times} \chi(\nu)\phi_f^{w_\infty}\left(\left[\frac{-\nu}{c}\right] - [\infty]\right)((cz+\nu)^j)$$

for $j = 0, \ldots, (k-2)/2$.

PROOF. See [**25**, Section 8].   □

Next, we briefly recall the construction of the p-adic L-function from [**25**]. Let

$$\mathbf{Z}_{p,c} := \varprojlim \mathbf{Z}/p^n c\mathbf{Z} \cong \mathbf{Z}_p \times \mathbf{Z}/c\mathbf{Z}.$$

For $x \in \mathbf{Z}_{p,c}$, let x_p denote the projection of x to $\mathbf{Z}_p$. For $a \in \mathbf{Z}_{p,c}^\times$, write $D(a,r) := a + cp^r \subset \mathbf{Z}_{p,c}^\times$.

The p-adic L-function is constructed from the (unique) distributions $\mu_{f,\mathrm{MTT}}^\pm$ on the space of locally analytic functions on $\mathbf{Z}_{p,c}^\times$ satisfying

$$(3.2.4) \quad \int_{D(a,r)} P(x_p)d\mu_{f,\mathrm{MTT}}^\pm(x) = \left(wp^{\frac{k-2}{2}}\right)^{-r}\phi_f^\pm\left([\infty] - \left[\frac{a}{p^r c}\right]\right)(P(p^r cz + a))$$

for all polynomials of P degree at most $(k-2)/2$. Here and in the sequel, the left side is shorthand notation for

$$\mu_{f,MTT}^\pm\left(\delta_{D(a,r)}(x)P(x_p)\right),$$

where $\delta_{D(a,r)}$ denotes the characteristic function of the open set $D(a,r)$.

The boundedness properties of the modular symbols imply that there is a unique distribution on locally analytic functions on $\mathbf{Z}_{p,c}$ that, restricted to locally polynomial functions, satisfies the condition in equation 3.2.4. This is another instance of the p-adic integration theory that we referred to in Theorem 2.3.2. For details of the construction, see [**25**, Section 11].

Let $\chi : \mathbf{Z}_{p,c}^\times \to \mathbf{C}_p^\times$, and define $\langle \cdot \rangle : \mathbf{Z}_{p,c}^\times \to 1 + p\mathbf{Z}_p$ by $\langle x \rangle := x_p/\omega_{\mathrm{Teich}}(x)$, where $\omega_{\mathrm{Teich}}(x)$ is the p-adic Teichmuller character. The p-adic L-functions attached to f and χ are defined as follows:

$$L_p^\pm(f,\chi,s) := \int_{\mathbf{Z}_{p,c}^\times} \chi(x)\langle x \rangle^s d\mu_{f,\mathrm{MTT}}^\pm(x).$$

If $\epsilon = \chi(-1) \cdot (-1)^{\frac{k-2}{2}} = \pm 1$, then $L_p^{-\epsilon}(f,\chi,s) = 0$ and $L_p^\epsilon(f,\chi,s)$ will be *a priori* non-trivial. Thus writing $L_p = L_p^+ + L_p^-$, we see that $L_p = L_p^\epsilon$.

3.2.2. *Modular symbols and the tree.* Darmon introduced the remarkable idea of blending p-adic integration and the p-adic upper half plane with classical modular forms in his paper [**11**]. Assume that the level M of the preceding section can be written $M = Np$ with $(N,p) = 1$. Define

$$\Gamma_0^p(N) = \left\{\begin{pmatrix} a & b \\ c & d \end{pmatrix} \in \mathrm{SL}_2(\mathbf{Z}[1/p]), \ c \equiv 0 \pmod{N}\right\}.$$

Similarly, let

$$\tilde{\Gamma}_0^p(N) = \left\{ \begin{pmatrix} a & b \\ c & d \end{pmatrix} \in \mathrm{GL}_2(\mathbf{Z}[1/p])^+, \ c \equiv 0 \pmod{N} \right\}$$

where $\mathrm{GL}_2(\mathbf{Q})^+$ is the group of invertible matrices with positive determinant.

DEFINITION 3.2.5. Let $\mathcal{H}$ be the classical upper half plane over $\mathbf{C}$. Define $S_k^0(X, \Gamma_0^p(N))$ to be the complex vector space of "harmonic, modular form-valued" functions

$$F : \mathrm{Edges}(X) \times \mathcal{H} \to \mathbf{C}$$

satisfying the following conditions:

(1) $F(\gamma e, z) = \gamma(F(e, z)) = F(e, z)|_{\gamma^t}$ for $\gamma = \begin{pmatrix} a & b \\ c & d \end{pmatrix} \in \Gamma_0^p(N)$.

(2) $F(e', z) = -F(e, z)$ where e' is the edge opposite to e.

(3) $\sum_{e \mapsto v} F(e, z) = 0$ where the sum is over the edges leaving v.

(4) Each $F(e, \cdot)$ is a cusp form of weight k on $\mathcal{H}$ for the group

$$\Gamma_e = \{ \gamma \in \Gamma_0^p(N) : \gamma e = e \}.$$

The group $\mathrm{GL}_2(\mathbf{Q})^+$ acts on the left on $S_k^0(X, \Gamma_0^p(N))$ via the formula

$$g(F)(e, z) = g(F(g^{-1}e, z)) = (bz + d)^{-k} \det(g)^{k/2-1} F\left(g^{-1}e, \frac{az + c}{bz + d}\right).$$

The space $S_k^0(X, \Gamma_0^p(N))$ is quite small. Notice that, if $F \in S_k^0(X, \Gamma_0^p(N))$, then the restriction $F(e_0, \cdot)$ of F to the basic edge e_0 satisfies

$$\gamma(F(e_0, z)) = F(e_0, z)$$

for all

$$\gamma = \begin{pmatrix} a & b \\ c & d \end{pmatrix} \in \Gamma_0^p(N) \cap \Gamma_{e_0} = \Gamma_0(Np).$$

In other words, $F(e_0, \cdot)$ is a cusp form for $\Gamma_0(M) = \Gamma_0(Np)$.

PROPOSITION 3.2.6. *The restriction map*

$$S_k^0(X, \Gamma_0^p(N)) \to S_k(\Gamma_0(Np), \mathbf{C})$$

is injective and has image equal to the subspace of forms that are "new at p."

PROOF. See [28, Section 2.1]. The point is that the harmonicity requirement amounts to the statement that the form has to be in the kernel of the trace map(s) from forms of level Np to forms of level N. □

Given a p-new form f, we can find a corresponding element of $S_k^0(X, \Gamma_0^p(N))$ by defining

$$F(ge_0, z) = g(f(z)) = w^{\mathrm{ord}(\det(ad - bc))} (bz + d)^{-k} \det(g)^{k/2-1} f\left(\frac{az + c}{bz + d}\right)$$

for $g \in \tilde{\Gamma}$, where w is the sign such that $W_p(f) = -wf$ for the Atkin-Lehner operator W_p.

3.2.3. *Modular symbols, harmonic cocycles, and distributions.* Let f be a cusp form of level M that is new at p, and let F be the element of $S_k^0(X, \Gamma_0^p(N))$ associated to f by Proposition 3.2.6. Define a harmonic function Φ_f with values in $\mathrm{Hom}(P_{k-2}(E), E)$ on the edges of X by the rule

$$\Phi_f^\pm([r] - [s])(e)(P) := \phi_{F(e,\cdot)}^\pm([r] - [s])(P).$$

PROPOSITION 3.2.7. *(Orton) The harmonic function $\Phi_f^\pm([r] - [s])$ is bounded.*

PROOF. We need to verify that

$$\omega(\Phi^\pm([r] - [s])(\gamma e)(\gamma P)) \geq N$$

for some fixed integer N and polynomials P with coefficients in R. But

$$\begin{aligned}
\phi_{F(\gamma e,\cdot)}^\pm([r] - [s])(P) &= \pm\left(\int_r^s \gamma(f)(z)\gamma(P)(z)dz\right)^\pm \\
&= \pm\left(\int_{\gamma(r)}^{\gamma(s)} f(z)P(z)dz\right)^\pm
\end{aligned}$$

and this is bounded by Theorem 3.2.1.   □

From this boundedness result, we obtain from $\Phi_f^\pm([r] - [s])$ a distribution $\lambda_f^\pm([r] - [s])$ on $C^{la}(K, 2 - k)$ that extends to $C^{an}(K, 2 - k)$ following the procedure discussed in Section 2.3.

Let $\mathcal{M} := \mathrm{Hom}(D_0, \mathrm{Hom}(P_{k-2}(\mathbf{C}_p), \mathbf{C}_p))$, the space of modular symbols valued in the dual of the space of polynomials of degree at most $k - 2$. Then choosing any $a \in \mathcal{X}$, we obtain maps

$$\{\text{Cusp forms of level } Np \text{ new at } p\} \to H^1(\Gamma_0^p(N), \mathcal{M})$$

defined by

$$(3.2.8) \qquad lc_f^\pm(\gamma)([r] - [s])(P) = \lambda_f^\pm([r] - [s])\left(P(x)\log\left(\frac{x - \gamma a}{x - a}\right)\right)$$

and

$$(3.2.9) \qquad oc_f^\pm(\gamma)([r] - [s])(P) = \lambda_f^\pm([r] - [s])\left(P(x)\,\mathrm{ord}\left(\frac{x - \gamma a}{x - a}\right)\right).$$

The cohomology classes of these maps are independent of the choice of a.

3.3. Orton's L-invariant. The difficulty with the invariant $\mathcal{L}_T$ is that it is only indirectly related to the Mazur-Swinnerton-Dyer p-adic L-function that plays a role in the exceptional zero conjecture. This is because the p-adic L-function is constructed using a modular form on the usual upper half plane corresponding to a usual modular curve, while the construction of the $\mathcal{L}_T$ invariant uses a Shimura curve. The connection between these two constructions comes from the Jacquet-Langlands lifting theorem, which asserts that there is a correspondence between modular forms on Shimura curves and certain modular forms on classical modular curves. In fact, not all forms on modular curves come from Shimura curves, and so the invariant $\mathcal{L}_T$ isn't even defined for a modular form on $\Gamma_0(N')$, with N' general.

The invariant $\mathcal{L}_O$ constructed by Darmon and Orton is a hybrid object that mixes p-adic uniformization with classical modular forms. It's construction has something of the same flavor as that of $\mathcal{L}_T$, but it is directly connected to both the

p-adic L-function and the classical L-function of a form f on $\Gamma_0(N)$. In this section we will construct Orton's invariant (see Definition 3.3.4) and relate it to L-values.

3.3.1. *Cohomology of Modular Symbols.* For each prime $\ell \nmid N$, we define an action of the Hecke operator T_ℓ on $H^1(\Gamma_0^p(N), \mathcal{M})$. Let $\{\delta_j\}_{j=0}^\ell$ be a set of matrices in $\mathrm{GL}_2(\mathbf{Q})$ such that

$$\Gamma_0^p(N) \begin{pmatrix} 1 & 0 \\ 0 & \ell \end{pmatrix} \Gamma_0^p(N) = \bigsqcup_{j=0}^\ell \Gamma_0^p(N)\delta_j.$$

For each $\gamma \in \Gamma_0^p(N)$ and $j = 0, \dots, \ell$, there exists a unique $\gamma_j \in \Gamma_0^p(N)$ and index $i(\gamma, j)$ such that $\delta_j \gamma = \gamma_j \delta_{i(\gamma,j)}$. Let $\tilde{c}$ be a cohomology class in $H^1(\Gamma_0^p(N), \mathcal{M})$ represented by a cocycle c. The cohomology class represented by the cocycle

$$T_\ell(c)(\gamma) := \ell^{\frac{k-2}{2}} \sum_{j=0}^\ell \delta_j^{-1} c(\gamma_j)$$

is independent of choices, and is defined to be $T_\ell(\tilde{c})$.

We now define an Atkin-Lehner involution "at infinity." Let $\alpha_\infty = \begin{pmatrix} -1 & 0 \\ 0 & 1 \end{pmatrix}$. The operator W_∞ on $H^1(\Gamma_0^p(N), \mathcal{M})$ is defined by

$$W_\infty(c)(\gamma) := \alpha_\infty c(\alpha_\infty \gamma \alpha_\infty).$$

DEFINITION 3.3.1. Let V be a space endowed with an action of the Hecke algebra $\mathbf{T}$. For a sign $w_\infty = \pm 1$, let V^{f,w_∞} denote the space of elements $v \in V$ such that $T_\ell(v) = a_\ell \cdot v$ for each $\ell \nmid N$ and $W_\infty(v) = w_\infty \cdot v$, where a_ℓ denotes the eigenvalue of f for the Hecke operator T_ℓ.

In the next section, we will prove:

LEMMA 3.3.2. *For each $w_\infty = \pm 1$, the cohomology group $H^1(\Gamma_0^p(N), \mathcal{M})^{f,w_\infty}$ is a 1-dimensional $\mathbf{C}_p$-vector space.*

LEMMA 3.3.3. *For each $w_\infty = \pm 1$, we have*

$$lc_f^{w_\infty}, oc_f^{w_\infty} \in H^1(\Gamma_0^p(N), \mathcal{M})^{f,w_\infty}.$$

PROOF. (Sketch; see [**28**, Lemma 5.3]) The fact that $lc_f^{w_\infty}, oc_f^{w_\infty}$ are in the f, w_∞-isotypic subspace of $H^1(\Gamma_0^p(N), \mathcal{M})$ follows from the corresponding fact for Φ. More precisely, one can show that

$$\ell^{(k-2)/2} \sum_{j=0}^\ell \Phi_f^{w_\infty}([\delta_j x] - [\delta_j y])(\delta_j e)(P|_{\delta_j^{-1}}) = a_\ell \Phi_f^{w_\infty}([x] - [y])(e)(P)$$

and

$$\Phi_f^{w_\infty}([\alpha_\infty x] - [\alpha_\infty y])(\alpha_\infty e)(P|_{\alpha_\infty^{-1}}) = w_\infty \Phi_f^{w_\infty}([x] - [y])(e)(P)$$

from the corresponding formulas for $\phi_f^{w_\infty}$. $\qquad\square$

In Corollary 3.3.22 we will show that $oc_f^{w_\infty} \neq 0$. In view of Lemmas 3.3.2 and 3.3.3, we therefore propose:

DEFINITION 3.3.4. For each $w_\infty = \pm 1$, define $\mathcal{L}_O^{w_\infty} \in \mathbf{C}_p$ by the equality

$$lc_f^{w_\infty} = \mathcal{L}_O^{w_\infty} \cdot oc_f^{w_\infty}.$$

The goal of the remainder of this section is to prove Lemma 3.3.2. To simplify the notation, let $V := \operatorname{Hom}(P_{k-2}(\mathbf{C}_p), \mathbf{C}_p)$. Applying $\operatorname{Hom}(-, V)$ to the short exact sequence

$$0 \to D_0 \to D \to \mathbf{Z}$$

defining D_0, we obtain

(3.3.5) $$0 \to V \to \mathcal{F} \to \mathcal{M} \to 0,$$

where $\mathcal{F} := \operatorname{Hom}(D, V)$ and $\mathcal{M} := \operatorname{Hom}(D_0, V)$.

Consider the long exact sequence arising from (3.3.5) by taking cohomology for $\Gamma_0(N)$. The first term $V^{\Gamma_0(N)}$ is trivial when $k > 2$ and equal to $\mathbf{C}_p$ when $k = 2$ [20, p. 165 Lemma 2]. Furthermore, $H^2(\Gamma_0(N), V) = 0$ [20, p. 162, Prop. 1].

For each cusp of $\Gamma_0(N)$, i.e. for each class in $\Gamma_0(N)\backslash\mathbf{P}^1(\mathbf{Q})$, we choose a representative $x \in \mathbf{P}^1(\mathbf{Q})$ and let $\Gamma_0(N)_x$ denote the stabilizer of x in $\Gamma_0(N)$. The module $\mathcal{F}$ is easily seen to be a sum of induced modules:

$$\mathcal{F} = \bigoplus_x \operatorname{Ind}_{\Gamma_0(N)_x}^{\Gamma_0(N)} V.$$

By Shapiro's Lemma, we therefore have

$$H^i(\Gamma_0(N), \mathcal{F}) = \bigoplus_x H^i(\Gamma_0(N)_x, V).$$

In the long exact sequence associated to (3.3.5), the map

$$H^1(\Gamma_0(N), V) \to H^1(\Gamma_0(N), \mathcal{F}) \cong \bigoplus_x H^1(\Gamma_0(N)_x, V)$$

is simply the direct sum of restriction maps; its kernel is called the parabolic cohomology group, and denoted $H^1_{\mathrm{par}}(\Gamma_0(N), V)$. We thus obtain two exact sequences:

(3.3.6) $$0 \to \left(\bigoplus_x V^{\Gamma_0(N)_x}\right)/V^{\Gamma_0(N)} \to \mathcal{M}^{\Gamma_0(N)} \to H^1_{\mathrm{par}}(\Gamma_0(N), V) \to 0$$

and

(3.3.7) $$0 \to H^1_{\mathrm{par}}(\Gamma_0(N), V) \to H^1(\Gamma_0(N), V) \to$$
$$\bigoplus_x H^1(\Gamma_0(N)_x, V) \to H^1(\Gamma_0(N), \mathcal{M}) \to 0.$$

The key results we will use to study these sequences are the classical Eichler-Shimura isomorphisms, which state [20, Section 6.2]:

(3.3.8) $$H^1_{\mathrm{par}}(\Gamma_0(N), V) \cong S_k(\Gamma_0(N)) \oplus \overline{S_k(\Gamma_0(N))},$$

(3.3.9) $$H^1(\Gamma_0(N), V) \cong S_k(\Gamma_0(N)) \oplus \overline{S_k(\Gamma_0(N))} \oplus E_k(\Gamma_0(N)).$$

The sequences (3.3.6) and (3.3.7) are Hecke equivariant, as are the Eichler-Shimura isomorphisms. The Hecke structure of the module on the left in (3.3.6) is given by the action of the Hecke operators on the cusps of $\Gamma_0(N)$; therefore it is not surprising that [28, §7.2]:

LEMMA 3.3.10. *We have an isomorphism of Hecke modules:*

$$\left(\bigoplus_x V^{\Gamma_0(N)_x}\right)/V^{\Gamma_0(N)} \cong E_k(\Gamma_0(N)).$$

We leave the proof to the reader, as well as that of:

LEMMA 3.3.11. *Given a Hecke equivariant short exact sequence of finite dimensional $\mathbf{C}_p$-vector spaces $0 \to V_1 \to V_2 \to V_3 \to 0$ with $V_1^f = 0$, the map $V_2 \to V_3$ induces an isomorphism $V_2^f \to V_3^f$. Alternatively, if $V_3^f = 0$, then the map $V_1 \to V_2$ induces an isomorphism $V_1^f \to V_2^f$.*

PROPOSITION 3.3.12. *We have $\mathcal{M}^{\Gamma_0(N),f} = 0$.*

PROOF. This follows from the previous two lemmas, sequence (3.3.6), and the Eichler-Shimura isomorphism (3.3.8), since f is a form of level Np which is not old at p. □

Arguing similarly for N replaced by Np, we find:

PROPOSITION 3.3.13. *For each $w_\infty = \pm 1$, the space $\mathcal{M}^{\Gamma_0(Np),f,w_\infty}$ is a 1-dimensional $\mathbf{C}_p$-vector space.*

Turning now to sequence (3.3.7), we note that each group $\Gamma_0(N)_x$ is infinite cyclic, generated by an element denoted π_x. Thus $H^1(\Gamma_0(N)_x, V) = V/(\pi_x - 1)V$. One checks (see [**20**, p. 166 (2a)], for example) that this is a 1-dimensional $\mathbf{C}_p$-vector space. A dimension count in (3.3.7) using the Eichler-Shimura isomorphisms shows that $H^1(\Gamma_0(N), \mathcal{M})$ is trivial when $k > 2$, and has dimension 1 when $k = 2$; in either case the module is Eisenstein, so we obtain:

PROPOSITION 3.3.14. *We have $H^1(\Gamma_0(N), \mathcal{M})^f = 0$.*

We are now in a position to prove Lemma 3.3.2. The group $\Gamma_0^p(N)$ is the amalgamation of the groups $\Gamma_0(N)$ and its conjugate $\Gamma_0(N)' := \begin{pmatrix} p & 0 \\ 0 & 1 \end{pmatrix}^{-1} \Gamma_0(N) \begin{pmatrix} p & 0 \\ 0 & 1 \end{pmatrix}$, with respect to their intersection $\Gamma_0(Np)$. This fact follows from the fact that a fundamental domain for the action of $\Gamma_0^p(N)$ on the tree X is given by the single edge e_0 with stabilizer $\Gamma_0(Np)$, and its two boundary vertices with stabilizers $\Gamma_0(N)$ and $\Gamma_0(N)'$. From this amalgamation property, one deduces an exact sequence (see [**43**, §2.6]):

$$0 \longrightarrow \mathcal{M}^{\Gamma_0^p(N)} \longrightarrow \mathcal{M}^{\Gamma_0(N)} \oplus \mathcal{M}^{\Gamma_0(N)'} \longrightarrow \mathcal{M}^{\Gamma_0(Np)}$$

$$H^1(\Gamma_0^p(N), \mathcal{M}) \longrightarrow H^1(\Gamma_0(N), \mathcal{M}) \oplus H^1(\Gamma_0(N)', \mathcal{M}).$$

By breaking this exact sequence into short exact sequences, we find from Lemma 3.3.11, Proposition 3.3.12 and Proposition 3.3.14 that:

$$(3.3.15) \qquad \mathcal{M}^{\Gamma_0^p(N),f} = 0 \quad \text{and} \quad (\mathcal{M}^{\Gamma_0(Np)})^{f,w_\infty} \xrightarrow{\sim} H^1(\Gamma_0^p(N), \mathcal{M})^{f,w_\infty}.$$

Proposition 3.3.13 then concludes the proof of Lemma 3.3.2.

3.3.2. *Specializations of the cohomology classes.* Fix a positive integer c and an integer ν relatively prime to c. The pair (c, ν) gives rise to a $\mathbf{Q}$-algebra embedding $\Psi : \mathbf{Q} \times \mathbf{Q} \to M_2(\mathbf{Q})$ via the formula

$$\Psi(1,0) = \begin{pmatrix} 1 & \nu/c \\ 0 & 0 \end{pmatrix}.$$

Let s be the order of p^2 in $(\mathbf{Z}/c\mathbf{Z})^\times$. The group $\Psi(\mathbf{Q}^\times \times \mathbf{Q}^\times) \cap \Gamma_0^p(N)$ is an infinite cyclic group, generated by

$$\gamma_\Psi := \begin{pmatrix} p^s & (p^s - p^{-s})\nu/c \\ 0 & p^{-s} \end{pmatrix}.$$

The fixed points of γ_Ψ are $x_\Psi = \infty$ and $y_\Psi = -\nu/c$, and the polynomial

$$P_\Psi(z) = (cz + \nu)^{\frac{k-2}{2}}$$

is fixed by γ_Ψ as well.

DEFINITION 3.3.16. Define

(3.3.17) $$LI_\Psi^\pm := lc_f^\pm(\gamma_\Psi)([x_\Psi] - [y_\Psi])(P_\Psi)$$

and

(3.3.18) $$W_\Psi^\pm := oc_f^\pm(\gamma_\Psi)([x_\Psi] - [y_\Psi])(P_\Psi).$$

Since γ_Ψ fixes x_Ψ, y_Ψ, and P_Ψ, one easily checks that $b(\gamma_\Psi)([x_\Psi] - [y_\Psi])(P_\Psi) = 0$ for a coboundary b; thus the equations for LI_Ψ and W_Ψ are well-defined. As we now explain, the values LI_Ψ and W_Ψ encode the central critical values of the p-adic and classical L-functions attached to f. As usual, let χ be a Dirichlet character of conductor c with $\chi(p) = w$ and $\chi(-1) = w_\infty$.

THEOREM 3.3.19. *With notation as above, we have*

(3.3.20) $$L^{alg}\left(f, \chi, \frac{k-2}{2}\right) = \frac{1}{2s} \sum_{\nu \in (\mathbf{Z}/c\mathbf{Z})^\times} \chi(\nu) W_\Psi^{w_\infty}$$

and

(3.3.21) $$L_p'\left(f, \omega^{\frac{k-2}{2}}\chi, \frac{k-2}{2}\right) = \frac{1}{2s} \sum_{\nu \in (\mathbf{Z}/c\mathbf{Z})^\times} \chi(\nu) LI_\Psi^{w_\infty}.$$

Consequently,

$$L_p'(f, \omega^{\frac{k-2}{2}}\chi, (k-2)/2) = \mathcal{L}_O^{w_\infty}(f)L(f, \chi, (k-2)/2)^{alg}.$$

(This is the "exceptional zero conjecture" (for weight $k \geq 2$) as originally posed in [25, Section 15]*).*

PROOF. The proofs of equations (3.3.20) and (3.3.21) involve calculations on the tree. This will take the next few sections. □

COROLLARY 3.3.22. $oc_f^{w_\infty} \neq 0$.

PROOF. A result of Rohrlich [30] implies that there is a Dirichlet character χ as desired such that $L(f, \chi, k/2) \neq 0$. □

3.3.3. *First part of proof of Orton's Theorem.* The first step in proving Orton's theorem is to evaluate W_Ψ. We begin with an explicit evaluation of the right side of equation (3.2.9), which defines a cocycle representing the cohomology class oc_f.

LEMMA 3.3.23. *Suppose that $a \in X$ reduces to a vertex v of the tree X. Then we have*

$$oc_f^\pm(\gamma)([r] - [s])(P) = \sum_{e \in (v \to \gamma v)} \Phi_f^\pm([r] - [s])(e)(P),$$

where $(v \to \gamma v)$ represents the unique path in X from the vertex v to the vertex γv, and the sum on the right side is indexed by the oriented edges e in this path.

PROOF. This is the same argument that we used in Lemma 3.1.1. □

For each $\nu \in (\mathbf{Z}/c\mathbf{Z})^\times$, let J_ν denote the coset $\nu\langle p \rangle \subset (\mathbf{Z}/c\mathbf{Z})^\times$. Let s' denote the order of p modulo c, so $s = s'$ if s' is odd and $s = s'/2$ if s' is even. For $a \in J_\nu$, denote by $j(a)$ the equivalence class mod s' such that $a \equiv \nu p^{j(a)} \pmod{c}$. Note that the expression $w^{j(a)}$ is well-defined if either $w = 1$ or s' is even.

PROPOSITION 3.3.24. *We have*
$$W_\Psi^\pm = \beta \sum_{a \in J_\nu} w^{j(a)} \phi_f^\pm \left(\left[-\frac{a}{c} \right] - [\infty] \right) \left((cz + a)^{(k-2)/2} \right),$$

where
$$\beta = \begin{cases} 1 & \text{if } s' \text{ is even,} \\ 2 & \text{if } s' \text{ is odd and } w = 1, \\ 0 & \text{if } s' \text{ is odd and } w = -1. \end{cases}$$

PROOF. Suppose we choose a in equation (3.2.9) to reduce to the central vertex v_0 of X. Then the definition of W_Ψ in (3.3.18) and Lemma 3.3.23 yield
$$W_\Psi^\pm = \sum_{e \in (v_0 \to \gamma_\Psi v_0)} \Phi_f^\pm \left([\infty] - \left[\frac{-\nu}{c} \right] \right) (e, P_\Psi).$$

The edges e in the sum may be written $e_j = \gamma_j^{-1} e_0$ where $\gamma_j = \begin{pmatrix} 1 & -\nu' \\ 0 & p^j \end{pmatrix}$, where ν' is an integer such that $\nu' \equiv -\nu/c \pmod{p^{2s}}$, and $j = 0, \ldots, 2s - 1$. We evaluate each term in the sum:

$$\Phi_f^\pm \left([\infty] - \left[-\frac{\nu}{c} \right] \right) (\gamma_j^{-1} e_0, P_\Psi) = w^{|\gamma_j|} \Phi_f^\pm \left([\gamma_j \infty] - \left[\gamma_j \left(-\frac{\nu}{c} \right) \right] \right) (e_0, P_\Psi|_{\gamma_j^{-1}})$$

$$(3.3.25) \qquad = w^j \Phi_f^\pm \left([\infty] - \left[\frac{(-\nu - c\nu')/p^j}{c} \right] \right) \left(e_0, \left(cz + \frac{c\nu' + \nu}{p^j} \right)^{\frac{k-2}{2}} \right).$$

From the invariance of f under the transformation $z \mapsto z + 1$, it is clear that the expression in (3.3.25) depends on the integer $(c\nu' + \nu)/p^j$ only up to its equivalence class modulo c. As $j = 0, \ldots, 2s - 1$, these integers run over the set J_ν: once if s' is even and twice if s' is odd. In the latter case, the coefficients w^j appear with opposite sign in the two occurrences when $w = -1$, and with the same sign when $w = 1$. The result follows. □

We may now prove the first half of theorem 3.3.19. Let χ be a Dirichlet character of conductor c with $\chi(p) = w$ and $\chi(-1) = w_\infty$. Note that $\beta \neq 0$, and hence $s' = 2s/\beta$. Then by Proposition 3.3.24 we have:
$$\frac{1}{2s} \sum_{\nu \in (\mathbf{Z}/c\mathbf{Z})^\times} \chi(\nu) W_\Psi^{w_\infty} =$$

$$(3.3.26) \qquad \frac{1}{s'} \sum_{\nu \in (\mathbf{Z}/c\mathbf{Z})^\times} \chi(\nu) \sum_{a \in J_\nu} w^{j(a)} \phi_f^{w_\infty} \left(\left[-\frac{a}{c} \right] - [\infty] \right) \left((cz + a)^{(k-2)/2} \right).$$

As ν ranges over $(\mathbf{Z}/c\mathbf{Z})^\times$, the sets J_ν cover $(\mathbf{Z}/c\mathbf{Z})^\times$ with each element repeated s' times. Furthermore, for $a \in J_\nu$ we have $\chi(\nu) w^{j(a)} = \chi(\nu \cdot p^{j(a)}) = \chi(a)$. We find that (3.3.26) equals

$$\sum_{a\in(\mathbf{Z}/c\mathbf{Z})^\times}\chi(a)\phi_f^{w_\infty}\left(\left[-\frac{a}{c}\right]-[\infty]\right)\left((cz+a)^{(k-2)/2}\right).$$

Equation (3.3.20) now follows from Lemma 3.2.3.

3.3.4. *Second part of the Proof of Orton's Theorem.* In this section, we relate the distribution λ_f to the p-adic L-function of f. Let us now compare the distribution $\mu_{f,MTT}^\pm$ on $\mathbf{Z}_{p,c}^\times$ to the Darmon-Orton modular symbol of distributions $\lambda_f^\pm$. Let Ψ be an embedding as in section 3.3.2. For each integer i, define

$$U(v_i):=\{t\in\mathbf{P}^1(\mathbf{Q}_p)-\{x_\Psi,y_\Psi\}:\mathrm{ord}(M_\Psi(t))=i\}.$$

The motivation for this notation is as follows. Let $(x_\Psi\to y_\Psi)$ denote the bi-infinite path from the end of X corresponding to x_Ψ to the end corresponding to y_Ψ. The vertices of $(x_\Psi\to y_\Psi)$ may be labeled $\{v_i\}$ in such a way that $U(v_i)$ is the set of points corresponding to ends of X that intersect $(x_\Psi\to y_\Psi)$ precisely at v_i. If e_i is the edge from v_{i-1} to v_i, then $U(v_i)=U(e_i)-U(e_{i+1})$. A fundamental region for the action of γ_Ψ on $\mathbf{P}^1(\mathbf{Q}_p)-\{x_\Psi,y_\Psi\}$ is given by:

$$\mathcal{F}_\Psi:=\bigcup_{i=0}^{2s-1}U(v_i).$$

For $z\in\mathcal{F}_\Psi$, write $i(z)=\mathrm{ord}(M_\Psi(z))$, i.e. the index i such that $z\in U(v_i)$. We also define

$$J_{\infty,\nu}=\{a\in\mathbf{Z}_{p,c}^\times:a\equiv\nu p^j\pmod{c}\text{ for some }j=j(a)\}.$$

PROPOSITION 3.3.27. *If F is a locally analytic function on $\mathbf{Z}_p^\times$, then*

$$\int_{\mathcal{F}_\Psi}p^{i(z)\cdot\frac{k-2}{2}}F\left(\frac{cz+\nu}{p^{i(z)}}\right)d\lambda_f^\pm([x_\Psi]-[y_\Psi])(z)=\beta\int_{J_{\infty,\nu}}w^{j(x)}F(x_p)d\mu_{f,MTT}^\pm(x).$$

PROOF. For $j=0,\ldots,2s-1$, write $J_{\infty,\nu,j}=\{a\in\mathbf{Z}_{p,c}^\times:b\equiv\nu p^j\pmod{c}\}$. We will show
(3.3.28)

$$w^jp^{j\cdot\frac{k-2}{2}}\int_{U(v_j)}F\left(\frac{cz+\nu}{p^j}\right)d\lambda_f^\pm([x_\Psi]-[y_\Psi])(z)=\int_{J_{\infty,\nu,j}}F(x_p)d\mu_{f,MTT}^\pm(x).$$

The result will then follow by summing from $j=0$ to $j=2s-1$; as j varies the $J_{\infty,\nu,j}$ cover $J_{\infty,\nu}$ once if s' is even, twice if s' is odd, and with opposite sign in the latter case when $w=-1$.

To prove (3.3.28), fix an integer $n>2s$. Refine each $U(v_j)$ by

$$U(v_j)=\bigcup_{a\in(\mathbf{Z}/p^n\mathbf{Z})}U_{j,a},\quad\text{where}\quad U_{j,a}=\{t\in U(v_j):(ct+\nu)/p^j\equiv a\pmod{p^n}\}$$

and correspondingly refine $J_{\infty,\nu,j}$ as

$$J_{\infty,\nu,j}=\bigcup_{a\in(\mathbf{Z}/p^n\mathbf{Z})^\times}D(b_{a,j},n).$$

where $b_{a,j} = (\nu + c\nu')/p^j + ac$ and $\nu' \in \mathbf{Z}$ satisfies $\nu' \equiv -\nu/c \pmod{n+2s}$. Then from the definition of the distribution $\mu_{f,\mathrm{MTT}}$, we have for a polynomial P of degree $\leq k-2$:

$$\int_{D(b,n)} P(x) d\mu_{f,\mathrm{MTT}}^\pm(x) = w^n p^{-n\left(\frac{k-2}{2}\right)} \phi_f^\pm \left([\infty] - \left[\frac{b}{p^n c}\right] \right) (P(p^n cz + b))$$

$$(3.3.29) \qquad\qquad = w^n p^{-n\left(\frac{k-2}{2}\right)} \Phi_f^\pm \left([\infty] - \left[\frac{b}{p^n c}\right] \right) (e_0, P(p^n cz + b)).$$

where $b = b_{a,j}$. Now if $\gamma = \begin{pmatrix} 1 & -\nu' - p^j a \\ 0 & p^{n+j} \end{pmatrix}$, then $U_{j,a} = \gamma^{-1} U(e_0)$. Thus using the transformation property of $\Phi_f^\pm$ under γ, the right side of (3.3.29) may be written:

$$w^j p^{-n\left(\frac{k-2}{2}\right)} \Phi_f^\pm \left([\gamma^{-1}(\infty)] - \left[\gamma^{-1}\left(\frac{b}{p^n c}\right)\right] \right) \left(\gamma^{-1} e_0, P(p^n cz + b)|_\gamma\right)$$

$$= w^j p^{j\left(\frac{k-2}{2}\right)} \Phi_f^\pm \left([\infty] - \left[\frac{\nu}{c}\right] \right) \left(\gamma^{-1} e_0, P\left(\frac{cz + \nu}{p^j}\right)\right)$$

$$= w^j p^{j\left(\frac{k-2}{2}\right)} \int_{U_{j,a}} P\left(\frac{cz + \nu}{p^j}\right) d\lambda_f^\pm ([x_\Psi] - [y_\Psi])(z).$$

Thus we have proven the result for polynomials of degree $\leq k-2$ on the arbitrarily small balls $D(b,n)$ and $U(j,a)$. By the uniqueness properties of the extensions of $\mu_{f,\mathrm{MTT}}^\pm$ and $\lambda_f^\pm$ from the distributions on P_{k-2} to the space of locally analytic functions, the result follows. $\square$

3.3.5. *End of Orton's Theorem.* In this section we conclude the proof of Theorem 3.3.19. Recall the definition:

$$LI_\Psi^\pm = \int_{\mathbf{P}^1(\mathbf{Q}_p)} \log\left(\frac{x - \gamma_\Psi z}{x - z}\right) P_\Psi(x) d\lambda_f^\pm ([x_\Psi] - [y_\Psi])(x).$$

Recall also how $\lambda_f^\pm([x_\Psi] - [y_\Psi])$ is applied to a locally analytic function such as $\log(\frac{x-\gamma_\Psi z}{x-z})P_\Psi(x)$: we cover $\mathbf{P}^1(\mathbf{Q}_p)$ by smaller and smaller open balls, write the function as a power series on each open ball, truncate the power series to a polynomial of degree $k-2$, evaluate $\lambda_f^\pm([x_\Psi] - [y_\Psi])$ on each of these polynomials on the open balls via $\Phi_f^\pm([x_\Psi] - [y_\Psi])$, and sum the results; the limit as the covers become uniformly finer is the desired value.

In the present case, we write

$$(3.3.30) \qquad \mathbf{P}^1(\mathbf{Q}_p) = \gamma_\Psi^{-n} U(\bar{e}_0) \sqcup \bigsqcup_{j=-n}^n \gamma_\Psi^j \mathcal{F}_\Psi \sqcup \gamma_\Psi^{n+1} U(e_0),$$

where $\bar{e}_0$ denotes the edge e_0 with the opposite orientation. We will refine the middle term of (3.3.30) later, but indicate first why, in the limit, the end divisions contribute nothing to the integral. Let $T_n(x)$ denote the truncation of the power series of $\log(\frac{x-\gamma_\Psi z}{x-z})P_\Psi(x)$ expanded around y_Ψ on the open set $\gamma_\Psi^{n+1} U(e_0)$ to a polynomial of degree $k-2$. From the invariance of Φ under Γ, and the fact that

γ_Ψ stabilizes x_Ψ, y_Ψ, and P_Ψ, we have

$$\lim_{n\to\infty} \int_{\gamma^{n+1}U(e_0)} \log\left(\frac{x - \gamma_\Psi z}{x - z}\right) P_\Psi(x) d\lambda_f^\pm([x_\Psi] - [y_\Psi])(x)$$

$$= \lim_{n\to\infty} \Phi_f^\pm([x_\Psi] - [y_\Psi])(\gamma^{n+1}e_0, T_n(x))$$

$$= \lim_{n\to\infty} \Phi_f^\pm([x_\Psi] - [y_\Psi])(e_0, V_n(x)),$$

where V_n is the truncation to a polynomial of degree $k - 2$ of the power series of

$$\log\left(\frac{\gamma_\Psi^{n+1}x - \gamma_\Psi z}{\gamma_\Psi^{n+1}x - z}\right) P_\Psi(x) = \log\left(\frac{x - \gamma_\Psi^{-n}z}{x - \gamma_\Psi^{-(n+1)}z}\right) P_\Psi(x)$$

on $U(e_0)$ expanded around y_Ψ. We leave it to the reader to verify (or consult [28, §6.4]) the explicit formula:

$$V_n(x) = \sum_{i=1}^{\frac{k-2}{2}} \left(-\left(\gamma_\Psi^{-n}z - y_\Psi\right)^{-i} + \left(\gamma_\Psi^{-(n+1)}z - y_\Psi\right)^{-i}\right) \frac{c^{\frac{k-2}{2}}}{i}(x - y_\Psi)^{i+\frac{k-2}{2}}.$$

As $n \to \infty$, $\gamma_\Psi^{-n}z \to x_\Psi = \infty$, so the coefficients above tend to zero. Thus we have proven that

$$\lim_{n\to\infty} \int_{\gamma^{n+1}U(e_0)} \log\left(\frac{x - \gamma_\Psi z}{x - z}\right) P_\Psi(x) d\lambda_f^\pm([x_\Psi] - [y_\Psi])(x) = 0,$$

and a similar result holds for the other end term in the decomposition (3.3.30).

Now we are left to analyze

$$LI_{\Psi,n}^\pm := \sum_{j=-n}^{n} \int_{\gamma_\Psi^j \mathcal{F}_\Psi} \log\left(\frac{x - \gamma_\Psi z}{x - z}\right) P_\Psi(t) d\lambda_f^\pm([x_\Psi] - [y_\Psi])(x).$$

For each term in the sum we invoke the change of variables $x \mapsto \gamma_\Psi^{-j}x$; using the $\Gamma_0^p(N)$-invariance of $\lambda_f^\pm$ we obtain

$$LI_{\Psi,n}^\pm = \sum_{j=-n}^{n} \int_{\mathcal{F}_\Psi} \log\left(\frac{\gamma_\Psi^j x - \gamma_\Psi z}{\gamma_\Psi^j x - z}\right) P_\Psi(t) d\lambda_f^\pm([x_\Psi] - [y_\Psi])(x)$$

$$= \sum_{j=-n}^{n} \int_{\mathcal{F}_\Psi} \log\left(\frac{x - \gamma_\Psi^{1-j}z}{x - \gamma_\Psi^{-j}z}\right) P_\Psi(t) d\lambda_f^\pm([x_\Psi] - [y_\Psi])(x)$$

$$(3.3.31) \qquad = \int_{\mathcal{F}_\Psi} \log\left(\frac{x - \gamma_\Psi^{1+n}z}{x - \gamma_\Psi^{-n}z}\right) P_\Psi(t) d\lambda_f^\pm([x_\Psi] - [y_\Psi])(x),$$

as the sum telescopes. Now in the limit as $n \to \infty$, we have $\gamma_\Psi^{1+n}z \to y_\Psi$ and $\gamma_\Psi^{-n}z \to x_\Psi$; thus we would like to say that in the limit, we can replace the argument of log in (3.3.31) by a linear fractional transformation taking y_Ψ to 0 and x_Ψ to ∞, namely, $M_\Psi(t) := t + \nu/c$. More precisely, let

$$M_n(t) = \frac{-y_\Psi \gamma_\Psi^{-n}z}{\gamma_\Psi^{n+1}z} \cdot \frac{x - \gamma_\Psi^{1+n}z}{x - \gamma_\Psi^{-n}z}.$$

Now since

$$\int_{\mathcal{F}_\Psi} P_\Psi(x) d\lambda_f^\pm([x_\Psi] - [y_\Psi])(x)$$
$$= \Phi_f^\pm([x_\Psi] - [y_\Psi])(e_{2s}, P_\Psi) - \Phi_f^\pm([x_\Psi] - [y_\Psi])(e_0, P_\Psi)$$
$$= \Phi_f^\pm([x_\Psi] - [y_\Psi])(\gamma_\Psi e_0, P_\Psi) - \Phi_f^\pm([x_\Psi] - [y_\Psi])(e_0, P_\Psi)$$
$$= 0,$$

it follows that

$$LI_{\Psi,n}^\pm = \int_{\mathcal{F}_\Psi} \log(M_n(x)) P_\Psi(x) d\lambda_f^\pm([x_\Psi] - [y_\Psi])(x).$$

Since $M_n(x) \to M_\Psi(x)$ as $n \to \infty$ for any $x \in \mathcal{F}_\Psi$, the continuity of λ gives

$$LI_\Psi^\pm = \int_{\mathcal{F}_\Psi} \log(M_\Psi(x)) P_\Psi(x) d\lambda_f^\pm([x_\Psi] - [y_\Psi])(x)$$
$$= \int_{\mathcal{F}_\Psi} \log(cx + \nu)(cx + \nu)^{\frac{k-2}{2}} d\lambda_f^\pm([x_\Psi] - [y_\Psi])(x)$$
$$= \int_{\mathcal{F}_\Psi} p^{i(z) \cdot \frac{k-2}{2}} \log\left(\frac{cx+\nu}{p^{i(z)}}\right) \left(\frac{cx+\nu}{p^{i(z)}}\right)^{\frac{k-2}{2}} d\lambda_f^\pm([x_\Psi] - [y_\Psi])(x)$$
$$= \beta \int_{J_{\infty,\nu}} w^{j(t)} t_p^{\frac{k-2}{2}} \log(t_p) d\mu_{f,\mathrm{MTT}}^\pm(t),$$

by Proposition 3.3.27. We are now in a position to conclude the proof of Theorem 3.3.19. Let χ be a Dirichlet character of conductor c with $\chi(p) = w$ and $\chi(-1) = w_\infty$. We evaluate:

$$\frac{1}{2s} \sum_{\nu \in (\mathbf{Z}/c\mathbf{Z})^\times} \chi(\nu) LI_\Psi^{w_\infty} = \frac{1}{s'} \sum_{\nu \in (\mathbf{Z}/c\mathbf{Z})^\times} \chi(\nu) \int_{J_{\infty,\nu}} w^{j(t)} t_p^{\frac{k-2}{2}} \log(t_p) d\mu_{f,\mathrm{MTT}}^{w_\infty}(t)$$

$$(3.3.32) \qquad = \frac{1}{s'} \sum_{\nu \in (\mathbf{Z}/c\mathbf{Z})^\times} \int_{J_{\infty,\nu}} \chi(t) t_p^{\frac{k-2}{2}} \log(t_p) d\mu_{f,\mathrm{MTT}}^{w_\infty}(t).$$

Now as ν ranges over $(\mathbf{Z}/c\mathbf{Z})^\times$, the sets $J_{\infty,\nu}$ cover $\mathbf{Z}_{p,c}^\times$, with each point being covered s' times. Thus (3.3.32) becomes

$$\int_{\mathbf{Z}_{p,c}^\times} \chi(t) t_p^{\frac{k-2}{2}} \log(t_p) d\mu_{f,\mathrm{MTT}}^{w_\infty} = \int_{\mathbf{Z}_{p,c}^\times} \chi(t)(\omega(t)\langle t \rangle)^{\frac{k-2}{2}} \log(t_p) d\mu_{f,\mathrm{MTT}}^{w_\infty}$$
$$= \frac{d}{ds}\left(\int_{\mathbf{Z}_{p,c}^\times} \chi(t)\omega(t)^{\frac{k-2}{2}} \langle t \rangle^s d\mu_{f,\mathrm{MTT}}^{w_\infty}\right)\Big|_{s=\frac{k-2}{2}}$$
$$= \frac{d}{ds} L_p^{w_\infty}(f, \omega^{\frac{k-2}{2}} \chi, s)\Big|_{s=\frac{k-2}{2}}.$$

Finally, we remark that

$$(\omega^{\frac{k-2}{2}} \chi)(-1) = w_\infty(-1)^{\frac{k-2}{2}},$$

so

$$L_p(f, \omega^{\frac{k-2}{2}} \chi, s) = L_p^{w_\infty}(f, \omega^{\frac{k-2}{2}} \chi, s).$$

This concludes the proof of Theorem 3.3.19.

4. Breuil duality and p-adic Langlands theory

4.1. Brief remarks on the p-adic Langlands program. In this lecture, we approach the Darmon-Orton theory developed earlier from the point of view of p-adic Langlands theory. In general terms, the classical Langlands program sets up a correspondence between Galois representations and classical automorphic forms. The p-adic Langlands program, which is currently in an early but exciting phase of development, seeks (at least in its local version) to relate classes of continuous p-adic representations of reductive groups to local p-adic Galois representations. Because p-adic representations are so much more complicated than complex representations, and continuous p-adic representations are more complicated than classical smooth representations, the p-adic Langlands program requires the introduction of many new concepts. For an overview of some of the ideas in the (local) p-adic Langlands program, see the introduction to Breuil's paper [**2**]. See also the paper [**14**], which adopts a representation theoretic perspective on many of the ideas we have discussed.

To get a taste of these new ideas in the situation of interest to us in these lectures, let us denote by $\sigma(f)$ the p-adic representation of $\mathrm{Gal}(\overline{\mathbf{Q}}/\mathbf{Q})$ attached to the newform f of weight $k \geq 2$ and level M. The philosophy of the p-adic Langlands program suggests that one should be able to recover this Galois representation from purely automorphic data associated to f. Similarly, the "local" p-adic Langlands program should relate the local representation $\sigma_p(f) := \sigma(f)|_{\mathrm{Gal}(\overline{\mathbf{Q}}_p/\mathbf{Q}_p)}$ to the local automorphic component $\pi_p(f)$ of the representation of GL_2 determined by f.

From the classical theory, we know that when the level of f is exactly divisible by p, the local automorphic component $\pi_p(f)$ does not contain enough information to isolate σ_p. Indeed, in this case, $\pi_p(f)$ is always the Steinberg representation and it is exactly the $\mathcal{L}$ invariant of f (and its weight) that provide the additional information necessary to identify the local Galois representation associated to f. In the papers [**2**] and [**3**] Breuil seeks to answer the question:

> How can we extract the $\mathcal{L}$-invariant of f from "automorphic" information?

Breuil's answer to this question begins with a certain p-adic completion $\hat{H}^1_c(N) \otimes E$ of the étale cohomology of the tower of modular curves of level Np^r. This completion is a p-adic Banach space with a Hecke action and $\mathrm{GL}_2(\mathbf{Q}_p)$ action that preserves its norm, defined over the finite extension E of $\mathbf{Q}_p$ generated by the Hecke eigenvalues of f. (This space is defined in more detail in the next section.) Roughly speaking, the part of the space $\hat{H}^1_c(N) \otimes E$ cut out by insisting that the Hecke algebra act through the eigenvalues of the form f contains the locally algebraic representation

$$\mathrm{Sym}^{k-2}(E^2) \otimes_E \pi_p(f) \text{``}\cong\text{''} C^{la}(\mathbf{Q}_p, 2-k)/P_{k-2} \otimes E$$

embedded $\mathrm{GL}_2(\mathbf{Q}_p)$–equivariantly, where the symbol "$\cong$" means "more or less the same as." For a thorough discussion of these ideas, see [**14**, Section 4], which in turn refers to [**13**].

Denote by $\hat{\pi}_p(f)$ the closure of $\mathrm{Sym}^{k-2} E^2 \otimes_E \pi_p(f)$ in $\hat{H}^1_c(N) \otimes E$. Breuil proves that:

(1) when $k > 2$, $\sigma_p(f)$ is absolutely irreducible, and $\hat{\pi}_p(f)$ indeed exactly determines $\mathcal{L}(f)$.

(2) When $k = 2$, $\sigma_p(f)$ is reducible and is not determined by $\hat{\pi}_p(f)$. However, Breuil shows that $\hat{H}_c^1(N) \otimes E$ contains a topologically reducible Banach space representation of length 2 with $\hat{\pi}_p(f)$ its unique sub-object, which determines $\mathcal{L}(f)$, and which depends only on $\sigma_p(f)$.

For the purposes of these notes, we will be mainly interested in how Breuil's point of view gives an alternate definition of the $\mathcal{L}$-invariant. As before, we assume that the level of f is $M = Np$, with $p \nmid N$. In this situation, the sign w occurring in the previous section is $w = a_p^{-1} \cdot p^{\frac{k-2}{2}}$. Let $\mathrm{nr}(w)$ denote the representation of $\mathrm{GL}_2(\mathbf{Q}_p)$ which sends $x \mapsto w^{\mathrm{ord}_p(\det(x))}$. For each $\mathcal{L} \in E$, Breuil uses the theory of modular symbol-valued measures on the upper half plane to define a Banach space representation $B(k, \mathcal{L})$ of $\mathrm{GL}_2(\mathbf{Q}_p)$. He then proves:

THEOREM 4.1.1. *There exists a (unique) $\mathcal{L}_B(f) \in E$ such that*

$$\mathrm{Hom}_{\mathrm{GL}_2(\mathbf{Q}_p)} \left(B(k, \mathcal{L}) \otimes \mathrm{nr}(w), (\hat{H}_c^1(N) \otimes E)^f \right) \cong \begin{cases} 0 & \text{if } \mathcal{L} \neq -\mathcal{L}_B(f), \\ E^2 & \text{if } \mathcal{L} = -\mathcal{L}_B(f). \end{cases}$$

Furthermore, we have $\mathcal{L}_B(f) = \mathcal{L}_O^+(f) = \mathcal{L}_O^-(f)$.

Here $(\hat{H}_c^1(N) \otimes E)^f$ denotes the f-isotypic component of $\hat{H}_c^1(N) \otimes E$, i.e. the subspace on which the Hecke algebra acts via the eigenvalues of f. In the remainder of this lecture, we discuss $B(k, \mathcal{L})$ and sketch a proof of Theorem 4.1.1.

4.2. Completed Étale cohomology. Let $K = \prod_\ell K_\ell$ be an open compact subgroup of $\mathrm{GL}_2(\hat{\mathbf{Z}})$. We denote by $Y(K)$ the open modular curve over $\mathbf{Q}$, whose complex points are given by:

$$Y(K)(\mathbf{C}) := \mathrm{GL}_2(\mathbf{Q}) \backslash \mathrm{GL}_2(\mathbf{A}_{\mathbf{Q}}) / \mathrm{SO}_2(\mathbf{R}) \mathbf{R}^\times K.$$

The number of geometric connected components of $Y(K)$ is the size of

$$(\mathbf{Q}^+)^\times \backslash \mathbf{A}_{\mathbf{Q},f}^\times / \det(K).$$

For a positive integer n, we write $H^1(Y(K), \mathbf{Z}/p^n\mathbf{Z})$ for the Betti cohomology of the complex space $Y(K)(\mathbf{C})$, or equivalently, the étale cohomology of the algebraic variety $Y(K)_{\overline{\mathbf{Q}}}$. Similarly, $H_c^1(Y(K), \mathbf{Z}/p^n\mathbf{Z})$ represents cohomology with compact supports. For $* = c$ or $* = $ empty, the $\mathbf{Z}/p^n\mathbf{Z}$-module $H_*^1(Y(K), \mathbf{Z}/p^n\mathbf{Z})$ is naturally endowed with an action of $\mathrm{Gal}(\overline{\mathbf{Q}}/\mathbf{Q})$. In the most concrete terms, the Galois action on $H_c^1(Y(K), \mathbf{Z}/p^n\mathbf{Z})$ can be understood by identifying this space with the p^n torsion of the jacobian of the closed curve $X(K)$, relative to its set of cusps.

In addition, $H_*^1(Y(K), \mathbf{Z}/p^n\mathbf{Z})$ is endowed with a Hecke action. First let

$$H_*^1(Y(K), \mathbf{Z}_p) = \varprojlim H_*^1(Y(K), \mathbf{Z}/p^n\mathbf{Z}).$$

Note that for $K' \subset K$, the group K acts on the right on $Y(K')$ and hence on the cohomology $H_*^1(Y(K'), \mathbf{Z}_p)$. Furthermore, the collection of such K' forms a direct system under inclusion, with an inclusion $K_1 \subset K_2 \subset K$ inducing a K-equivariant map

$$H_*^1(Y(K_2), \mathbf{Z}_p) \to H_*^1(Y(K_1), \mathbf{Z}_p).$$

If we take the direct limit of this system, it is clear that $H^1_*(Y(K), \mathbf{Z}_p)$ maps to this direct limit, and lies in the subspace which is invariant under the action of K:

$$H^1_*(Y(K), \mathbf{Z}_p) \subset (\varinjlim_{K'} H^1_*(Y(K'), \mathbf{Z}_p))^K.$$

Now we may define the Hecke action. For each ℓ with $K_\ell = \mathrm{GL}_2(\mathbf{Z}_\ell)$, write

$$K \begin{pmatrix} 1 & 0 \\ 0 & \ell \end{pmatrix} K = \bigsqcup \delta_i K$$

for the matrix

$$\begin{pmatrix} 1 & 0 \\ 0 & \ell \end{pmatrix} \in \mathrm{GL}_2(\mathbf{Q}_\ell) \subset \mathrm{GL}_2(\mathbf{A_Q}).$$

For $x \in H^1_*(Y(K), \mathbf{Z}_p)$, define $T_\ell(x) := \sum \delta_i^{-1} x$ in $(\varinjlim_{K'} H^1_*(Y(K'), \mathbf{Z}_p))^K$; one must check that the image $T_\ell(x)$ again lies in $H^1_*(Y(K), \mathbf{Z}_p)$. This action induces the action on the quotients $H^1_*(Y(K), \mathbf{Z}/p^n\mathbf{Z}) = H^1_*(Y(K), \mathbf{Z}_p)/p^n$ as well. The Hecke operator S_ℓ is defined similarly via the matrix $\begin{pmatrix} \ell & 0 \\ 0 & \ell \end{pmatrix}$. The Galois and Hecke actions commute.

Now let $K^p = \prod_{\ell \neq p} K_\ell$ be an open compact subgroup of $\mathrm{GL}_2(\hat{\mathbf{Z}})$, with trivial component at p. Let $K(p^r) := \ker(\mathrm{GL}_2(\mathbf{Z}_p) \to \mathrm{GL}_2(\mathbf{Z}/p^r\mathbf{Z}))$ and define

$$H^1_*(K^p) := \varinjlim_r H^1_*(Y(K^p K(p^r)), \mathbf{Z}_p).$$

The transition maps in the inductive limit for $* = $ empty are the usual contravariant maps on cohomology induced by the projection maps $Y(K^p K(p^{r+1})) \to Y(K^p K(p^r))$. For $* = c$, the transition maps are the duals of the trace maps $H^1_*(Y(K^p K(p^{r+1})), \mathbf{Z}_p) \to H^1_*(Y(K^p K(p^r)), \mathbf{Z}_p)$ induced by the projections. The $\mathbf{Z}_p$-module $H^1_*(K^p)$ is torsion-free. Furthermore, it is endowed with a smooth left $\mathrm{GL}_2(\mathbf{Q}_p)$-action which we now describe.

For each $\gamma \in \mathrm{GL}_2(\mathbf{Q}_p)$, right multiplication by γ^{-1} induces a map $Y(\gamma^{-1} K \gamma) \mapsto Y(K)$, which in turn gives a map $H^1_*(Y(K), \mathbf{Z}_p) \to H^1_*(Y(\gamma^{-1} K \gamma), \mathbf{Z}_p)$. Given any element $x \in H^1_*(K^p)$, we may choose an index r large enough so that x is represented by an element in $H^1_*(K^p K(p^r), \mathbf{Z}_p)$, and such that $\gamma^{-1} K^p K(p^r) \gamma \subset K^p K(p^s)$ for some $s > 1$. Then the image of x under the composition of maps

$$H^1_*(K^p K(p^r), \mathbf{Z}_p) \to H^1_*(\gamma^{-1} K^p K(p^r) \gamma, \mathbf{Z}_p) \to H^1_*(K^p K(p^s), \mathbf{Z}_p)$$

represents an element of $H^1_*(K^p)$, which is defined to be γx.

Finally, we define

$$\hat{H}^1_*(K^p) := \varprojlim_n \left(\varinjlim_r H^1_*(Y(K^p K(p^r)), \mathbf{Z}/p^n\mathbf{Z}) \right) \cong \varprojlim_n H^1_*(K^p)/p^n.$$

For the remainder of this lecture, we will be interested in particular in the case

$$K^p = \left\{ \begin{pmatrix} a & b \\ c & d \end{pmatrix} \in \mathrm{GL}_2 \left(\prod_{\ell \neq p} \mathbf{Z}_\ell \right) : c \equiv 0 \pmod{N}, \ a \equiv 1 \pmod{N} \right\}.$$

In this case, $Y(K^p K(p^r)) = Y(N, p^r) := Y_1(N) \times_{Y(1)} Y(p^r)$, the open modular curve whose connected geometric component $Y^0(N, p^r)$ can be identified with $\Gamma_1(N) \cap \Gamma(p^r) \backslash \mathcal{H}$.

The modules $H^1_*(K^p)$ and $\hat{H}^1(K^p)$ are endowed with $\mathbf{Z}_p$-linear actions of $\mathrm{Gal}(\overline{\mathbf{Q}}/\mathbf{Q})$, the Hecke operators T_ℓ, S_ℓ for $\ell \nmid Np$, and $\mathrm{GL}_2(\mathbf{Q}_p)$. This last action endows $\hat{H}_c^1(K^p) \otimes \mathbf{Q}_p$ with the structure of an admissible unitary Banach space representation of $\mathrm{GL}_2(\mathbf{Q}_p)$. We will simply write $H_c^1 = H_c^1(N)$ and $\hat{H}_c^1 = \hat{H}_c^1(N)$ for $H_c^1(K^p)$ and $\hat{H}_c^1(K^p)$, respectively.

4.3. $\mathrm{GL}_2(\mathbf{Q}_p)$ representations and modular symbols. In order to connect the theory of p-adic Banach representations of $\mathrm{GL}_2(\mathbf{Q}_p)$ with the work of Orton, Breuil gave a reinterpretation of the space of $\mathrm{GL}_2(\mathbf{Q}_p)$-equivariant maps from an arbitrary Banach space into $\hat{H}_c^1$, in terms of modular symbols. Let E denote a finite extension of $\mathbf{Q}_p$.

THEOREM 4.3.1 ([**2**], Théorème 2.4.2). *Let B be a unitary p-adic Banach space representation of $\mathrm{GL}_2(\mathbf{Q}_p)$ with coefficients in E, and let $B^* := \mathrm{Hom}_E(B, E)$ be its $\mathrm{GL}_2(\mathbf{Q}_p)$ unitary Banach dual. We have a canonical Hecke equivariant isomorphism*

$$(4.3.2) \qquad \mathrm{Hom}_{\mathrm{GL}_2(\mathbf{Q}_p)}(B, \hat{H}_c^1 \otimes_{\mathbf{Z}_p} E) \cong \mathrm{Hom}_{\tilde{\Gamma}_1^p(N)}(D_0, B^*),$$

where the left hand side denotes continuous E-linear $\mathrm{GL}_2(\mathbf{Q}_p)$-equivariant maps between the two indicated unitary Banach space representations, and the right side denotes B^*-valued modular symbols ϕ such that

$$\phi([\gamma r_1] - [\gamma r_2])(b) = \phi([r_1] - [r_2])(\gamma^{-1}b)$$

for all $\gamma \in \tilde{\Gamma}_1^p(N)$. (Note: a unitary representation means that the norm is G-invariant).

We state this theorem as Breuil does, for the groups

$$\tilde{\Gamma}_1^p(N) = \left\{ \begin{pmatrix} a & b \\ c & d \end{pmatrix} \in \mathrm{GL}_2(\mathbf{Z}[1/p])^+ : \ c \equiv 0 \ (\mathrm{mod} \ N), \ a \equiv 1 \ (\mathrm{mod} \ N) \right\}$$

and

$$\Gamma_1^p(N) = \left\{ \begin{pmatrix} a & b \\ c & d \end{pmatrix} \in \mathrm{SL}_2(\mathbf{Z}[1/p]) : \ c \equiv 0 \ (\mathrm{mod} \ N), \ a \equiv 1 \ (\mathrm{mod} \ N) \right\}.$$

This is a slightly more general setting then that in the previous section, where we worked with Γ_0^p. No doubt the results in the previous section hold in this more general setting, since the difference between Γ_1^p and Γ_0^p is "away from p," but we have not attempted to make this generalization and we content ourselves here with following Breuil.

The Hecke equivariance in Thoerem 4.3.1 is with respect to the operators T_ℓ and S_ℓ for $\ell \nmid Np$, and w_∞. These operators act on the left of (4.3.2) via their action on $\hat{H}_c^1$, and on the right via the usual action on modular symbols.

To prove Theorem 4.3.1, let M be the closed unit ball in B. We will show that there is a Hecke equivariant isomorphism
$$(4.3.3)$$
$$\mathrm{Hom}_{\mathrm{GL}_2(\mathbf{Q}_p)}\left(M, H_c^1 \otimes \mathcal{O}_E/\pi_E^n \mathcal{O}_E\right) \cong \mathrm{Hom}_{\tilde{\Gamma}_1^p(N)}\left(D_0, \mathrm{Hom}_{\mathcal{O}_E}(M, \mathcal{O}_E/\pi_E^n \mathcal{O}_E)\right).$$

Theorem 4.3.1 follows by passing to the limit over n and tensoring with E.

To prove (4.3.3), we begin by providing an alternate description of the left hand side. Denote by $\mathrm{Ind}_1^{\mathrm{GL}_2(\mathbf{Z}/p^r\mathbf{Z})} 1_{\mathbf{Z}/p^n\mathbf{Z}}$ the $\mathbf{Z}/p^n\mathbf{Z}$-module of functions $f : \mathrm{GL}_2(\mathbf{Z}/p^r\mathbf{Z}) \to \mathbf{Z}/p^n\mathbf{Z}$. Denote by

$$(4.3.4) \qquad \mathrm{Hom}_{\Gamma_1(N)}(D_0, \mathrm{Ind}_1^{\mathrm{GL}_2(\mathbf{Z}/p^r\mathbf{Z})} 1_{\mathbf{Z}/p^n\mathbf{Z}})$$

the $\mathbf{Z}/p^n\mathbf{Z}$-module of group homomorphisms $\phi : D_0 \to \mathrm{Ind}_1^{\mathrm{GL}_2(\mathbf{Z}/p^r\mathbf{Z})} 1_{\mathbf{Z}/p^n\mathbf{Z}}$ such that

$$\phi([r_1] - [r_2])(x\gamma) = \phi([\gamma r_1] - [\gamma r_2])(x)$$

for all $[r_1] - [r_2] \in D_0$, $x \in \mathrm{GL}_2(\mathbf{Z}/p^r\mathbf{Z})$, and $\gamma \in \Gamma_1(N)$. The $\mathbf{Z}/p^n\mathbf{Z}$-module in (4.3.4) is endowed with a left $\mathrm{GL}_2(\mathbf{Z}/p^r\mathbf{Z})$-module action via:

$$(4.3.5) \qquad (g(\phi))([r_1] - [r_2])(x) := \phi([r_1] - [r_2])(g^{-1}x).$$

We then have:

LEMMA 4.3.6. *For each pair of integers $n > 0$ and $r > 1$, there is a canonical $\mathrm{GL}_2(\mathbf{Z}/p^r\mathbf{Z})$-equivariant isomorphism*

$$(4.3.7) \qquad H_c^1(Y(N, p^r), \mathbf{Z}/p^n\mathbf{Z}) \cong \mathrm{Hom}_{\Gamma_1(N)}(D_0, \mathrm{Ind}_1^{\mathrm{GL}_2(\mathbf{Z}/p^r\mathbf{Z})} 1_{\mathbf{Z}/p^n\mathbf{Z}}).$$

Furthermore, this isomorphism is equivariant with respect to the Hecke operators T_ℓ, S_ℓ for $\ell \nmid Np$, and transforms the action of complex conjugation on the left to that of w_∞ on the right.

PROOF. We will give the geometric intuition behind the slightly weaker isomorphism

$$(4.3.8) \qquad H_c^1(Y^0(N, p^r), \mathbf{Z}/p^n\mathbf{Z}) \cong \mathrm{Hom}_{\Gamma_1(N)}(D_0, \mathrm{Ind}_1^{\mathrm{SL}_2(\mathbf{Z}/p^r\mathbf{Z})} 1_{\mathbf{Z}/p^n\mathbf{Z}}).$$

The curve $Y^0(N, p^r)$ may be thought of as the many holed torus $X^0(N, p^r)$, minus a certain finite number of cusps. By Poincaré duality, the left side of (4.3.8) is

$$\mathrm{Hom}(H_1(X^0(N, p^r), \mathrm{cusps}; \mathbf{Z}), \mathbf{Z}/p^n\mathbf{Z}),$$

the $\mathbf{Z}/p^n\mathbf{Z}$-dual of the homology of the torus $X^0(N, p^r)$ relative to its set of cusps. Meanwhile, the right side of (4.3.8) is $\mathrm{Hom}_{\Gamma_1(N) \cap \Gamma(p^r)}(D_0, \mathbf{Z}/p^n\mathbf{Z})$ by Shapiro's Lemma. It thus remains to prove:

$$(4.3.9) \quad \mathrm{Hom}(H_1(X^0(N, p^r), \mathrm{cusps}; \mathbf{Z}), \mathbf{Z}/p^n\mathbf{Z}) \cong \mathrm{Hom}_{\Gamma_1(N) \cap \Gamma(p^r)}(D_0, \mathbf{Z}/p^n\mathbf{Z}).$$

Given an element $[x] - [y] \in D_0$, consider any path in $\mathcal{H} \cup \mathbf{P}^1(\mathbf{Q})$ starting at x and ending at y. The image of this path in $\Gamma_1(N) \cap \Gamma(p^r) \backslash (\mathcal{H} \cup \mathbf{P}^1(\mathbf{Q}))$ yields a well-defined element of $H_1(X^0(N, p^r), \mathrm{cusps}; \mathbf{Z})$. The theory of modular symbols states that this identification induces an isomorphism as in (4.3.9).

The compatibility of the isomorphism with the Hecke algebra involves computations with the group action. The isomorphism (4.3.7) follows from similar, but more involved reasoning. For details of all of this, see [2, Lemme 2.3.2]. $\qquad\square$

Passing to the inductive limit over r in (4.3.7), we obtain a Hecke equivariant isomorphism

$$(4.3.10) \qquad H_c^1/p^n \cong \mathrm{Hom}_{\Gamma_1(N)}(D_0, \mathrm{Ind}_1^{\mathrm{GL}_2(\mathbf{Z}_p)} 1_{\mathbf{Z}/p^n\mathbf{Z}}),$$

where $\mathrm{Ind}_1^{\mathrm{GL}_2(\mathbf{Z}_p)} 1_{\mathbf{Z}/p^n\mathbf{Z}}$ denotes the $\mathbf{Z}/p^n\mathbf{Z}$-module of locally constant functions $\mathrm{GL}_2(\mathbf{Z}_p) \to \mathbf{Z}/p^n\mathbf{Z}$. The right hand side of (4.3.10) is endowed with a left $\mathrm{GL}_2(\mathbf{Z}_p)$ action, as given in equation (4.3.5). This action may be extended to $\mathrm{GL}_2(\mathbf{Q}_p)$ as follows. For any $g \in \mathrm{GL}_2(\mathbf{Q}_p)$ and $x \in \mathrm{GL}_2(\mathbf{Z}_p)$, we may write $g^{-1}x = ba$ with $b \in \mathrm{GL}_2(\mathbf{Z}_p)$ and $a \in \tilde{\Gamma}_1^p(N)$. We then define

$$(4.3.11) \qquad (g(\phi))([r_1] - [r_2])(x) := \phi([ar_1] - [ar_2])(b).$$

One easily checks that this definition is independent of the choice of a and b, and yields a well-defined $\mathrm{GL}_2(\mathbf{Q}_p)$-action extending the $\mathrm{GL}_2(\mathbf{Z}_p)$-action defined in (4.3.5). Furthermore, with this action, the isomorphism (4.3.10) is $\mathrm{GL}_2(\mathbf{Q}_p)$-equivariant.

We are now in a position to prove (4.3.3). In view of (4.3.10), we must construct an isomorphism

$$\varphi : \mathrm{Hom}_{\mathrm{GL}_2(\mathbf{Q}_p)}\left(M, \mathrm{Hom}_{\Gamma_1(N)}(D_0, \mathrm{Ind}_1^{\mathrm{GL}_2(\mathbf{Z}_p)} 1_{\mathcal{O}_E/\pi_E^n \mathcal{O}_E})\right)$$

$$\downarrow$$

$$\mathrm{Hom}_{\bar{\Gamma}_1^p(N)}\left(D_0, \mathrm{Hom}_{\mathcal{O}_E}(M, \mathcal{O}_E/\pi_E^n \mathcal{O}_E)\right).$$

Such a map φ is given by

$$\varphi(F)([r_1] - [r_2])(m) := F(m)([r_1] - [r_2])(1)$$

for all $m \in M$. Its inverse ψ is given by

$$\psi(G)(m)([r_1] - [r_2])(h) := G([r_2] - [r_1])(h^{-1}m)$$

for all $m \in M$ and $h \in \mathrm{GL}_2(\mathbf{Z}_p)$. We leave it to the reader to verify (or consult [**3**, Proposition 2.4.1]) that the functions φ and ψ indeed map the indicated spaces to one another, that the two maps are mutually inverse, and that they are Hecke equivariant. This concludes the proof of Theorem 4.3.1.

4.4. Breuil Duality. Breuil Duality is a generalization of the Morita Duality that we discussed in Section 2.2.4. For details on this duality, see the proof of [**3**, Theorem 3.2.3]. In this section, we continue to assume that the ground field (over which $\mathfrak{X}$ is defined) is the field $\mathbf{Q}_p$.

Let $\mathcal{O}_E(k)$ be the space of rigid analytic functions $H : \mathfrak{X} \to E$ with a left $G = \mathrm{GL}_2(\mathbf{Q}_p)$ action given by

$$\left(\begin{pmatrix} a & b \\ c & d \end{pmatrix} H\right)(z) := \frac{\det(g)^{k/2}}{(bz+d)^k} H\left(\frac{az+c}{bz+d}\right).$$

Let $|\cdot|_p$ be the usual p-adic absolute value, and let ϵ be the character $\epsilon : \mathbf{Q}_p^* \to \mathbf{Z}_p^*$ defined by

$$\epsilon(x) = x|x|_p.$$

We view ϵ as a character of G through the determinant:

$$\epsilon(g) = \epsilon(\det(g)).$$

Let $\mathcal{O}_{E,\epsilon}(k)$ be the space obtained by adjusting the action of the center of G

$$\mathcal{O}_{E,\epsilon}(k) = \epsilon^{(2-k)/2} \otimes \mathcal{O}_E(k).$$

We now define the space $\mathcal{O}_E(k, \mathcal{L})$ for each $\mathcal{L} \in E$. This space consists of the integrals of functions in $\mathcal{O}_{E,\epsilon}(k)$. These integrals are not rigid analytic, but involve a fixed "branch" of the p-adic logarithm determined by the number $\mathcal{L}$. (See the partial fractions expansion in Lemma 2.2.7 and consider what's involved in integrating it formally).

DEFINITION 4.4.1. Let $\log_{\mathcal{L}} : \mathbf{C}_p^\times \to \mathbf{C}_p$ be the branch of the p-adic logarithm which satisfies $\log_{\mathcal{L}}(p) = \mathcal{L}$.

Let

$$U = \bigsqcup_{i=0}^{s} U_i$$

be a covering of $\mathbf{Q}_p$ in $\mathbf{C}_p$ by pairwise disjoint opens U_i, such that $U_0 := \{z \in \mathbf{C}_p : |z| > r_0\}$ and $U_i := \{z \in \mathbf{C}_p : |z - z_i| < r_i\}$ for $1 \le i \le s$, with $r_i \in |E^\times|$ and $z_i \in \mathbf{Q}_p$. Define $\mathcal{O}(2 - k, \mathcal{L})$ to be the space of functions $H : \mathcal{X} \to \mathbf{C}_p$ such that the restriction to each affinoid $\mathcal{X}_U := \mathbf{C}_p - U \subset \mathcal{X}$ with U as above, has the form:

$$H|_{\mathcal{X}_U} = H_U + \sum_{i=1}^{s} \sum_{n=0}^{k-2} c_{i,n} z^n \log_{\mathcal{L}}(z - z_i),$$

with $c_{i,n} \in E$ and H_U an E-rational rigid analytic function on $\mathcal{X}_U$. The space $\mathcal{O}(2 - k, \mathcal{L}, E)$ is endowed with the left $\mathrm{GL}_2(\mathbf{Q}_p)$ action given by

$$\left(\begin{pmatrix} a & b \\ c & d \end{pmatrix} H \right)(z) := \epsilon(ad - bc)^{-\frac{k-2}{2}} \frac{(bz + d)^{k-2}}{(ad - bc)^{(k-2)/2}} H\left(\frac{az + c}{bz + d} \right).$$

As Lemma 2.2.7 suggests, the $(k - 1)$st derivative map induces a short exact sequence of $\mathrm{GL}_2(\mathbf{Q}_p)$-representations:

$$(4.4.2) \qquad 0 \to P_{k-2}(E) \otimes \epsilon^{\frac{2-k}{2}} \to \mathcal{O}(2 - k, \mathcal{L}, E) \to \mathcal{O}_{E,\epsilon}(k) \to 0.$$

Let $C_{\log}(\mathbf{Q}_p, 2 - k, \mathcal{L}, E)$ be the space of locally analytic functions h on $\mathbf{Q}_p$ such that in a neighborhood of ∞ we have,

$$(4.4.3) \qquad h(z) = -2P(z) \log_{\mathcal{L}}(z) + z^{k-2} \sum_{n=0}^{\infty} \frac{a_n}{z^n}$$

with $a_n \in E$ and $P(z) \in P_{k-2}(E)$. The space $C_{\log}(\mathbf{Q}_p, 2 - k, \mathcal{L}, E)$ has a direct limit topology similar to that on $C^{an}(\mathbf{Q}_p, 2 - k, E)$ defined in Definition 2.1.4. To be precise, given a covering U of $\mathbf{Q}_p$ in $\mathbf{C}_p$ as above, a collection of power series f_i on U_i, along with a function f_0 as in Equation (4.4.3) on the set U_0, gives an element f in $C_{\log}$. The functions defined relative to a fixed covering form a Banach space, and the full space $C_{\log}(\mathbf{Q}_p, 2 - k, \mathcal{L}, E)$ is the direct limit of these Banach spaces as the coverings are refined.

The group action on $C_{\log}(\mathbf{Q}_p, 2 - k, \mathcal{L}, E)$ is given by the formula

$$g_*(h)(z) = \epsilon^{\frac{k-2}{2}}(g) \frac{(bz + d)^{k-2}}{\det(g)^{(k-2)/2}} \left[h\left(\frac{az + c}{bz + d} \right) + P\left(\frac{az + c}{bz + d} \right) \log_{\mathcal{L}}\left(\frac{ad - bc}{(bz + d)^2} \right) \right].$$

Define

$$(4.4.4) \qquad \Sigma(\mathbf{Q}_p, 2 - k, \mathcal{L}, E) = C_{\log}(\mathbf{Q}_p, 2 - k, \mathcal{L}, E)/P_{k-2}(E).$$

Let us now consider the dual exact sequence to (4.4.2). For even integer $k \ge 2$, Morita duality (Theorem 2.2.1) identifies the dual of the rightmost term $\mathcal{O}_{E,\epsilon}(k)$ in (4.4.2) with the space

$$C_\epsilon^{an}(\mathbf{Q}_p, 2 - k, E)/P_{k-2} := \epsilon^{\frac{k-2}{2}} \otimes (C^{an}(\mathbf{Q}_p, 2 - k, E)/P_{k-2}(E)),$$

where $C^{an}(\mathbf{Q}_p, 2 - k, E)$ is the space of E-valued locally analytic functions on $\mathbf{Q}_p$ with poles of the correct order at ∞. Breuil duality (Theorem 4.4.5 below) identifies the dual of the larger space $\mathcal{O}(2-k, \mathcal{L}, E)$ with the space $\Sigma(\mathbf{Q}_p, 2-k, \mathcal{L}, E)$ of (4.4.4). Thus the dual exact sequence to (4.4.2) may be written:

$$0 \longrightarrow C_\epsilon^{an}(\mathbf{Q}_p, 2-k, E)/P_{k-2} \longrightarrow \Sigma(\mathbf{Q}_p, 2-k, \mathcal{L}, E)$$

$$\epsilon^{(k-2)/2} \otimes \mathrm{Hom}(P_{k-2}(E), E) \longrightarrow 0.$$

The map on the right side of this series picks out the logarithmic part at infinity of the function h, taking into account the group action.

THEOREM 4.4.5. *There exists a unique G-invariant pairing*

$$\langle\, \cdot\, ,\, \cdot\, \rangle_B : C_{\log}(\mathbf{Q}_p, 2-k, \mathcal{L}, E)/P_{k-2}(E) \times \mathcal{O}(2-k, \mathcal{L}, E) \to E$$

satisfying:

(1) *for $z \in \mathfrak{X}$, $\langle \frac{(x-z)^{k-2}}{(k-2)!} \log_\mathcal{L}(x-z), G\rangle = G(z)$;*
(2) *if $f \in C_\epsilon^{an}(\mathbf{Q}_p, 2-k, E)/P_{k-2}$ then*

$$\langle f, G\rangle_B = \langle f, G^{(k-1)}\rangle_M;$$

(3) *if $G \in \mathcal{O}_{E,\epsilon}(2-k)$, then*

$$\langle f, G\rangle_B = (-1)^{k-1}\langle f^{(k-1)}, G\rangle_M,$$

where $\langle\, \cdot\, ,\, \cdot\, \rangle_M$ is the Morita pairing.

PROOF. See [**2**, Section 3]. $\qquad\qquad\qquad\qquad\qquad\qquad\qquad\qquad\square$

Of particular importance in Breuil's work are two Banach spaces, $B(2-k, \mathcal{L})$ and $B(2-k)$ that contain the spaces $\Sigma(\mathbf{Q}_p, 2-k, \mathcal{L}, E)$ and $C_\epsilon^{an}(\mathbf{Q}_p, 2-k, E)/P_{k-2}$ respectively as dense subspaces. One way to think of the space $B(2-k)$ is as the continuous dual of the space $\mathcal{O}_{\epsilon, E}(k)^b$ of functions $f \in \mathcal{O}_{\epsilon, E}(k)$ having bounded residues, defined as in Corollary 2.3.4 or Section 2.3 — and, in fact, this is literally true, provided one equips $\mathcal{O}_{\epsilon, E}(k)^b$ with the proper topology. It is also true that $B(2-k)$ is the completion of the space $C^{la}(\mathbf{Q}_p, 2-k, E)/P_{k-2}(E)$ of locally polynomial functions on $\mathbf{Q}_p$ with respect to a certain norm. When $k=2$, this is just the sup-norm.

To avoid too many functional analytic complications, we content ourselves with listing some key properties of $B(2-k)$:

(1) $B(2-k)$ is a Banach space.
(2) When $k > 2$, it is the completion of the locally polynomial functions $C_\epsilon^{la}(\mathbf{Q}_p, 2-k, E)/P_{k-2}$ in a certain G-invariant norm.
(3) Let $C_{har}(k, \epsilon, E)$ denote the space of harmonic cocycles with values in $\mathrm{Hom}(P_{k-2}(E), E)$ and group action twisted by $\epsilon^{\frac{k-2}{2}}$. There is a continuous duality between the space $\mathcal{O}_{\epsilon, E}(k)^b \xrightarrow{\sim} C_{har}(k, \epsilon, E)^b$ of bounded rigid functions and $B(2-k)$; with the proper topology on the harmonic cocycles, each is the continuous dual of the other.
(4) The space $C_{\epsilon, E}^{an}(\mathbf{Q}_p, 2-k, E)/P_{k-2}$ is dense in $B(2-k)$.

Similarly, one can identify a space of bounded functions $\mathcal{O}(2-k, \mathcal{L}, E)^b$ in the space $\mathcal{O}(2-k, \mathcal{L}, E)$. In simple terms, a function is bounded if it, and all of its translates by the G-action, are bounded on a fixed affinoid domain in $\mathfrak{X}$. This space of bounded functions is then given a topology so that its dual is a Banach space $B(2 - k, \mathcal{L})$.

The relevant properties of $B(2 - k, \mathcal{L})$ are:

(1) $B(2 - k, \mathcal{L})$ is a Banach space with a G-invariant norm.
(2) There is a continuous duality between the "bounded" elements $\mathcal{O}(2 - k, \mathcal{L}, E)^b$ of $\mathcal{O}_{\epsilon,E}(2 - k, \mathcal{L})$ and $B(2 - k, \mathcal{L})$.
(3) When $k > 2$, the surjection $\mathcal{O}(\mathbf{Q}_p, 2 - k, \mathcal{L}) \to \mathcal{O}_{\epsilon,E}(k)$ becomes an isomorphism.

To avoid too many functional-analytic complications, we will not give a precise definition of the Banach spaces $B(2 - k)$ and $B(2 - k, \mathcal{L})$. See [**3**, Section 3.1-3.3] for the full definition, which relies on the duality described in [**41**].

One important remark: the fact that $\mathcal{O}(2 - k, \mathcal{L}, E)^b$ is non-zero is highly nontrivial! The paper [**2**] proves this when $\mathcal{L} = \mathcal{L}(f)$ for some cusp for f of level Np; see the papers of Colmez for the general case.

4.5. Orton's $\mathcal{L}$-invariant from Breuil's viewpoint.

In this section we combine the ideas of Breuil and Morita duality with the Darmon picture of modular symbols on the tree to approach Breuil's interpretation of the $\mathcal{L}$-invariant.

We first assemble a few facts. First, recall from Proposition 3.2.7 that, given a p-new cusp form f on $\Gamma_1(N)$ we have a function

$$\Phi_f^{\pm} : D_0 \times \mathrm{Edges}(X) \times P_{k-2} \to E$$

that, holding the D_0-variable fixed, is a bounded harmonic cocycle in $C_{har}(k)^b$ on X.

At the same time, the cohomological calculations from Section 3.3.1 tell us that

LEMMA 4.5.1. *We have* $w_\infty \Phi_f^{\pm} = \pm \Phi_f^{\pm}$, *and*

$$\mathrm{Hom}_{\Gamma_1^p(N)}(D_0, C_{har}(k))^f = E\Phi_f^+ \oplus E\Phi_f^-.$$

PROOF. This is essentially the content of Proposition 3.3.13. Since the map is Hecke equivariant, and the group $\Gamma_1^p(N)$ acts transitively on the edges (at least up to orientation) the function Φ_f is determined by its value on e_0, where it must lie in a one dimensional subspace. $\qquad\square$

THEOREM 4.5.2. *The residue map* $\mathcal{O}_{\epsilon,E}(k) \to C_{har}(k, \epsilon, E)$ *induces isomorphisms*

$$\mathrm{Hom}_{\Gamma_1^p(N)}(D_0, \mathcal{O}_{\epsilon,E}(k)^b) \cong \mathrm{Hom}_{\Gamma_1^p(N)}(D_0, \mathcal{O}_{\epsilon,E}(k))$$
$$\cong \mathrm{Hom}_{\Gamma_1^p(N)}(D_0, C_{har}(k, \epsilon, E)^b).$$

Furthermore,

$$\mathrm{Hom}_{\Gamma_1^p(N)}(D_0, \mathcal{O}(2 - k, \mathcal{L})^b) \cong \mathrm{Hom}_{\Gamma_1^p(N)}(D_0, \mathcal{O}(2 - k, \mathcal{L})).$$

In words: an invariant modular symbol with values in harmonic cocycles automatically takes values in bounded harmonic cocycles.

PROOF. We will prove the first result; the second is proved similarly, but since we avoided giving a precise definition of the norm on $\mathcal{O}(2 - k, \mathcal{L})$, we won't give details. Suppose $\phi : D_0 \to \mathbf{C}_{har}(k, \epsilon, E)$ is $\Gamma_1^p(N)$ invariant. To compute the value

$\phi(m)(e) \in \text{Hom}(P_{k-2}(E), E)$, first use transitivity of the $\Gamma_1^p(N)$ action on the edges to find γ so that $e = \gamma e_0$. Then

$$\phi(m)(e) = \gamma(\phi(\gamma^{-1}(m))(e_0)).$$

Next use the fact that the stabilizer of e_0 in $\Gamma_1^p(N)$ is $\Gamma_1(pN)$ and that $D_0/\Gamma_1(pN)$ is finitely generated; in other words, there are finitely many $m_i \in D_0$, $\tau_i \in \Gamma_1(pN)$, and integers a_i so that

$$\gamma^{-1}(m) = \sum a_i \tau_i m_i.$$

Therefore

$$\omega(\gamma^{-1}\phi(m)(\gamma e_0)) = \omega(\phi(\gamma^{-1}(m))(e_0)) = \omega(\phi(\sum a_i \tau_i m_i)(e_0)) \leq \inf_i \omega(\phi(m_i)(e_0)).$$

Thus the cocycle is automatically bounded. Now use the Poisson integral to integrate the associated measure (using Corollary 2.3.4) to obtain, for each $m \in D_0$, an element in $\mathcal{O}_{\epsilon,E}(k)$ which is, of necessity, bounded. $\qquad\square$

Let $\Phi_F^{\pm} \in \text{Hom}_{\Gamma_1^p(N)}(D_0, \mathcal{O}(k))$ denote the elements which map to

$$\Phi_f^{\pm} \in \text{Hom}_{\Gamma_1^p(N)}(D_0, C_{har}(k)^b)$$

under the isomorphism of Theorem 4.5.2. Using the Poisson integral, we may write

$$\Phi_F^{\pm}(m)(z) = \int_{\mathbf{P}^1(K)} \frac{1}{z - x} d\Phi_f^{\pm}(m)$$

viewing $\Phi_f^{\pm}$ as a bounded distribution.

LEMMA 4.5.3. *We have* $w_\infty \Phi_F^{\pm} = \pm\Phi_F^{\pm}$, *and*

$$\text{Hom}_{\Gamma_1^p(N)}(D_0, \mathcal{O}(k))^f = E\Phi_F^+ \oplus E\Phi_F^-.$$

PROOF. This follows directly from Lemma 4.5.1 and Theorem 4.5.2. $\qquad\square$

Before we proceed, we introduce some extra notation. Fix $\mathcal{L} \in E$. If H is a rigid analytic E-rational function on Ω, we may view H as an element of $\mathcal{O}(2)$ and choose a lift $\tilde{H}$ of H via the surjection $\mathcal{O}(2, \mathcal{L}) \to \mathcal{O}(2)$; i.e. $\tilde{H}$ is a $\log_{\mathcal{L}}$-rigid anti-derivative of H. The function $\tilde{H}$ is determined up to a constant, so for $z_1, z_2 \in \Omega$, the value

$$\int_{z_1}^{z_2} H(z)dz := \tilde{H}(z_1) - \tilde{H}(z_2)$$

is well-defined, and called a *Coleman line integral* relative to the choice of $\mathcal{L}$.

For each $\mathcal{L} \in E$ and $Q \in \Omega$, we define a 1-cocyle

$$c_{\mathcal{L},Q}^{\pm} \in Z^1(\Gamma_1^p(N), \text{Hom}(D_0, \text{Hom}(P_{k-2}(E), E)))$$

by the rule

$$\gamma \mapsto c_{\mathcal{L},Q}^{\pm}(\gamma)([r_1] - [r_2])(P(z)) := \int_Q^{\gamma Q} \Phi_F^{\pm}([r_1] - [r_2])(z)P(z)dz.$$

The class of $c_{\mathcal{L},Q}^{\pm}$ in $H^1(\Gamma_1^p(N), \text{Hom}(D_0, \text{Hom}(P_{k-2}(E), E)))$ is independent of Q and is denoted $c_{\mathcal{L}}^{\pm}$.

PROPOSITION 4.5.4. *We have*

$$c_{\mathcal{L},Q}^{\pm}(\gamma)([r_1] - [r_2])(P(z)) = \Phi_f^{\pm}([r_1] - [r_2]) \left(\log_{\mathcal{L}} \left(\frac{z - \gamma Q}{z - Q} \right) P(z) \right)$$

and hence

$$c_{\mathcal{L}}^{\pm} = l c_f^{\pm} + \mathcal{L} \cdot o c_f^{\pm}.$$

PROOF. We give a completely formal, but essentially correct proof. Use the representation of Φ_F as a Poisson integral to write $c_{\mathcal{L},Q}^{\pm}$ as a "double integral":

$$c_{\mathcal{L},Q}^{\pm}(\gamma)([r_1] - [r_2])(P(z)) = \int_Q^{\gamma Q} \int_{\mathbf{P}^1(K)} \frac{P(z)}{z - x} d\Phi_f^{\pm}([r_1] - [r_2]).$$

Interchanging the order of integration, using the selected branch of the logarithm, and taking into account the fact that $d\Phi_f^{\pm}$ vanishes on polynomials, yields

$$
\begin{aligned}
c_{\mathcal{L},Q}^{\pm}(\gamma)([r_1] - [r_2])(P(z)) &= \int_{\mathbf{P}^1(\mathbf{Q}_p)} \int_Q^{\gamma Q} \frac{P(z)}{z - x} d\Phi_f^{\pm}([r_1] - [r_2]) \\
&= \int_{\mathbf{P}^1(\mathbf{Q}_p)} \log_{\mathcal{L}} \left(\frac{x - \gamma Q}{x - Q} \right) P(x) d\Phi_f^{\pm}([r_1] - [r_2])
\end{aligned}
$$

as claimed. The last statement then follows from the fact that

$$\log_{\mathcal{L}}(z) = \log(z) + \mathcal{L} \operatorname{ord}(z).$$

$\square$

THEOREM 4.5.5. *As usual, $k \geq 2$ and even. For every $\mathcal{L} \in E$, the surjection $\mathcal{O}(2 - k, \mathcal{L}) \to \mathcal{O}(k)$ induces an injection:*

$$(4.5.6) \qquad \operatorname{Hom}_{\Gamma_1^p(N)}(D_0, \mathcal{O}(2 - k, \mathcal{L}))^f \hookrightarrow \operatorname{Hom}_{\Gamma_1^p(N)}(D_0, \mathcal{O}_{\epsilon, E}(k))^f,$$

and

$$\Phi_F^{\pm} \in \operatorname{Hom}_{\Gamma_1^p(N)}(D_0, \mathcal{O}(2 - k, \mathcal{L}))^f \Leftrightarrow \mathcal{L} = \mathcal{L}_O^{\pm}(f).$$

PROOF. The exact sequence (4.4.2) induces a Hecke- and $\mathrm{GL}_2(\mathbf{Q})$-equivariant sequence

$$(4.5.7) \qquad 0 \to \operatorname{Hom}\left(D_0, P_{k-2}(E) \otimes \epsilon^{-\frac{k-2}{2}}\right) \to \operatorname{Hom}(D_0, \mathcal{O}(2 - k, \mathcal{L}))$$
$$\to \operatorname{Hom}(D_0, \mathcal{O}_{\epsilon, E}(k)) \to 0.$$

Take $\Gamma_1^p(N)$-invariants and f-isotypic components. We have

$$\operatorname{Hom}_{\Gamma_1^p(N)}\left(D_0, P_{k-2} \otimes \epsilon^{-\frac{k-2}{2}}\right)^f = 0$$

by an argument as in Lemma 3.3.15.

Let

$$\delta_{\mathcal{L}} : \operatorname{Hom}_{\Gamma_1^p(N)}(D_0, \mathcal{O}(k)) \to H^1\left(\Gamma_1^p(N), \operatorname{Hom}\left(D_0, P_{k-2} \otimes \epsilon^{-\frac{k-2}{2}}\right)\right)$$

denote the coboundary map in the long exact sequence associated to (4.5.7). Recall that $\Phi_F^{\pm} \in \operatorname{Hom}_{\Gamma_1^p(N)}(D_0, \mathcal{O}(k))^f$. We will show that

$$(4.5.8) \qquad \delta_{\mathcal{L}}(\Phi_F^{\pm}) = 0 \Leftrightarrow \mathcal{L} = -\mathcal{L}_O^{\pm}(f).$$

Thus if $\mathcal{L} \neq -\mathcal{L}_O^\pm(f)$, then $\Phi_F^\pm \notin \mathrm{Hom}_{\Gamma_1^p(N)}(D_0, \mathcal{O}(2-k, \mathcal{L}))$. Suppose on the other hand that $\mathcal{L} = -\mathcal{L}_O^\pm(f)$. We have an exact sequence of *finite dimensional* E-vector spaces:

$$(4.5.9) \quad 0 \to \mathrm{Hom}_{\Gamma_1^p(N)}\left(D_0, P_{k-2} \otimes \epsilon^{-\frac{k-2}{2}}\right) \to \mathrm{Hom}_{\Gamma_1^p(N)}(D_0, \mathcal{O}(2-k, \mathcal{L}))$$
$$\to \ker(\delta_\mathcal{L}) \to 0.$$

The f-isotypic component of the leftmost non-trivial term in (4.5.9) is trivial; this implies that the rightmost nontrivial arrow induces an isomorphism on f-isotypic components. Since $\Phi_F^\pm \in (\ker \delta_\mathcal{L})^f$, we have that $\Phi_F^\pm \in \mathrm{Hom}_{\Gamma_1^p(N)}(D_0, \mathcal{O}(2-k, \mathcal{L}))^f$ as desired.

It remains to prove (4.5.8). Let $\tilde{\Phi}$ denote a lift of $\Phi_F^\pm$ via the surjection

$$\mathrm{Hom}(D_0, \mathcal{O}(2-k, \mathcal{L})) \to \mathrm{Hom}(D_0, \mathcal{O}(k)).$$

By definition, $\delta_\mathcal{L}(\Phi_F^\pm)$ is the class of the 1-cocycle $\Delta_\mathcal{L}$ defined by

$$\gamma \mapsto \Delta_\mathcal{L}(\gamma) := \gamma(\tilde{\Phi}) - \tilde{\Phi} \in \mathrm{Hom}\left(D_0, P_{k-2}(E) \otimes \epsilon^{-\frac{k-2}{2}}\right).$$

(Recall that the map from $\mathcal{O}(2-k, \mathcal{L})$ to $\mathcal{O}(k)$ is the $(k-1)^{st}$ derivative.)

For $m \in D_0$ and $\gamma \in \Gamma_1^p(N)$, the functions $\gamma\tilde{\Phi}(m)$ and $\tilde{\Phi}(m)$ are $\log_\mathcal{L}$-rigid functions of a variable $T \in \mathcal{X}$, and the difference is a polynomial in T of degree at most $k-2$; around each point $z_0 \in \mathcal{X}$ we therefore have the representation

$$\gamma\tilde{\Phi}(\gamma^{-1}(m)) - \tilde{\Phi}(m) = \sum_{i=0}^{k-2} \gamma\tilde{\Phi}(\gamma^{-1}m)^{(i)}(z_0)\frac{(T-z_0)^i}{i!}$$

$$(4.5.10) \qquad\qquad\qquad - \sum_{i=0}^{k-2} \tilde{\Phi}(m)^{(i)}(z_0)\frac{(T-z_0)^i}{i!},$$

where the exponent (i) represents the ith derivative.

Now one checks that

$$(4.5.11)$$
$$\sum_{i=0}^{k-2} \gamma\tilde{\Phi}(\gamma^{-1}m)^{(i)}(z_0)\frac{(T-z_0)^i}{i!} = \gamma\left(\sum_{i=0}^{k-2} \tilde{\Phi}(\gamma^{-1}m)^{(i)}(\gamma^{-1}z_0)\frac{(T-\gamma^{-1}z_0)^i}{i!}\right).$$

Now we will correct $\Delta_\mathcal{L}$ by a coboundary in order to make it possible to finish the computation. Fix $Q \in \mathcal{X}$. Define $\psi \in \mathrm{Hom}(D_0, P_{k-2} \otimes \epsilon^{-\frac{k-2}{2}})$ by the formula

$$\psi(m) := \sum_{i=0}^{k-2} \tilde{\Phi}(m)^{(i)}(Q)\frac{(T-Q)^i}{i!},$$

and let $d\psi$ be the coboundary $d\psi(\gamma) = \gamma\psi - \psi$. Using (4.5.10) and (4.5.11) applied to $z_0 = \gamma Q$, one calculates

$$(4.5.12)$$
$$(d\psi(\gamma) - \Delta_\mathcal{L}(\gamma))(m) = \sum_{i=0}^{k-2} \tilde{\Phi}(m)^{(i)}(\gamma Q)\frac{(T-\gamma Q)^i}{i!} - \sum_{i=0}^{k-2} \tilde{\Phi}(m)^{(i)}(Q)\frac{(T-Q)^i}{i!}.$$

Now we apply $\left(\frac{d}{dz}\right)$ to the function

$$\sum_{i=0}^{k-2} \tilde{\Phi}(m)^{(i)}(z)\frac{(T-z)^i}{i!}$$

and obtain

$$\tilde{\Phi}(m)^{(k-1)}(z)\frac{(T-z)^{k-2}}{(k-2)!} = \Phi_F^\pm(m)\frac{(T-z)^{k-2}}{(k-2)!}.$$

We can therefore express the right side of (4.5.12) as a Coleman line integral relative to $\mathcal{L}$:

$$d\psi(\gamma) - \Delta_\mathcal{L}(\gamma) = \frac{1}{(k-2)!}\int_Q^{\gamma Q}\Phi_F^\pm(m)(z)(T-z)^{k-2}dz.$$

To compare this with $c_{\mathcal{L},Q}^\pm$, we must first recall the fact that $P_{k-2}(E)$ and $\mathrm{Hom}(P_{k-2}(E), E))$ are isomorphic irreducible G-representations. Up to scalar multiplication, there is a unique isomorphism between these spaces. In fact, expanding the representative for $\Delta_\mathcal{L}(\gamma)$ we have computed above, we find

$$\Delta_\mathcal{L}(\gamma) \equiv \frac{1}{(k-2)!}\sum_{j=0}^{k-2}(-1)^j\binom{k-2}{j}T^j\int_Q^{\gamma Q}z^{k-2-j}\Phi_F^\pm(m).$$

In the isomorphism between $P_{k-2}(E)$ and $\mathrm{Hom}(P_{k-2}(E), E)$, This polynomial corresponds (up to multiples) to the cocycle

$$\delta_\mathcal{L}(\gamma)(x^j) = \int_Q^{\gamma Q}z^j\Phi_F^\pm(m)$$

which is nothing but $c_{\mathcal{L},Q}^\pm$. Consequently $\Delta_\mathcal{L}$ is cohomologous to zero precisely when $c_{\mathcal{L},Q}^\pm = 0$; the equivalence (4.5.8) follows from Proposition 4.5.4. $\qquad\square$

Note that, taking into account Theorem 4.5.2, we have in fact proven:

COROLLARY 4.5.13. *Let $\mathcal{L} \in E$. If $\mathcal{L} \neq -\mathcal{L}_O^\pm(f)$, then*

$$\mathrm{Hom}_{\Gamma_1^p(N)}(D_0, \mathcal{O}(2-k, \mathcal{L})^b)^{f,\pm} = 0.$$

If $\mathcal{L} = -\mathcal{L}_O^\pm(f)$, then we have isomorphisms

$$\mathrm{Hom}_{\Gamma_1^p(N)}(D_0, \mathcal{O}(2-k, \mathcal{L})^b)^{f,\pm} \cong \mathrm{Hom}_{\Gamma_1^p(N)}(D_0, \mathcal{O}(k)^b)^{f,\pm} = E\Phi_F^\pm.$$

4.6. Conclusion of proof of Breuil's Theorem.

LEMMA 4.6.1. *For any $\mathcal{L} \in E$, we have*

$$\mathrm{Hom}_{\tilde{\Gamma}_1^p(N)}(D_0, \mathcal{O}(2-k, \mathcal{L})^b \otimes \mathrm{nr}(w^{-1}))^f = \mathrm{Hom}_{\Gamma_0^p(N)}(D_0, \mathcal{O}(2-k, \mathcal{L})^b)^f.$$

PROOF. In view of Corollary 4.5.13, we must verify that $W_p\Phi_F^\pm = \lambda\Phi_F^\pm$ where W_p is a matrix in $\tilde{\Gamma}_1^p(N)$ which does not lie in $\Gamma_1^p(N)$. Such a matrix is given by $\begin{pmatrix} pu & v \\ Nps & t \end{pmatrix}$ where u, v, s, t are integers with $ut - Nsv = 1$. The desired result then follows from

$$F(W_p^{-1}\alpha, z) = \frac{p^{k-1}}{a_p}F(\alpha, W_pz)(pNsz+pt)^{-k},$$

which itself follows from $f|_{W_p} = -a_pf$. For the details, see [**2**, Proposition 5.1.1] $\quad\square$

Combining Theorem 4.3.1, Theorem 4.5.2, Lemma 4.6.1, and the duality between $B(2-k, \mathcal{L})$ and $\mathcal{O}(2-k, \mathcal{L})^b$, we find that
(4.6.2)
$$\mathrm{Hom}_{\mathrm{GL}_2(\mathbf{Q}_p)}(B(k, \mathcal{L}) \otimes \mathrm{nr}(w), \hat{H}_c^1(K_1^p(N)) \otimes E)^f \cong \mathrm{Hom}_{\Gamma_1^p(N)}(D_0, \mathcal{O}(2-k, \mathcal{L}))^f$$

for all $\mathcal{L} \in E$.

We may now deduce Breuil's Theorem:

THEOREM 4.6.3. *Let $\mathcal{L} \in E$. We have $\mathcal{L}_O^+ = \mathcal{L}_O^-$ and if we let $\mathcal{L}_B$ denote this common value we have:*

$$H = \mathrm{Hom}_{\mathrm{GL}_2(\mathbf{Q}_p)} \left(B(k, \mathcal{L}) \otimes \mathrm{nr}(w), (\hat{H}_c^1(K_1^p(N)) \otimes E)^f \right)$$

satisfies

$$H \cong \begin{cases} 0 & \text{if } \mathcal{L} \neq -\mathcal{L}_B(f), \\ E\Phi_F^+ + E\Phi_F^- & \text{if } \mathcal{L} = -\mathcal{L}_B(f). \end{cases}$$

PROOF. The equality of the $\mathcal{L}_O^\pm$ invariants is a consequence of the fact that the space H in the statement of the theorem carries an action by the Hecke algebra for f. Given one homomorphism h in this space, one can construct another by taking, for example, $T_\ell(h)$ for some ℓ prime to Np. These two homomorphisms are then independent (because, by the Eichler-Shimura relations, f picks out a two-dimensional subspace of the target). Thus H cannot be one-dimensional, and therefore the two $\mathcal{L}$ invariants must agree. The full result then follows from (4.6.2) and Corollary 4.5.13.   □

Breuil shows further that $E\Phi_F^+ + E\Phi_F^- \cong \sigma_p(f)^*$. He also proves that for all $\mathcal{L} \in E$, we have

$$\mathrm{Hom}_{\mathrm{GL}_2(\mathbf{Q}_p)} \left(B(k, \mathcal{L}) \otimes \mathrm{nr}(w), \hat{\pi}_p(f) \right) = 0$$

if $\mathcal{L} \neq -\mathcal{L}_B(f)$ and that if $k > 2$,

$$\mathrm{Hom}_{\mathrm{GL}_2(\mathbf{Q}_p)} \left(B(k, -\mathcal{L}_B(f)) \otimes \mathrm{nr}(w), \hat{\pi}_p(f) \right) = E.$$

In the introduction to this lecture, we asked Breuil's question: can we extract the $\mathcal{L}$-invariant of f from automorphic information? This result shows that this is, indeed, the case. This result is, essentially, a "formula" for $\mathcal{L}_B(f)$ that uses only representation theoretic information. Indeed, the space $\hat{\pi}_p(f)$ is born inside a large p-adic Banach space representation constructed from global automorphic data – the cohomology of modular curves. One might view it as the global p-adic automorphic representation attached to the modular form f. Breuil then shows that the $\mathcal{L}$ invariant of f identifies exactly which member of the family $B(k, -\mathcal{L})$ of Banach spaces occurs in this large representation. We should view $B(k, -\mathcal{L}_B(f))$ as the local representation at p associated to f in the big representation $\hat{\pi}_p(f)$.

Bibliography

[1] M. Bertolini, H. Darmon, A. Iovita. Families of automorphic forms on definite quaternion algebras and Teitelbaum's conjecture, preprint, 2006.

[2] C. Breuil, Serie spéciale p-adique et cohomologie étale complétée, preprint, 2006.

[3] C. Breuil, Invariant $\mathcal{L}$ et série spéciale p-adique. Ann. Sci. École Norm. Sup. (4) 37 (2004), no. 4, 559–610.

[4] J.-F. Boutot and H. Carayol, Uniformisation p-adique des courbes de Shimura: les théorèmes de Čerednik et de Drinfel'd, Astérisque No. 196-197 (1991), 7, 45–158 (1992)

[5] R. Coleman and A. Iovita, The Frobenius and monodromy operators for curves and abelian varieties. Duke Math. J. 97 (1999), no. 1, 171–215.

[6] P. Colmez, Fonctions d'une variable p-adique, Preprint, 2005. Available at http://www.math.jussieu.fr/~colmez/publications.html.

[7] P. Colmez, Zèros supplémentaires de fonctions L p-adiques de formes modulaires, Algebra and Number Theory, edited by R. Tandon, Hindustan book agency (2005), 193-210. See also http://www.math.jussieu.fr/~colmez/publications.html, #22

[8] P. Colmez, Une correspondance de Langlands locale p-adique pour les représentations semi-stables de dimension 2, preprint, 2004. See
http://www.math.jussieu.fr/~colmez/publications.html, #4.

[9] P. Colmez, La conjecture de Birch et Swinnerton-Dyer p-adique, Sém. Bourbaki 2002-03, exp. 919, Astèrisque 294 (2004), 251-319.

[10] E. de Shalit, Eichler cohomology and periods of modular forms on p-adic Schottky groups. J. Reine Anew. Math. **400**, 1989, 3–31.

[11] H. Darmon, Integration on $\mathcal{H}_p \times \mathcal{H}$ and arithmetic applications, Ann. of Math. (2) 154 (2001), no. 3, 589–639.

[12] V. G. Drinfel'd, Coverings of p-adic symmetric domains, Funkcional. Anal. i Priložen. **10** (1976), no. 2, 29–40.

[13] M. Emerton, On the interpolation of systems of eigenvalues attached to automorphic Hecke eigenforms, Invent. Math. 164 (2006), no. 1, 1–84.

[14] M. Emerton, p-adic L-functions and unitary completions of representations of p-adic reductive groups. Duke Math. J. 130 (2005), no. 2, 353–392.

[15] Féaux de Lacroix, C. T., Einige Resultate über die topologischen Darstellungen p-adischer Liegruppen auf unendlich dimensionalen Vektorräumen über einem p-adischen Körper. Schriftenreihe des Mathematischen Instituts der Universität Münster. 3. Serie, Heft 23, x+111 pp., Univ. Münster, Münster, 1999.

[16] O. Forster, Zur Theorie der Steinschen Algebren und Moduln, Math. Z. **97** (1967), 376–405.

[17] L. Gerritzen and M. van der Put, *Schottky groups and Mumford curves*, Lecture Notes in Math., 817, Springer, Berlin, 1980.

[18] O. Goldman and N. Iwahori, The space of **p**-adic norms, Acta Math. **109** (1963), 137–177.

[19] R. Greenberg and G. Stevens, p-adic L-functions and p-adic periods of modular forms. Invent. Math. 111 (1993), no. 2, 407–447.

[20] H. Hida, *Elementary theory of L-functions and Eisenstein series*. London Mathematical Society Student Texts, 26. Cambridge University Press, Cambridge, 1993.

[21] A. Iovita and M. Spiess. Derivatives of p-adic L-functions, Heegner cycles and monodromy modules attached to modular forms. Invent. Math. 154 (2003), no. 2, 333–384.

[22] R. Kiehl, Theorem A und Theorem B in der nichtarchimedischen Funktionentheorie, Invent. Math. **2** (1967), 256–273.

[23] H. Komatsu, Projective and injective limits of weakly compact sequences of locally convex spaces, J. Math. Soc. Japan, **19**, (1967), 366–383.

[24] B. Mazur and P. Swinnerton-Dyer. Arithmetic of Weil curves. Invent. Math. 25 (1974), 1–61.

[25] B. Mazur, J. Tate and J. Teitelbaum, On p-adic analogues of the conjectures of Birch and Swinnerton-Dyer, Invent. Math. **84** (1986), no. 1, 1–48.

[26] Y. Morita, Analytic representations of SL$_2$ over a **p**-adic number field. II, in *Automorphic forms of several variables (Katata, 1983)*, 282–297, Progr. Math., 46, Birkhäuser, Boston, MA, 1984.

[27] D. Mumford, An analytic construction of degenerating curves over complete local rings, Compositio Math. **24** (1972), 129–174.

[28] L. Orton, An elementary proof of a weak exceptional zero conjecture. Canad. J. Math. 56 (2004), no. 2, 373–405.

[29] A. Pizer, On the arithmetic of quaternion algebras. Acta Arith. 31 (1976), no. 1, 61–89.

[30] D. Rohrlich, Non-vanishing of L-functions for GL$_2$, Invent. Math. **97**, 1989, no. 2, 381–403.

[31] P. Schneider, Rigid-analytic L-transforms. Number theory, Noordwijkerhout 1983 (Noordwijkerhout, 1983), 216–230, Lecture Notes in Math., 1068, Springer, Berlin, 1984.

[32] P. Schneider, The cohomology of local systems on p-adically uniformized varieties, Math. Ann. **293** (1992), no. 4, 623–650.

[33] P. Schneider, Gebäude in der Darstellungstheorie über lokalen Zahlkörpern, Jahresber. Deutsch. Math.-Verein. **98** (1996), no. 3, 135–145.

[34] P. Schneider, *Nonarchimedean functional analysis*, Springer , Berlin, 2002.

[35] P. Schneider and U. Stuhler, The cohomology of p-adic symmetric spaces, Invent. Math. **105** (1991), no. 1, 47–122.

[36] P. Schneider and J. Teitelbaum, An integral transform for p-adic symmetric spaces, Duke Math. J. **86** (1997), no. 3, 391–433.

[37] P. Schneider and J. Teitelbaum, $U(\mathfrak{g})$-finite locally analytic representations, Represent. Theory **5** (2001), 111–128 (electronic).

[38] P. Schneider and J. Teitelbaum, p-adic boundary values, Astérisque No. 278 (2002), 51–125.

[39] P. Schneider and J. Teitelbaum, Locally analytic distributions and p-adic representation theory, with applications to GL_2, J. Amer. Math. Soc. **15** (2002), no. 2, 443–468 (electronic).

[40] P. Schneider and J. Teitelbaum, Correction to: "p-adic boundary values", Astérisque No. 295 (2004), 291–299.

[41] P. Schneider and J. Teitelbaum, Banach space representations and Iwasawa theory. Israel J. Math. 127 (2002), 359–380.

[42] P. Schneider and J. Teitelbaum, Algebras of p-adic distributions and admissible representations, Invent. Math. **153** (2003), no. 1, 145–196.

[43] J. P. Serre, *Trees.* Translated from the French original by John Stillwell. Corrected 2nd printing of the 1980 English translation. Springer Monographs in Mathematics. Springer-Verlag, Berlin, 2003. x+142 pp. ISBN: 3-540-44237-5

[44] Shimura, G. *Introduction to the arithmetic theory of automorphic functions.* (reprint of 1971 edition) Princeton University Press, Princeton, NJ 1994.

[45] J. T. Teitelbaum, Values of p-adic L-functions and a p-adic Poisson kernel, Invent. Math. **101** (1990), no. 2, 395–410.

[46] M. F. Vigneras, *Arithmétique des Algèbres de Quaternions*, Lecture Notes in Math. 800, Springer, Berlin, 1980.

[47] A. Werner, Compactification of the Bruhat-Tits building of PGL by seminorms, Math. Z. **248** (2004), no. 3, 511–526.

CHAPTER 3

An introduction to Berkovich analytic spaces and non-archimedean potential theory on curves

Matthew Baker

Introduction

This is an expository set of lecture notes meant to accompany the author's lectures at the 2007 Arizona Winter School on p-adic geometry. It is partially adapted from the author's monograph with R. Rumely [**BR08**], and also draws on ideas from V. Berkovich's monograph [**Ber90**], A. Thuillier's thesis [**Thu05**], and Rumely's book [**Rum89**]. We have purposely chosen to emphasize examples, pictures, discussion, and the intuition behind various constructions rather than emphasizing formal proofs and rigorous arguments. (Indeed, there are very few proofs given in these notes!) Once the reader has acquired a basic familiarity with the ideas in the present survey, it should be easier to understand the material in the sources cited above.

We now explain the three main goals of these notes.

First, we hope to provide the reader with a concrete and "visual" introduction to Berkovich's theory of analytic spaces, with a particular emphasis on understanding the Berkovich projective line. Berkovich's general theory, which is nicely surveyed in [**Con07**], requires quite a bit of machinery to understand, and it is not always clear from the definitions what the objects being defined "look like". However, the Berkovich projective line can in fact be defined, visualized, and analyzed with little technical machinery, and already there are interesting applications of the theory in this case. Furthermore, general Berkovich curves (i.e., one-dimensional Berkovich analytic spaces) can be studied by combining the semistable reduction theorem and a good understanding of the local structure of the Berkovich projective line.

Second, we hope to give an accessible introduction to the monograph [**BR08**], which develops the foundations of potential theory on the Berkovich projective line, including the construction of a Laplacian operator and a theory of harmonic and subharmonic functions. Using Berkovich spaces one obtains surprisingly precise analogues of various classical results concerning subdomains of the Riemann

The author was supported by NSF grants DMS-0602287 and DMS-0600027 during the writing of these lecture notes. The author would like to thank Xander Faber, David Krumm, and Christian Wahle for sending comments and corrections on an earlier version of the notes, Michael Temkin and the anonymous referees for their helpful comments, and Clay Petsche for his help as the author's Arizona Winter School project assistant. Finally, the author would like to express his deep gratitude to Robert Rumely for his inspiration, support, and collaboration.

sphere $\mathbb{P}^1(\mathbb{C})$, including a Poisson formula, a maximum modulus principle for subharmonic functions, and a version of Harnack's principle. A more general theory (encompassing both arbitrary curves and more general base fields) has been worked out by Thuillier in his recent Ph.D. thesis [**Thu05**]. One of the reasons for wanting to develop potential theory on Berkovich spaces is purely aesthetic. Indeed, one of the broad unifying principles within number theory is the idea that all completions of a global field (e.g., a number field) should be treated in a symmetric way. This has been a central theme in number theory for almost a hundred years, as illustrated for example by Chevalley's idelic formulation of class field theory and Tate's development of harmonic analysis on adeles. In the traditional formulation of Arakelov intersection theory, this symmetry principle is violated: one uses analytic methods from potential theory at the archimedean places, and intersection theory on schemes at the non-archimedean places, which at first glance appear to be completely different tools. But thanks to the recent work of Thuillier, one can now formulate Arakelov theory for curves in a completely symmetric fashion at all places. A higher-dimensional synthesis of Berkovich analytic spaces, pluripotential theory, and Arakelov theory has yet to be accomplished, but achieving such a synthesis should be viewed as an important long-term goal.

Third, we hope to give the reader a small sampling of some of the applications of potential theory on Berkovich curves. The theories developed in [**BR08**] and [**Thu05**], as well as the related theory developed in [**FJ04**], have already found applications to subjects as diverse as non-archimedean equidistribution theorems [**BR06, FRL04, FRL06, CL06**], canonical heights over function fields [**Bak07**], Arakelov theory [**Thu05**], p-adic dynamics [**BR08, FRL06, RL03b**], *complex* dynamics [**FJ07**], and complex pluripotential theory [**FJ05**]. However, because these notes have been written with beginning graduate students in mind, and because of space and time constraints, we will not be able to explore in detail any of these applications. Instead, we content ourselves with some "toy" applications, such as Berkovich space interpretations of the product formula (see Example 3.1.2) and the theory of valuation polygons (see §4.7), as well as new proofs of some known results from p-adic analysis (see §4.8). We also briefly explain a connection between potential theory on Berkovich elliptic curves and Néron canonical local height functions (see Example 5.4.5). Although these are not the most important applications of the general theory being described, they are hopefully elegant enough to convince the reader that Berkovich spaces provide an appealing and useful way to approach certain natural problems.

The organization of these notes is as follows. In §1, we will provide an elementary introduction to the Berkovich projective line $\mathbb{P}^1_{\mathrm{Berk}}$, and we will explore in some detail its basic topological and metric properties. In §2, we will discuss more general Berkovich spaces, including the Berkovich analytic space $\mathcal{M}(\mathbb{Z})$ associated to $\mathbb{Z}$. We will then give a brief overview of the topological structure of general Berkovich analytic curves. In §3, we will explore the notion of a *harmonic function* in the context of the spaces $\mathcal{M}(\mathbb{Z})$ and $\mathbb{P}^1_{\mathrm{Berk}}$, formulating analogues of classical results such as the Poisson formula, and we will briefly discuss the related notions of subharmonic and superharmonic functions. In §4, we will define a Laplacian operator on the Berkovich projective line which is analogous in many ways to the classical Laplacian operator on subdomains of the Riemann sphere. Finally, in §5, we will describe how to generalize some of our constructions and "visualization techniques" from $\mathbb{P}^1_{\text{Berk}}$ to more general Berkovich curves.

We set the following notation, which will be used throughout:

K an algebraically closed field which is complete with respect to a non-archimedean absolute value.

$|\ |$ the absolute value on K, which we assume to be nontrivial.

$|K^*|$ the value group of K, i.e., $\{|\alpha| : \alpha \in K^*\}$.

$\tilde{K}$ the residue field of K.

q_v a fixed real number greater than 1, chosen so that $z \mapsto -\log_{q_v}(z)$ is a suitably normalized valuation v on K.

$\log_v$ shorthand for $\log_{q_v}$.

$B(a, r)$ the closed disk $\{z \in K : |z - a| \leq r\}$ of radius r about a in K. Here r is any positive real number, and sometimes we allow the degenerate case $r = 0$ as well. If $r \in |K^*|$ we call the disk *rational*, and if $r \notin |K^*|$ we call it *irrational*.

$B(a, r)^-$ the open disk $\{z \in K : |z - a| < r\}$ of radius r about a in K.

$\mathbb{A}^1_{\text{Berk}}$ the Berkovich affine line over K.

$\mathbb{P}^1_{\text{Berk}}$ the Berkovich projective line over K.

$\mathbf{H}_{\text{Berk}}$ the "Berkovich hyperbolic space" $\mathbb{P}^1_{\text{Berk}} \backslash \mathbb{P}^1(K)$.

$\mathbb{Q}_p$ the field of p-adic numbers.

$\mathbb{C}_p$ the completion of a fixed algebraic closure of $\mathbb{Q}_p$ for some prime number p.

1. The Berkovich projective line

In this lecture, we will introduce in a concrete way the Berkovich affine and projective lines, and we will explore in detail their topological properties. We will also define a subset of the Berkovich projective line called "Berkovich hyperbolic space", which is equipped with a canonical metric. Finally, we define and describe various special functions (e.g. the "canonical distance") on the Berkovich projective line.

1.1. Motivation. Let K be an *algebraically closed field* which is *complete* with respect to a nontrivial non-archimedean absolute value. (These conventions will hold throughout unless explicitly stated otherwise.) The topology on K induced by the given absolute value is Hausdorff, but it is also totally disconnected and not locally compact. This makes it difficult to define in a satisfactory way a good notion of an *analytic function* on K. Tate dealt with this problem by developing the subject which is now known as *rigid analysis*, in which one works with a certain Grothendieck topology on K. This gives a satisfactory theory of analytic functions on K, but the underlying topological space is unchanged, so problems remain for certain other applications. For example, even with Tate's theory in hand it is not at all obvious how to define a Laplacian operator on K analogous to the classical Laplacian on $\mathbb{C}$, or to formulate a natural notion of harmonic and subharmonic functions on K.

However, these difficulties, and many more, can be resolved in an extremely satisfactory way using Berkovich's theory. The Berkovich affine line $\mathbb{A}^1_{\text{Berk}}$ over K is a locally compact, Hausdorff, and path-connected topological space which contains K (with the topology induced by the given absolute value) as a dense subspace.

One obtains the Berkovich projective line $\mathbb{P}^1_{\mathrm{Berk}}$ by adjoining to $\mathbb{A}^1_{\mathrm{Berk}}$ in a suitable manner a point at infinity; the resulting space $\mathbb{P}^1_{\mathrm{Berk}}$ is a compact, Hausdorff, and path-connected topological space which contains $\mathbb{P}^1(K)$ (with its natural topology) as a dense subspace. In fact, $\mathbb{A}^1_{\mathrm{Berk}}$ and $\mathbb{P}^1_{\mathrm{Berk}}$ are better than just path-connected: they are *uniquely* path-connected, in the sense that any two distinct points can be joined by a unique arc. The unique path-connectedness is closely related to the fact that $\mathbb{A}^1_{\mathrm{Berk}}$ and $\mathbb{P}^1_{\mathrm{Berk}}$ are endowed with a natural *profinite $\mathbb{R}$-tree* structure. The profinite $\mathbb{R}$-tree structure on $\mathbb{A}^1_{\mathrm{Berk}}$ (resp. $\mathbb{P}^1_{\mathrm{Berk}}$) can be used to define a *Laplacian operator* in terms of the classical Laplacian on a finite graph. This in turn leads to a good theory of harmonic and subharmonic functions which closely parallels the classical theory over $\mathbb{C}$.

1.2. Multiplicative seminorms. The definition of $\mathbb{A}^1_{\mathrm{Berk}}$ is quite simple, and makes sense with K replaced by an arbitrary field k endowed with a (possibly archimedean or even trivial) absolute value. A *multiplicative seminorm* on a ring A is a function $|\ |_x : A \to \mathbb{R}_{\geq 0}$ satisfying:

- $|0|_x = 0$ and $|1|_x = 1$.
- $|fg|_x = |f|_x \cdot |g|_x$ for all $f, g \in A$.
- $|f + g|_x \leq |f|_x + |g|_x$ for all $f, g \in A$.

As a set, $\mathbb{A}^1_{\mathrm{Berk},k}$ consists of all multiplicative seminorms on the polynomial ring $k[T]$ which extend the usual absolute value on k. By an aesthetically desirable abuse of notation, we will identify seminorms $|\ |_x$ with points $x \in \mathbb{A}^1_{\mathrm{Berk},k}$, and we will usually omit explicit reference to the field k, writing $\mathbb{A}^1_{\mathrm{Berk}}$ and assuming that we are working over a complete and algebraically closed non-archimedean field K. The Berkovich topology on $\mathbb{A}^1_{\mathrm{Berk},k}$ is defined to be the weakest one for which $x \mapsto |f|_x$ is continuous for every $f \in k[T]$.

To motivate the definition of $\mathbb{A}^1_{\mathrm{Berk}}$, we can observe that in the classical setting, every multiplicative seminorm on $\mathbb{C}[T]$ which extends the usual absolute value on $\mathbb{C}$ is of the form $f \mapsto |f(z)|$ for some $z \in \mathbb{C}$. (This is a consequence of the well-known Gelfand-Mazur theorem from functional analysis.) It is then easy to see that $\mathbb{A}^1_{\mathrm{Berk},\mathbb{C}}$ is homeomorphic to $\mathbb{C}$ itself, and also to the Gelfand space of all maximal ideals in $\mathbb{C}[T]$.

1.3. Berkovich's classification theorem. In the non-archimedean world, K can once again be identified with the Gelfand space of maximal ideals in $K[T]$, but now there are many more multiplicative seminorms on $K[T]$ than just the ones given by evaluation at a point of K. The prototypical example is if we fix a closed disk $B(a, r) = \{z \in K \ : \ |z - a| \leq r\}$ in K, and define $|\ |_{B(a,r)}$ by

$$|f|_{B(a,r)} = \sup_{z \in B(a,r)} |f(z)| \ .$$

It is an elementary consequence of the well-known "Gauss lemma" that $|\ |_{B(a,r)}$ is *multiplicative*, and the other axioms for a seminorm are trivially satisfied. Thus each disk $B(a, r)$ gives rise to a point of $\mathbb{A}^1_{\mathrm{Berk}}$. Note that this includes disks for which $r \notin |K^*|$, i.e., *irrational disks* for which the set $\{z \in K \ : \ |z - a| = r\}$ is empty. Also, we may consider the point a itself as a "degenerate" disk of radius zero, in which case we set $|f|_{B(a,0)} = |f(a)|$.

It is not hard to see that distinct disks $B(a, r)$ with $r \geq 0$ give rise to distinct multiplicative seminorms on $K[T]$, and therefore the set of all such disks embeds naturally into $\mathbb{A}^1_{\text{Berk}}$. In particular, K embeds naturally into $\mathbb{A}^1_{\text{Berk}}$ as the set of disks of radius zero, and it is not hard to show that K is in fact *dense* in $\mathbb{A}^1_{\text{Berk}}$ in the Berkovich topology.

Suppose $x, x' \in \mathbb{A}^1_{\text{Berk}}$ are distinct points corresponding to the (possibly degenerate) disks $B(a, r), B(a', r')$, respectively. The unique path in $\mathbb{A}^1_{\text{Berk}}$ between x and x' has the following very intuitive description. If $B(a, r) \subset B(a', r')$, this path consists of all points of $\mathbb{A}^1_{\text{Berk}}$ corresponding to disks containing $B(a, r)$ and contained in $B(a', r')$. The collection of such "intermediate disks" is totally ordered by containment, and if $a = a'$ it is just $\{B(a, t) : r \le t \le r'\}$, which is homeomorphic to the closed interval $[r, r']$ in $\mathbb{R}$. If $B(a, r)$ and $B(a', r')$ are disjoint, the unique path between x and x' consists of all points of $\mathbb{A}^1_{\text{Berk}}$ corresponding to disks of the form $B(a, t)$ with $r \le t \le |a - a'|$ or $B(a', t')$ with $r' \le t' \le |a - a'|$. The disk $B(a, |a - a'|)$ is the smallest one containing both $B(a, r)$ and $B(a', r')$, and if $x \vee x'$ denotes the corresponding point of $\mathbb{A}^1_{\text{Berk}}$, then the unique path from x to x' is just the path from x to $x \vee x'$ followed by the path from $x \vee x'$ to x'.

In particular, if a, a' are distinct points of K, then one can visualize the unique path in $\mathbb{A}^1_{\text{Berk}}$ from a to a' as follows: Start increasing the "radius" of the degenerate disk $B(a, 0)$ until we have a disk $B(a, r)$ which also contains a'. This disk can also be written as $B(a', s)$ with $r = s = |a - a'|$. Now decrease s until the radius reaches zero and we have the degenerate disk $B(a', 0)$. In this way we have "connected up" the totally disconnected space K by adding points corresponding to closed disks in K!

In order to obtain a *compact* space from this construction, it is usually necessary to add even more points. This is because K may not be *spherically complete*.[1] Intuitively, we need to add points corresponding to such sequences in order to obtain a space which has a chance of being compact. More precisely, if we return to the definition of $\mathbb{A}^1_{\text{Berk}}$ in terms of multiplicative seminorms, it is easy to see that if $\{B(a_n, r_n)\}$ is any decreasing nested sequence of closed disks, then the map

$$f \mapsto \lim_{n \to \infty} |f|_{B(a_n, r_n)}$$

defines a multiplicative seminorm on $K[T]$ extending the usual absolute value on K. Two such sequences of disks with empty intersection define the same seminorm if and only if the sequences are *cofinal* (or *interlacing*). This yields a large number of additional points of $\mathbb{A}^1_{\text{Berk}}$ which we are forced to throw into the mix. According to *Berkovich's classification theorem*, we have now described all points of $\mathbb{A}^1_{\text{Berk}}$. More precisely:

THEOREM 1.3.1 (Berkovich's Classification Theorem). *Every point $x \in \mathbb{A}^1_{\text{Berk}}$ corresponds to a nested sequence $B(a_1, r_1) \supseteq B(a_2, r_2) \supseteq B(a_3, r_3) \supseteq \cdots$ of closed disks, in the sense that*

$$|f|_x = \lim_{n \to \infty} |f|_{B(a_n, r_n)}.$$

Two such nested sequences define the same point of $\mathbb{A}^1_{\text{Berk}}$ if and only if
  (a) *each has a nonempty intersection, and their intersections are the same; or*
  (b) *both have empty intersection, and the sequences are cofinal.*

---

[1] A field is called *spherically complete* if there are no decreasing sequences of closed disks having empty intersection. For example, the field $\mathbb{C}_p$ is not spherically complete. In a complete field which is not spherically complete, any decreasing sequence of closed disks with empty intersection must have radii tending towards a strictly positive real number.

Consequently, we can categorize the points of $\mathbb{A}^1_{\text{Berk}}$ into four types according to the nature of $B = \bigcap B(a_n, r_n)$:

Type I: B is a point of K.

Type II: B is a closed disk with radius belonging to $|K^*|$.

Type III: B is a closed disk with radius not belonging to $|K^*|$.

Type IV: $B = \emptyset$.

The set of points of Type III can be either infinite or empty, and similarly for the Type IV points. The set of points of Type I is always infinite, as is the set of points of Type II.

The description of points of $\mathbb{A}^1_{\text{Berk}}$ in terms of closed disks is very useful, because it allows us to visualize quite concretely the abstract space of multiplicative seminorms which we started out with.

As a set[2], the Berkovich projective line $\mathbb{P}^1_{\text{Berk}}$ is obtained from $\mathbb{A}^1_{\text{Berk}}$ by adding a type I point at infinity, denoted ∞. The topology on $\mathbb{P}^1_{\text{Berk}}$ is that of the one-point compactification.

REMARK 1.3.2. Both $\mathbb{P}^1(K)$ and $\mathbb{P}^1_{\text{Berk}} \backslash \mathbb{P}^1(K)$ are dense in the Berkovich topology on $\mathbb{P}^1_{\text{Berk}}$.

We will denote by $\zeta_{a,r}$ the point of $\mathbb{A}^1_{\text{Berk}}$ of type II or III corresponding to the closed or irrational disk $B(a,r)$. Allowing degenerate disks (i.e., $r = 0$), we can extend this notation to points of type I: we let $\zeta_{a,0}$ (or simply a, depending on the context) denote the point of $\mathbb{A}^1_{\text{Berk}}$ corresponding to $a \in K$.

Following the terminology introduced by Chambert-Loir in [**CL06**], the distinguished point $\zeta_{0,1}$ in $\mathbb{A}^1_{\text{Berk}}$ corresponding to the Gauss norm on $K[T]$ will be called the *Gauss point*. We will usually write ζ_{Gauss} instead of $\zeta_{0,1}$.

1.4. Visualizing $\mathbb{P}^1_{\text{Berk}}$.

1.4.1. *A partial order on $\mathbb{P}^1_{\text{Berk}}$.* The space $\mathbb{A}^1_{\text{Berk}}$ is endowed with a natural partial order, defined by saying that $x \leq y$ if and only if $|f|_x \leq |f|_y$ for all $f \in K[T]$. In terms of (possibly degenerate) disks, if $x, y \in \mathbb{A}^1_{\text{Berk}}$ are points of type I, II, or III, we have $x \leq y$ if and only if the disk corresponding to x is contained in the disk corresponding to y. (We leave it to the reader to extend this description of the partial order to points of type IV.) For each pair of points $x, y \in \mathbb{A}^1_{\text{Berk}}$, there is a unique least upper bound $x \vee y \in \mathbb{A}^1_{\text{Berk}}$ with respect to this partial order. Concretely, if $x = \zeta_{a,r}$ and $y = \zeta_{b,s}$ are points of type I, II or III, then $x \vee y$ is the point of $\mathbb{A}^1_{\text{Berk}}$ corresponding to the smallest disk containing both $B(a,r)$ and $B(b,s)$.

We can extend the partial order to $\mathbb{P}^1_{\text{Berk}}$ by declaring that $x \leq \infty$ for all $x \in \mathbb{A}^1_{\text{Berk}}$. Writing

$$[x, x'] = \{z \in \mathbb{P}^1_{\text{Berk}} \ : \ x \leq z \leq x'\} \cup \{z \in \mathbb{P}^1_{\text{Berk}} \ : \ x' \leq z \leq x\} \, ,$$

---

[2]One should also define a sheaf of analytic functions on $\mathbb{A}^1_{\text{Berk}}$ and $\mathbb{P}^1_{\text{Berk}}$ and view them as locally ringed spaces endowed with the extra structure of a (maximal) K-affinoid atlas, but we will not emphasize the formalism of Berkovich's theory of K-analytic spaces in these lectures; see instead B. Conrad's article [**Con07**].

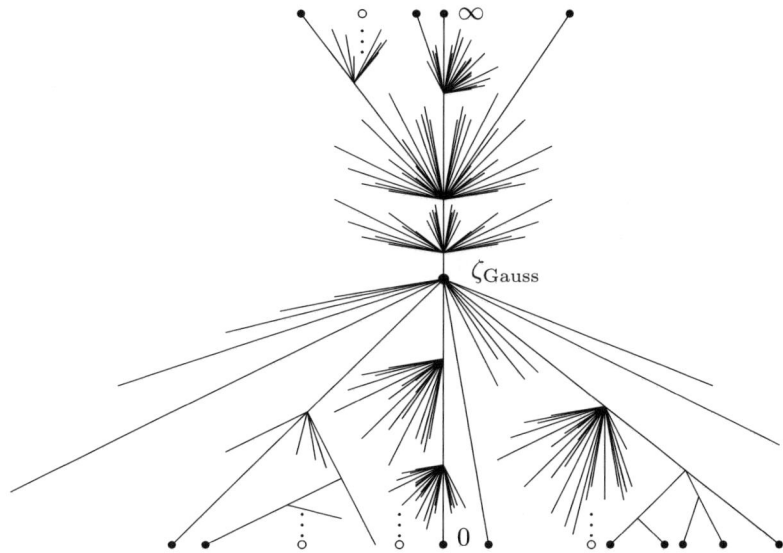

FIGURE 1. The Berkovich projective line (adapted from an illustration of Joe Silverman)

it is easy to see that the unique path between $x, y \in \mathbb{P}^1_{\mathrm{Berk}}$ is just

$$\ell_{x,y} := [x, x \vee y] \cup [x \vee y, y] \ .$$

1.4.2. *Navigating* $\mathbb{P}^1_{\mathrm{Berk}}$. You can visualize "navigating" the Berkovich projective line in the following way (cf. Figure 1). Starting from the Gauss point $\zeta_{0,1}$, there are infinitely many branches in which you can travel, one for each element of the residue field $\tilde{K}$ plus a branch leading up towards infinity. Having chosen a direction in which to move, at each point of type II along the chosen branch there are infinitely many new branches to choose from, and each subsequent branch behaves in the same way. This dizzying collection of densely splitting branches forms a configuration which Robert Rumely has christened a "witch's broom". However, the witch's broom has some structure:

- There is branching *only* at the points of type II, not those of type III.
- The branches emanating from a type II point $\zeta_{a,r}$ are in one-to-one correspondence with elements of $\mathbb{P}^1(\tilde{K})$: there is one branch going "up" to infinity, with the other branches corresponding to open disks $B(a', r)^-$ of radius r contained in $B(a, r)$.
- Some of the branches extend all the way to the bottom (terminating in points of type I), while others are "cauterized off" earlier and terminate at points of type IV. In any case, every branch terminates either at a point of type I or type IV.

1.4.3. *Tangent spaces and directional derivatives.* Let $x \in \mathbb{P}^1_{\mathrm{Berk}}$. We define the space T_x of *tangent directions* at x to be the set of equivalence classes of paths $\ell_{x,y}$ emanating from x, where y is any point of $\mathbb{P}^1_{\mathrm{Berk}}$ not equal to x and two paths $\ell_{x,y_1}, \ell_{x,y_2}$ are *equivalent* if they share a common initial segment. There is a natural bijection between elements $\vec{v} \in T_x$ and connected components of $\mathbb{P}^1_{\mathrm{Berk}} \backslash \{x\}$. We denote by $U(x; \vec{v})$ the connected component[3] of $\mathbb{P}^1_{\mathrm{Berk}} \backslash \{x\}$ corresponding to $\vec{v} \in T_x$. It is not hard to show that the open sets $U(x; \vec{v})$ for $x \in \mathbb{P}^1_{\mathrm{Berk}}$ and $\vec{v} \in T_x$ form a sub-base for the topology on $\mathbb{P}^1_{\mathrm{Berk}}$, so that finite intersections of such open sets form a neighborhood base for this topology.[4]

For example, consider the Gauss point ζ_{Gauss}. The different tangent directions $\vec{v} \in T_{\zeta_{\mathrm{Gauss}}}$ correspond bijectively to elements of $\mathbb{P}^1(\tilde{K})$, the projective line over the residue field of K. Equivalently, elements of $T_{\zeta_{\mathrm{Gauss}}}$ correspond to the open disks of radius 1 contained in the closed unit disk $B(0, 1)$, together with the open disk

$$B(\infty, 1)^- := \mathbb{P}^1(K) \backslash B(0, 1).$$

The correspondence between elements of $T_{\zeta_{\mathrm{Gauss}}}$ and open disks is given explicitly by $\vec{v} \mapsto U(\zeta_{\mathrm{Gauss}}; \vec{v}) \cap \mathbb{P}^1(K)$.

More generally, for each point $x = \zeta_{a,r}$ of type II, the set T_x of tangent directions at x is (non-canonically) isomorphic to $\mathbb{P}^1(\tilde{K})$: there is one tangent direction going "up" to infinity, and the other tangent directions correspond to open disks $B(a', r)^-$ of radius r contained in $B(a, r)$, which (after choosing a Möbius transformation sending $B(a, r)$ to $B(0, 1)$) correspond bijectively to elements of $\tilde{K}$.

For points $x = \zeta_{a,r}$ of type III, there are only two possible tangent directions: one leading "up" towards infinity, and one going "down" towards a. Similarly, since points of type I or IV are "endpoints" of $\mathbb{P}^1_{\mathrm{Berk}}$, the set T_x of tangent directions at a point $x \in \mathbb{P}^1_{\mathrm{Berk}}$ of type I or IV consists of just one element.

In particular, for $x \in \mathbb{P}^1_{\mathrm{Berk}}$, we have:

$$|T_x| = \begin{cases} |\mathbb{P}^1(\tilde{K})| & x \text{ of type II,} \\ 2 & x \text{ of type III,} \\ 1 & x \text{ of type I or type IV.} \end{cases}$$

Finally, we explain how to interpret the sets $U(x; \vec{v})$ as "open Berkovich disks". For $a \in K$ and $r > 0$, write

$$\mathcal{B}(a, r)^- = \{x \in \mathbb{A}^1_{\mathrm{Berk}} : |T - a|_x < r\},$$
$$\mathcal{B}(a, r) = \{x \in \mathbb{A}^1_{\mathrm{Berk}} : |T - a|_x \leq r\}.$$

We call a set of the form $\mathcal{B}(a, r)^-$ an *open Berkovich disk* in $\mathbb{A}^1_{\mathrm{Berk}}$, and a set of the form $\mathcal{B}(a, r)$ a *closed Berkovich disk* in $\mathbb{A}^1_{\mathrm{Berk}}$.

Similarly, we can define open and closed Berkovich disks in $\mathbb{P}^1_{\mathrm{Berk}}$: an open (resp. closed) Berkovich disk in $\mathbb{P}^1_{\mathrm{Berk}}$ is either an open (resp. closed) Berkovich disk in $\mathbb{A}^1_{\mathrm{Berk}}$ or the complement of a closed (resp. open) Berkovich disk in $\mathbb{A}^1_{\mathrm{Berk}}$.

It follows from the definitions that the intersection of a Berkovich open (resp. closed) disk in $\mathbb{P}^1_{\mathrm{Berk}}$ with $\mathbb{P}^1(K)$ is an open (resp. closed) disk in $\mathbb{P}^1(K)$.

We have the following result, whose proof is left as an exercise for the reader:

---

[3]It is not difficult to show that a subset of $\mathbb{P}^1_{\mathrm{Berk}}$ is connected if and only if it is path-connected, and in particular that the path-connected components of $\mathbb{P}^1_{\mathrm{Berk}}$ coincide with the connected components.

[4]This has been called the "observer's topology" (see [**CHL07**]), since a fundamental system of open neighborhoods at x is given by the set of points which can simultaneously be 'seen' by a finite number of "observers" $x_1, \ldots, x_n$ looking in the direction of x.

LEMMA 1.4.1. *Every open set $U(x; \vec{v})$ with x of type II or III and $\vec{v} \in T_x$ is a Berkovich open disk in $\mathbb{P}^1_{\mathrm{Berk}}$, and conversely.*

REMARK 1.4.2. A fundamental system of open neighborhoods for the topology on $\mathbb{P}^1_{\mathrm{Berk}}$ is given by the finite intersections of Berkovich open disks in $\mathbb{P}^1_{\mathrm{Berk}}$ (cf. Lemma 2.2.3 below).

1.5. The Berkovich hyperbolic space $\mathbf{H}_{\mathrm{Berk}}$ and its canonical metric. Following notation introduced by Juan Rivera-Letelier, we write $\mathbf{H}_{\mathrm{Berk}}$ for the subset of $\mathbb{P}^1_{\mathrm{Berk}}$ consisting of all points of type II, III, or IV, and call $\mathbf{H}_{\mathrm{Berk}}$ "Berkovich hyperbolic space". We also write $\mathbf{H}^{\mathbb{Q}}_{\mathrm{Berk}}$ for the set of type II points, and $\mathbf{H}^{\mathbb{R}}_{\mathrm{Berk}}$ for the set of points of type II or III.

The subset $\mathbf{H}^{\mathbb{Q}}_{\mathrm{Berk}}$ is dense in $\mathbb{P}^1_{\mathrm{Berk}}$ (and therefore $\mathbf{H}^{\mathbb{R}}_{\mathrm{Berk}}$ and $\mathbf{H}_{\mathrm{Berk}}$ are also dense).

There is a canonical metric ρ on $\mathbf{H}_{\mathrm{Berk}}$, which we refer to as the *path metric*, that is of great importance for potential theory. To define this metric, we first define the *diameter function* $\mathrm{diam} : \mathbb{A}^1_{\mathrm{Berk}} \to \mathbb{R}_{\geq 0}$ by setting $\mathrm{diam}(x) = \lim r_i$ if x corresponds to the nested sequence $\{B(a_i, r_i)\}$. This is well-defined independent of the choice of nested sequence. If $x \in \mathbf{H}^{\mathbb{R}}_{\mathrm{Berk}}$, then $\mathrm{diam}(x)$ is just the diameter ($=$ radius) of the corresponding closed disk. In terms of multiplicative seminorms, we have

$$\mathrm{diam}(x) = \inf_{a \in K} |T - a|_x \ .$$

Because K is complete, it is not hard to see that if x is of type IV, then necessarily $\mathrm{diam}(x) > 0$ (see footnote 1). Thus $\mathrm{diam}(x) = 0$ for $x \in \mathbb{A}^1_{\mathrm{Berk}}$ of type I, and $\mathrm{diam}(x) > 0$ for $x \in \mathbf{H}_{\mathrm{Berk}}$.

If $x, y \in \mathbf{H}_{\mathrm{Berk}}$ with $x \leq y$, we define

$$\rho(x, y) = \log_v \frac{\mathrm{diam}(y)}{\mathrm{diam}(x)} \ ,$$

where $\log_v$ denotes the logarithm to the base q_v, with $q_v > 1$ a fixed real number chosen so that $z \mapsto -\log_v |z|$ is a suitably normalized valuation on K. For example, if $K = \mathbb{C}_p$, endowed with the standard absolute value $|\ |_p$ for which $|p|_p = 1/p$, then we set $q_v = p$ in order to have

$$\{\log_v |z|_p \ : \ z \in \mathbb{C}_p^*\} = \mathbb{Q} \ .$$

More generally, for $x, y \in \mathbf{H}_{\mathrm{Berk}}$ arbitrary, we define

$$\rho(x, y) = \rho(x, x \vee y) + \rho(y, x \vee y)$$
$$= 2 \log_v \mathrm{diam}(x \vee y) - \log_v \mathrm{diam}(x) - \log_v \mathrm{diam}(y) \ .$$

It is not hard to verify that ρ defines a *metric* on $\mathbf{H}_{\mathrm{Berk}}$. One can extend ρ to a singular metric on $\mathbb{P}^1_{\mathrm{Berk}}$ by declaring that if $x \in \mathbb{P}^1(K)$ and $y \in \mathbb{P}^1_{\mathrm{Berk}}$, we have $\rho(x, y) = +\infty$ if $x \neq y$ and 0 if $x = y$. However, we will usually only consider ρ as being defined on $\mathbf{H}_{\mathrm{Berk}}$.

Intuitively, $\rho(x, y)$ is just the "length" of the unique path $\ell_{x,y}$ between x and y, which for closed disks $B(a, r) \subseteq B(a, R)$ is just $\log_v R - \log_v r$.

REMARK 1.5.1. It is important to note that the topology on $\mathbf{H}_{\mathrm{Berk}}$ defined by the metric ρ is *not* the subspace topology induced from the Berkovich topology on $\mathbb{P}^1_{\mathrm{Berk}}$. However, the inclusion map $i : \mathbf{H}_{\mathrm{Berk}} \hookrightarrow \mathbb{P}^1_{\mathrm{Berk}}$ is continuous with respect to these topologies.

The group $\mathrm{PGL}(2,K)$ of *Möbius transformations* acts continuously on $\mathbb{P}^1_{\mathrm{Berk}}$ in a natural way compatible with the usual action on $\mathbb{P}^1(K)$, and this action preserves $\mathbf{H}_{\mathrm{Berk}}, \mathbf{H}^{\mathbb{Q}}_{\mathrm{Berk}}$, and $\mathbf{H}^{\mathbb{R}}_{\mathrm{Berk}}$. (The action of $\mathrm{PGL}(2,K)$ on $\mathbb{P}^1_{\mathrm{Berk}}$ can be described quite concretely in terms of Berkovich's classification theorem, using the fact that each $M \in \mathrm{PGL}(2,K)$ takes closed disks to closed disks, but it can also be described more intrinsically in terms of multiplicative seminorms; see §2.1 for further details.)

An important observation (see Chapter 2 of [**BR08**]) is that $\mathrm{PGL}(2,K)$ acts *via isometries* on $\mathbf{H}_{\mathrm{Berk}}$, i.e.,

$$\rho(M(x), M(y)) = \rho(x,y)$$

for all $x,y \in \mathbf{H}_{\mathrm{Berk}}$ and all $M \in \mathrm{PGL}(2,K)$. This shows that the metric ρ is canonical and does not depend on a choice of coordinates for $\mathbb{P}^1$.

1.6. The canonical distance.

1.6.1. *The canonical distance relative to infinity.* The diameter function diam introduced in §1.5 can be used to extend the usual distance function $|x - y|$ on K to $\mathbb{A}^1_{\mathrm{Berk}}$ in a natural way. We call this extension the *canonical distance* (relative to infinity), and denote it by $[x,y]_\infty$.

Formally, for $x,y \in \mathbb{A}^1_{\mathrm{Berk}}$ we have

$$(1.6.1) \qquad\qquad [x,y]_\infty = \mathrm{diam}(x \vee y) \ .$$

It is easy to see that if $x,y \in K$ then $[x,y]_\infty = |x - y|$. More generally (see Chapter 4 of [**BR08**]), one has the formula

$$[x,y]_\infty = \limsup_{(x_0,y_0) \to (x,y)} |x_0 - y_0| \ ,$$

where $(x_0, y_0) \in K \times K$ and the lim sup is taken with respect to the product topology on $\mathbb{P}^1_{\mathrm{Berk}} \times \mathbb{P}^1_{\mathrm{Berk}}$. The canonical distance $[x,y]_\infty$ satisfies all of the axioms for an ultrametric except for the fact that $[x,x]_\infty > 0$ for $x \in \mathbf{H}_{\mathrm{Berk}}$.

REMARK 1.6.2. In [**BR08**], $[x,y]_\infty$ is written as $\delta(x,y)_\infty$, and is called the *Hsia kernel*.

1.6.2. *The canonical distance relative to an arbitrary point.* In this section, we describe a function $[x,y]_z$ which extends the canonical distance (relative to a point z) on $\mathbb{P}^1(K)$, as introduced by Rumely in [**Rum89**], to the Berkovich projective line. When $z = \infty$, it coincides with the canonical distance relative to infinity as defined in the previous section.

Let x,y,z be points of $\mathbb{P}^1_{\mathrm{Berk}}$, not all equal. Following the terminology introduced by Favre and Rivera-Letelier in [**FRL06, FRL04**], we define the *Gromov product* $(x|y)_z$ by

$$(x|y)_z = \rho(w,z),$$

where w is the first point where the unique paths from x to z and y to z intersect. By convention, we set $(x|y)_z = +\infty$ if $x = y$ and x is a point of type I, and we set $(x|y)_z = 0$ if $x = z$ or $y = z$.

REMARK 1.6.3. If $x,y,z \in \mathbf{H}_{\mathrm{Berk}}$, then one checks easily that

$$(x|y)_z = \frac{1}{2}(\rho(x,z) + \rho(y,z) - \rho(x,y)).$$

This is the usual definition of the Gromov product in Gromov's theory of δ-hyperbolic spaces, with $\mathbf{H}_{\text{Berk}}$ being an example of a 0-hyperbolic space.

REMARK 1.6.4. In [**BR08**], the function $(x|y)_z$ is written $j_z(x, y)$.

Next, define the *fundamental potential kernel relative to* z, written $\kappa_z(x, y)$, and the *canonical distance relative to* z, written $[x, y]_z$, by setting

$$(1.6.5) \qquad \kappa_z(x, y) = -\log_v [x, y]_z = (x|y)_\zeta - (x|z)_\zeta - (y|z)_\zeta,$$

where $\zeta = \zeta_{\text{Gauss}}$ is the Gauss point of $\mathbb{P}^1_{\text{Berk}}$. One can define $\kappa_z(x, y)$ as an extended-real number for all $x, y, z \in \mathbb{P}^1_{\text{Berk}}$ by setting $\kappa_z(z, y) = \kappa_z(x, z) = -\infty$ if z is a point of type I.

REMARK 1.6.6. 1. In Chapter 4 of [**BR08**], the notation $\delta(x, y)_z$ is used instead of $[x, y]_z$, and $\delta(x, y)_z$ is referred to as the *generalized Hsia kernel*. For $x, y, z \in \mathbb{P}^1(K)$, our definition of $[x, y]_z$ agrees with Rumely's definition of the canonical distance in [**Rum89**].

2. If ζ, ζ' are arbitrary points of $\mathbf{H}_{\text{Berk}}$, one can show that

$$(x|y)_\zeta - (x|z)_\zeta - (y|z)_\zeta = (x|y)_{\zeta'} - (x|z)_{\zeta'} - (y|z)_{\zeta'} + C_{\zeta,\zeta'}$$

for some constant $C_{\zeta,\zeta'}$ independent of x, y, z. Thus a different choice of ζ in (1.6.5) would only change the definition of $\kappa_z(x, y)$ by an additive constant. Our choice $\zeta = \zeta_{\text{Gauss}}$ is just a convenient normalization.

3. After defining the Laplacian operator Δ on $\mathbb{P}^1_{\text{Berk}}$, we will see in Example 4.5.4 below that for y, z fixed, the function $f(x) = \kappa_z(x, y)$ satisfies the *Laplace equation* $\Delta(f) = \delta_y - \delta_z$, and up to an additive constant f is in fact the *unique* such function.

Since the definition of $[x, y]_z$ takes some getting used to, we will attempt to orient the reader with the following illustrative examples:

EXAMPLE 1.6.7. If $z = \infty$, it is straightforward (but not completely trivial) to verify that the definitions of $[x, y]_\infty$ given in (1.6.5) and (1.6.1) coincide. Thus our notation is consistent, and $[x, y]_\infty$ extends the distance function $|x - y|$ on $K \times K$.

EXAMPLE 1.6.8. If x, y are written in homogeneous coordinates as $x = (x_1 : x_2)$ and $y = (y_1 : y_2)$, the *spherical metric* on $\mathbb{P}^1(K)$ is given by

$$\|x, y\| = \frac{|x_1 y_2 - x_2 y_1|}{\max(|x_1|, |x_2|) \cdot \max(|y_1|, |y_2|)}.$$

If $z = \zeta_{\text{Gauss}}$, then for x and y in $\mathbb{P}^1_{\text{Berk}}$, the function $-\log_v [x, y]_{\zeta_{\text{Gauss}}}$ coincides with the Gromov product $(x|y)_{\zeta_{\text{Gauss}}}$, and the restriction of $[x, y]_{\zeta_{\text{Gauss}}}$ to $x, y \in \mathbb{P}^1(K)$ coincides with the spherical metric $\|x, y\|$ on $\mathbb{P}^1(K)$.

We will sometimes write $\|x, y\|$ for the extended function $[x, y]_{\zeta_{\text{Gauss}}}$ on $\mathbb{P}^1_{\text{Berk}} \times \mathbb{P}^1_{\text{Berk}}$.

REMARK 1.6.9. 1. Note that unlike $[x, y]_\infty$, which is singular at infinity, the function $\|x, y\| = [x, y]_{\zeta_{\text{Gauss}}}$ is bounded and real-valued on all of $\mathbb{P}^1_{\text{Berk}} \times \mathbb{P}^1_{\text{Berk}}$.

2. By (1.6.5), we have the identity

$$[x, y]_z = \frac{\|x, y\|}{\|x, z\| \, \|y, z\|}.$$

The following result (see Chapter 4 of [**BR08**]) describes some of the main properties possessed by the canonical distance $[x, y]_z$ on $\mathbb{P}^1_{\text{Berk}}$. Recall that if X is a topological space, a real-valued function $f : X \to [-\infty, \infty)$ is called *upper semicontinuous* if for each $x_0 \in X$,

$$\limsup_{x \to x_0} f(x) \ \leq \ f(x_0) \ .$$

This is equivalent to requiring that $f^{-1}([-\infty, b))$ be open for each $b \in \mathbb{R}$.

PROPOSITION 1.6.10.

(1) *For each $z \in \mathbb{P}^1_{\text{Berk}}$, the canonical distance $[x, y]_z$ is nonnegative, symmetric, and continuous in each variable separately. If $z \in \mathbf{H}_{\text{Berk}}$, then $[x, y]_z$ is bounded. For $z \in \mathbb{P}^1(K)$ it is unbounded, and extends the canonical distance $[x, y]_z$ from [**Rum89**].*

(2) *As a function of x and y, the canonical distance $[x, y]_z$ is upper semicontinuous. It is continuous off the diagonal, and is continuous at (x_0, x_0) for each point $x_0 \in \mathbb{P}^1(K)$ of type I, but is discontinuous at (x_0, x_0) for each point $x_0 \in \mathbf{H}_{\text{Berk}}$.*

(3) *For each $x, y \in \mathbb{P}^1_{\text{Berk}}$,*

$$[x, y]_z \ = \ \limsup_{\substack{(a,b) \to (x,y) \\ a,b \in \mathbb{P}^1(K)}} [a, b]_z \ .$$

(4) *For all $x, y, w \in \mathbb{P}^1_{\text{Berk}}$, the ultrametric inequality*

$$[x, y]_z \ \leq \ \max([x, w]_z, [y, w]_z)$$

holds, with equality if $[x, w]_z \neq [y, w]_z$.

(5) *If f is a nonzero meromorphic function on $\mathbb{P}^1$ with divisor $\text{Div}(f) = \sum m_i(a_i)$, then for any $z \in \mathbb{P}^1_{\text{Berk}}$, there is a constant C (depending on z and f) such that*

$$|f(x)| = C \cdot \prod [x, a_i]_z^{m_i}$$

for all $x \in \mathbb{P}^1_{\text{Berk}}$.

2. Further examples of Berkovich analytic spaces

In this lecture, we will explore further properties and an alternative definition of the Berkovich projective line, and then we discuss some more general Berkovich spaces. For example, after defining the Berkovich analytic space $\mathcal{M}(\mathcal{A})$ associated to an arbitrary normed ring $\mathcal{A}$, we will describe in detail the topological structure of $\mathcal{M}(\mathbb{Z})$. We will then give a brief overview of the topological structure of Berkovich analytic curves. (A more detailed description will be given in §5.)

All rings throughout these notes will be commutative rings with an identity element 1.

2.1. The Berkovich "Proj" construction. As a topological space, we have defined the Berkovich projective line $\mathbb{P}^1_{\text{Berk}, K}$ to be the one-point compactification of the locally compact Hausdorff space $\mathbb{A}^1_{\text{Berk}, K}$. However, this description depends on a choice of coordinates, and is often awkward to use. For example, it is not immediately clear from this definition how a rational function $\varphi \in K(T)$ induces a natural map from $\mathbb{P}^1_{\text{Berk}}$ to itself. We therefore introduce the following alternate construction of $\mathbb{P}^1_{\mathrm{Berk},K}$, analogous to the "Proj" construction in algebraic geometry.[5]

Let S denote the set of multiplicative seminorms $[\ ]$ on the two-variable polynomial ring $K[X,Y]$ which extend the absolute value on K, and which are not identically zero on the maximal ideal (X,Y) of $K[X,Y]$. It is easy to see that $[\ ]$ is automatically non-archimedean, and that $[\ ]$ is identically zero on (X,Y) if and only if $[X] = [Y] = 0$.

We put an equivalence relation on S by declaring that $[\ ]_1 \sim [\ ]_2$ if and only if there exists a constant $C > 0$ such that $[G]_1 = C^d[G]_2$ for all homogeneous polynomials $G \in K[X,Y]$ of degree d.

As a set, define $\mathbb{P}^1_{\mathrm{Berk}}$ to be the equivalence classes of elements of S.

Define the point ∞ in $\mathbb{P}^1_{\mathrm{Berk}}$ to be the equivalence class of the seminorm $[\ ]_\infty$ defined by $[G]_\infty = |G(1,0)|$. More generally, if $P \in \mathbb{P}^1(K)$ has homogeneous coordinates $(a : b)$, the equivalence class of the evaluation seminorm $[G]_P = |G(a,b)|$ is independent of the choice of homogeneous coordinates, and therefore $[\ ]_P$ is a well-defined point of $\mathbb{P}^1_{\mathrm{Berk}}$. This furnishes an embedding of $\mathbb{P}^1(K)$ into $\mathbb{P}^1_{\mathrm{Berk}}$.

We say that a seminorm $[\ ]$ in S is *normalized* if $\max\{[X],[Y]\} = 1$. Every equivalence class of elements of S contains at least one normalized seminorm. From the definition of the equivalence relation on S, it is clear that all the normalized seminorms in a given class take the same value on homogeneous polynomials. Explicitly, if $[\ ]_z$ is any representative of the equivalence class of $z \in \mathbb{P}^1_{\mathrm{Berk}}$, then any normalized seminorm $[\ ]^*_z$ representing z satisfies

$$[G]^*_z = [G]_z / \max\{[X]_z, [Y]_z\}^d$$

for all homogeneous polynomials $G \in K[X,Y]$ of degree d.

The topology on $\mathbb{P}^1_{\mathrm{Berk}}$ is defined to be the weakest one such that $z \mapsto [G]^*_z$ is continuous for all homogeneous polynomials $G \in K[X,Y]$. One readily verifies:

LEMMA 2.1.1. *This definition of $\mathbb{P}^1_{\mathrm{Berk}}$ as a topological space agrees with the previous one.*

Let $\varphi \in K(T)$ be a rational function of degree $d \geq 1$. To conclude this section, we explain how to extend the usual action of φ on $\mathbb{P}^1(K)$ to a continuous map $\varphi : \mathbb{P}^1_{\mathrm{Berk}} \to \mathbb{P}^1_{\mathrm{Berk}}$.

Choose a homogeneous lifting $F = (F_1, F_2)$ of φ, where $F_i \in K[X,Y]$ are homogeneous of degree d and have no common zeros in K. (Recall that the field K is assumed to be algebraically closed.) The condition that F_1 and F_2 have no common zeros is equivalent to requiring that the homogeneous resultant $\mathrm{Res}(F) = \mathrm{Res}(F_1, F_2)$ is nonzero.

We define the action of φ on $\mathbb{P}^1_{\mathrm{Berk}}$ as follows: Let $G \in K[X,Y]$, and define

(2.1.2) $$[G]_{\varphi(z)} := [G(F_1(X,Y), F_2(X,Y))]_z.$$

It is readily verified that the right-hand side of (2.1.2) is independent of the lifting F of φ, up to equivalence of seminorms. As it is clear that the right-hand side of (2.1.2) gives a continuous multiplicative seminorm on $K[X,Y]$, to see that (2.1.2) induces a map from $\mathbb{P}^1_{\mathrm{Berk}}$ to itself, it suffices to note that $[X]_{\varphi(z)} = [F_1(X,Y)]_z$

---

[5]The alternate construction presented here is adapted from Berkovich's paper [**Ber95**].

and $[Y]_{\varphi(z)} = [F_2(X, Y)]_z$ cannot both be zero; this can be proved using standard properties of resultants (see Chapter 2 of [**BR08**] for details).

In particular, we see that the group $\mathrm{PGL}(2, K)$ acts naturally on $\mathbb{P}^1_{\mathrm{Berk}}$ via automorphisms, as mentioned in §1.5.

REMARK 2.1.3. One can show that $\varphi : \mathbb{P}^1_{\mathrm{Berk}} \to \mathbb{P}^1_{\mathrm{Berk}}$ is an open surjective mapping, and that every point $z \in \mathbb{P}^1_{\mathrm{Berk}}$ has at most d preimages under φ (see Chapter 9 of [**BR08**], §3 of [**RL03b**], and Lemma 3.2.4 of [**Ber90**]).

REMARK 2.1.4. Note that if $z \in \mathbf{H}_{\mathrm{Berk}}$ then $\varphi(z) \in \mathbf{H}_{\mathrm{Berk}}$ as well, because the seminorm $[G]_{\varphi(z)}$ has trivial kernel (i.e., is a norm), whereas for each $a \in \mathbb{P}^1(K)$, the corresponding seminorm has nonzero kernel.

More generally, one can verify that φ takes type I points to type I points, type II points to type II points, type III points to type III points, and type IV points to type IV points.

2.2. $\mathbb{P}^1_{\mathrm{Berk}}$ as an inverse limit of $\mathbb{R}$-trees. We now come to an important description of $\mathbb{P}^1_{\mathrm{Berk}}$ as a *profinite $\mathbb{R}$-tree*. We will need the following definitions.

Let X be a metric space, and let $x, y \in X$. A *geodesic* in X is the image of a one-to-one isometry from a real interval $[a, b]$ into X. An *arc* from x to y is a continuous one-to-one map $f : [a, b] \to X$ with $f(a) = x$ and $f(b) = y$. An *$\mathbb{R}$-tree* is a metric space T such that for each distinct pair of points $x, y \in T$, there is a unique arc from x to y, and this arc is a geodesic.

A topological space homeomorphic to an $\mathbb{R}$-tree (but which is not necessarily endowed with a distinguished metric) will be called a *topological tree*. A *branch point* of a topological tree is a point $x \in T$ for which $T \backslash \{x\}$ has either fewer than or more than two connected components. A *finite $\mathbb{R}$-tree* (resp. topological tree) is an $\mathbb{R}$-tree (resp. topological tree) with only finitely many branch points. Intuitively, a finite $\mathbb{R}$-tree is just a finite tree in the usual graph-theoretic sense, but where the edges are thought of as line segments having specific lengths. Finally, a *profinite $\mathbb{R}$-tree* is an inverse limit of finite $\mathbb{R}$-trees.

Here's how these definitions play out in the case of $\mathbb{P}^1_{\mathrm{Berk}}$. If $S \subset \mathbb{P}^1_{\mathrm{Berk}}$, define the *convex hull* of S to be the smallest path-connected subset of $\mathbb{P}^1_{\mathrm{Berk}}$ containing S. (This is the same as the union of all paths between points of S.) By a *finite subgraph* of $\mathbb{P}^1_{\mathrm{Berk}}$, we will mean the convex hull of a finite subset $S \subset \mathbf{H}^{\mathbb{R}}_{\mathrm{Berk}}$. Every finite subgraph Γ can be thought of as a finite $\mathbb{R}$-tree, with the metric induced by the path-distance ρ on $\mathbf{H}_{\mathrm{Berk}}$. By construction, a finite subgraph of $\mathbb{P}^1_{\mathrm{Berk}}$ is both finitely branched and of finite total length with respect to ρ.[6] We define the *locally metric topology* on $\mathbf{H}_{\mathrm{Berk}}$ to be the topology generated by the open subsets of Γ (endowed with its metric topology) as Γ varies over all finite subgraphs of $\mathbf{H}_{\mathrm{Berk}}$.

The collection of all finite subgraphs of $\mathbb{P}^1_{\mathrm{Berk}}$ is a directed set under inclusion. Moreover, if $\Gamma \leq \Gamma'$, then by a basic property of $\mathbb{R}$-trees, there is a continuous *retraction map* $r_{\Gamma', \Gamma} : \Gamma' \twoheadrightarrow \Gamma$. The following result can be thought of as a topological reformulation of Berkovich's classification theorem:

---

[6]We have chosen to require in addition that $\partial \Gamma \subset \mathbf{H}^{\mathbb{R}}_{\mathrm{Berk}}$, but this could be relaxed by allowing a finite subgraph to have boundary points of type IV without creating any major differences in the resulting theory. We could equally well impose the more stringent requirement that $\partial \Gamma \subset \mathbf{H}^{\mathbb{Q}}_{\mathrm{Berk}}$, and again, the resulting theory would be basically the same.

THEOREM 2.2.1. $\mathbb{P}^1_{\mathrm{Berk}}$ *is homeomorphic to the inverse limit* $\varprojlim \Gamma$ *over all finite subgraphs* $\Gamma \subset \mathbb{P}^1_{\mathrm{Berk}}$.

This description of $\mathbb{P}^1_{\mathrm{Berk}}$ as a profinite $\mathbb{R}$-tree provides a convenient way to visualize the topology on $\mathbb{P}^1_{\mathrm{Berk}}$: two points are "close" if they retract to the same point of a "large" finite subgraph.

We also have the following fact:

LEMMA 2.2.2. *The* direct limit *of all finite subgraphs* Γ *of* $\mathbb{P}^1_{\mathrm{Berk}}$ *with respect to inclusion is homeomorphic to the space* $\mathbf{H}^{\mathbb{R}}_{\mathrm{Berk}}$ *endowed with the locally metric topology.*

Let r_Γ be the natural map from $\mathbb{P}^1_{\mathrm{Berk}}$ to Γ coming from the universal property of the inverse limit. A fundamental system of open neighborhoods for the topology on $\mathbb{P}^1_{\mathrm{Berk}}$ is given by the *connected open affinoids*, or *simple domains*, which are subsets of the form $r_\Gamma^{-1}(V)$ for Γ a finite subgraph of $\mathbb{P}^1_{\mathrm{Berk}}$ and V a connected open subset of Γ (see Figure 4 below).

LEMMA 2.2.3. *For a subset* $U \subseteq \mathbb{P}^1_{\mathrm{Berk}}$, *the following are equivalent:*

(1) U *is a simple domain.*
(2) U *is a finite intersection of Berkovich open disks.*
(3) U *is a connected open set whose boundary is a finite subset of* $\mathbf{H}^{\mathbb{R}}_{\mathrm{Berk}}$.

2.3. The Berkovich spectrum of a normed ring. In this section, we explain a general construction which associates a Berkovich analytic space to an arbitrary normed ring.

2.3.1. *Seminorms and norms.* A *seminorm* on a ring $\mathcal{A}$ is a function $| \ | : \mathcal{A} \to \mathbb{R}_{\geq 0}$ with values in the set of nonnegative reals such that for every $f, g \in \mathcal{A}$, we have

(S1) $|0| = 0, |1| = 1$.
(S2) $|f + g| \leq |f| + |g|$.
(S3) $|f \cdot g| \leq |f| \cdot |g|$.

A seminorm $| \ |$ defines a topology on $\mathcal{A}$ in the usual way, and this topology is Hausdorff if and only if $| \ |$ is a *norm*, meaning that $|f| = 0$ if and only if $f = 0$.

A *normed ring* is a pair $(\mathcal{A}, \| \ \|)$ consisting of a ring $\mathcal{A}$ and a norm $\| \ \|$. It is called a *Banach ring* if $\mathcal{A}$ is complete with respect to this norm. Any ring may be regarded as a Banach ring with respect to the *trivial norm*, for which $\|0\| = 0$ and $\|f\| = 1$ for $f \neq 0$.

A seminorm $| \ |$ on a ring $\mathcal{A}$ is called *multiplicative* if for all $f, g \in \mathcal{A}$, we have

(S3)$'$ $|f \cdot g| = |f| \cdot |g|$,

and it is called *non-archimedean* if

(S2)$'$ $|f + g| \leq \max\{|f|, |g|\}$.

A multiplicative norm on a ring $\mathcal{A}$ is also called an *absolute value* on $\mathcal{A}$.

A seminorm $| \ |$ on a normed ring $(\mathcal{A}, \| \ \|)$ is called *bounded* if

(S4) there exists a constant $C > 0$ such that $|f| \leq C\|f\|$ for all $f \in \mathcal{A}$.

LEMMA 2.3.1. *If* $| \ |$ *is a multiplicative seminorm, then condition (S4) is equivalent to:*

(S4)$'$ $|f| \leq \|f\|$ *for all* $f \in \mathcal{A}$.

PROOF. Since $|f^n| \leq C\|f^n\| \leq C\|f\|^n$, we have $|f| \leq \sqrt[n]{C}\|f\|$ for all $n \geq 1$. Passing to the limit as n tends to infinity yields the desired result. $\square$

2.3.2. *The Berkovich spectrum of a normed ring.* Let $(\mathcal{A}, \| \ \|)$ be a normed ring. We define a topological space $\mathcal{M}(\mathcal{A})$, called the *Berkovich spectrum* of $\mathcal{A}$, as follows. As a set, $\mathcal{M}(\mathcal{A})$ consists of all bounded multiplicative seminorms on $\mathcal{A}$. The topology on $\mathcal{M}(\mathcal{A})$ (which we will call the *Berkovich topology*[7]) is defined to be the weakest one for which all functions of the form $| \ | \mapsto |f|$ for $f \in \mathcal{A}$ are continuous.

It is useful from a notational standpoint to denote points of $X = \mathcal{M}(\mathcal{A})$ by a letter such as x, and the corresponding bounded multiplicative seminorm by $| \ |_x$. With this notation, a sub-base of open neighborhoods for the topology on X is given by the collection

$$U(f, \alpha, \beta) \ = \ \{x \in X : \alpha < |f|_x < \beta\}$$

for all $f \in \mathcal{A}$ and all $\alpha < \beta$ in $\mathbb{R}$.

Equivalently, one may define the topology on $\mathcal{M}(\mathcal{A})$ as the *topology of pointwise convergence*: a net[8] $\langle x_\alpha \rangle$ in $\mathcal{M}(\mathcal{A})$ converges to $x \in \mathcal{M}(\mathcal{A})$ if and only if $|f|_{x_\alpha}$ converges to $|f|_x$ in $\mathbb{R}$ for all $f \in \mathcal{A}$.

THEOREM 2.3.2. *If $\mathcal{A}$ is a nonzero Banach ring, then the spectrum $\mathcal{M}(\mathcal{A})$ is a non-empty compact Hausdorff space.*

PROOF. This is proved in Theorem 1.2.1 of [**Ber90**]. The fact that $\mathcal{M}(\mathcal{A})$ is Hausdorff is an easy exercise. The proof that $\mathcal{M}(\mathcal{A})$ is always non-empty is rather subtle, though. (In many cases of interest, however, such as when the norm on $\mathcal{A}$ is multiplicative, the fact that $\mathcal{M}(\mathcal{A})$ is non-empty is obvious.)

Here is a quick proof, different from the one in [**Ber90**], of the compactness of $\mathcal{M}(\mathcal{A})$. It suffices by general topology to prove that every net in $X = \mathcal{M}(\mathcal{A})$ has a convergent subnet. Let T be the space $\prod_{f \in \mathcal{A}} [0, \|f\|]$ endowed with the product topology. By Tychonoff's theorem, T is compact. By Lemma 2.3.1, there is a natural map $\iota : X \to T$ sending $x \in X$ to $(|f|_x)_{f \in \mathcal{A}}$, and ι is clearly injective and continuous.

Let $\langle x_\alpha \rangle$ be a net in X. Since T is compact, $\langle \iota(x_\alpha) \rangle$ has a subnet $\langle \iota(y_\beta) \rangle$ converging to an element $(\alpha_f)_{f \in \mathcal{A}} \in T$. Define a function $| \cdot |_y : \mathcal{A} \to \mathbb{R}_{\geq 0}$ by $|f|_y = \alpha_f$. It is easily verified that $| \cdot |_y$ is a bounded multiplicative seminorm on $\mathcal{A}$, and thus defines a point $y \in X$. By construction, we have $\iota(y) = \lim_\beta \iota(y_\beta)$. This implies that $\lim_\beta |f|_{y_\beta} = |f|_y$ for all $f \in \mathcal{A}$, i.e., $y_\beta \to y$. Thus $\langle x_\alpha \rangle$ has a convergent subnet as desired. $\square$

2.4. The analytification of an algebraic variety. As discussed in [**Ber90**, §3.4.1] and [**Ber93**, §2.6] (see also [**Duc06**, §1.4] and [**Con07**]), one can associate in a functorial way to every algebraic variety X/K a locally ringed topological space X_{Berk} called the *Berkovich K-analytic space* associated to X. A Berkovich

---

[7]This topology is also referred to as the *Gelfand topology.*

[8]Recall that a *net* in a topological space X is a mapping from a directed set I to X, with a sequence being the special case where $I = \mathbb{N}$. For non-metrizable topological spaces, nets are much better than sequences for describing the interplay between concepts like convergence and continuity. The space $\mathbb{A}^1_{\text{Berk}, K}$ (or equivalently $\mathbb{P}^1_{\text{Berk}, K}$) is metrizable if and only if the residue field $\tilde{K}$ of K is countable.

K-analytic space is also endowed with an additional structure, called a K-*affinoid atlas*, which is crucial for gluing constructions, and for defining the general concept of a *morphism* in the category of K-analytic spaces.

We will refer to the functor from algebraic varieties over K to Berkovich K-analytic spaces as the *Berkovich analytification functor*. When $X = \mathrm{Spec}(A)$ is affine, the underlying topological space of X_{Berk} is the set of multiplicative semi-norms on A which extend the given absolute value on K, equipped with the weakest topology for which all functions of the form $| \ | \mapsto |f|$ for $f \in A$ are continuous. When $X = \mathbb{A}^1$ (resp. $\mathbb{P}^1$), we recover the definition of $\mathbb{A}^1_{\mathrm{Berk}}$ (resp. $\mathbb{P}^1_{\mathrm{Berk}}$) given above. The space X_{Berk} is locally compact and Hausdorff, and if X is proper then X_{Berk} is compact. Moreover, if X is connected in the Zariski topology then X_{Berk} is path-connected. Finally, there is a canonical embedding of $X(K)$ (endowed with its totally disconnected analytic topology) as a dense subspace of X_{Berk}.

As a concrete example, let X be a smooth, proper, and geometrically integral algebraic curve over K. We briefly describe the topological structure of X_{Berk}; further details will be given in §5.1.

A *finite topological graph* is just a finite connected graph whose edges are thought of as line segments; this is essentially the same thing as a connected one-dimensional CW-complex with finitely many cells. If the genus of X is at least one, there is a canonically defined subset $\Sigma \subset X_{\mathrm{Berk}}$, called the *skeleton* of X_{Berk}, which is homeomorphic to a finite topological graph. Moreover, the entire space X_{Berk} admits a deformation retraction r onto Σ. (In the case $X = \mathbb{P}^1$, there is no canonical skeleton, but after choosing coordinates, we can if we like think of the skeleton of $\mathbb{P}^1_{\mathrm{Berk}}$ as the Gauss point ζ_{Gauss}.) A useful fact which will be discussed in §5 is that the skeleton of X_{Berk} can be equipped with a canonical metric.

For each $x \in \Sigma$, the fiber $r^{-1}(x)$ is homeomorphic to a compact, connected subset of $\mathbb{P}^1_{\mathrm{Berk}}$, and in particular is a topological tree. Using this, one can define a notion of a "finite subgraph" of X_{Berk} in such a way that X_{Berk} is homeomorphic to the inverse limit of its finite subgraphs (see §5.1 below).

More generally, Berkovich proves in [**Ber99**] and [**Ber04**] that every smooth K-analytic space (for example, the analytification of a smooth projective variety over K) is locally contractible. This is a very difficult result which relies, among other things, on de Jong's theory of *alterations*, and we will not discuss the higher-dimensional case any further in these notes. See [**Duc06**, §2] for a nice overview of this and many other aspects of Berkovich's theory.

2.5. The Berkovich space $\mathcal{M}(\mathbb{Z})$. We now consider a simple but interesting example of the construction from (2.3.2): the Berkovich analytic space $\mathcal{M}(\mathbb{Z})$ associated to the normed ring $(\mathbb{Z}, | \ |_\infty)$, where $| \ |_\infty$ denotes the usual archimedean absolute value on $\mathbb{Z}$.

A famous result of Ostrowski asserts that every non-trivial absolute value on $\mathbb{Q}$ is equivalent to either $| \ |_\infty$, or to the standard p-adic absolute value $| \ |_p$ for some prime number p. (We normalize $| \ |_p$ in the usual way so that $|p|_p = \frac{1}{p}$.)

Thus, if we let $M_\mathbb{Q}$ denote the set of *places* (equivalence classes of non-trivial absolute values) of $\mathbb{Q}$, then there is a bijection

$$M_\mathbb{Q} \leftrightarrow \{\text{prime numbers } p\} \cup \{\infty\}.$$

With this notation, the *product formula* states that if $\alpha \in \mathbb{Q}^*$ is a non-zero rational number, then

$$\prod_{v \in M_{\mathbb{Q}}} |\alpha|_v = 1.$$

Following Berkovich[9], one can classify all *multiplicative seminorms* on $\mathbb{Z}$ as follows:

(Z1) The *p-trivial seminorms* $|\ |_{p,\infty}$ defined by

$$|n|_{p,\infty} = \left\{ \begin{array}{ll} 0 & p \mid n, \\ 1 & p \nmid n. \end{array} \right.$$

(Z2) The *trivial seminorm* $|\ |_0$ defined by

$$|n|_0 = \left\{ \begin{array}{ll} 0 & n = 0, \\ 1 & n \neq 0. \end{array} \right.$$

(Z3) The *p-adic absolute values* $|\ |_{p,\epsilon}$ for $0 < \epsilon < \infty$ defined by

$$|n|_{p,\epsilon} = |n|_p^\epsilon.$$

(Z4) The *archimedean absolute values* $|\ |_{\infty,\epsilon}$ for $0 < \epsilon \leq 1$ defined by

$$|n|_{\infty,\epsilon} = |n|_\infty^\epsilon.$$

All of the seminorms in (Z1)-(Z4) are clearly bounded. Moreover:

LEMMA 2.5.1. *For all $n \in \mathbb{Z}$, we have:*

(1) $\lim_{\epsilon \to 0} |n|_{\infty,\epsilon} = |n|_0$.
(2) $\lim_{\epsilon \to \infty} |n|_{p,\epsilon} = |n|_{p,\infty}$.
(3) $\lim_{\epsilon \to 0} |n|_{p,\epsilon} = |n|_0$.

We will therefore write $|\ |_{\infty,0}$ or $|\ |_{p,0}$ instead of $|\ |_0$ when convenient.

This leads to the following visual representation of the Berkovich analytic space $\mathcal{M}(\mathbb{Z})$ associated to $\mathbb{Z}$:

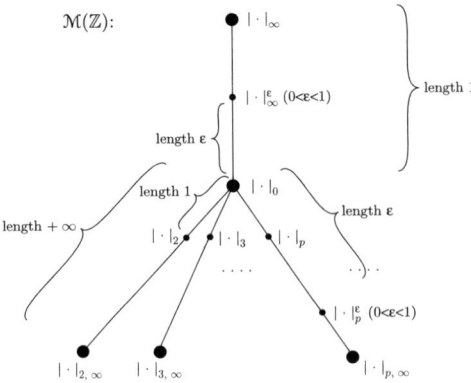

FIGURE 2. The space $\mathcal{M}(\mathbb{Z})$.

---

[9]See [**Ber90**, Example 1.4.1], although our notation is slightly different

Note that the different "tangent directions" emanating from the trivial semi-norm $|\ |_0$ are in one-to-one correspondence with the places of $\mathbb{Q}$. We will return to this observation later when we discuss harmonic functions and Laplacians.

Recall that the Berkovich topology on $\mathcal{M}(\mathbb{Z})$ is defined to be the weakest one for which the function $x \mapsto |n|_x$ is continuous for all $n \in \mathbb{Z}$. This can be described concretely as follows: each of the subsets

$$\ell_\infty = \{|\ |_0\} \cup \{|\ |_{\infty,\epsilon}\}_{0<\epsilon\leq 1} = \{|\ |_{\infty,\epsilon}\}_{0\leq\epsilon\leq 1}$$

and

$$\ell_p = \{|\ |_{p,\infty}\} \cup \{|\ |_{p,\epsilon}\}_{0<\epsilon<\infty} \cup \{|\ |_0\} = \{|\ |_{p,\epsilon}\}_{0\leq\epsilon\leq\infty}$$

is homeomorphic to a real interval, and the open neighborhoods of the trivial semi-norm $|\ |_0$ are the subsets U of $\mathcal{M}(\mathbb{Z})$ containing $|\ |_0$ for which:

(1) $U \cap \ell_v$ is open in ℓ_v for all $v \in M_\mathbb{Q}$.
(2) $U \cap \ell_v = \ell_v$ for all but finitely many $v \in M_\mathbb{Q}$.

It is a simple exercise to verify directly using this description of the topology that $\mathcal{M}(\mathbb{Z})$ is path-connected, compact, and Hausdorff.

If we identify the segment ℓ_∞ with the real interval $[0,1]$ via the association

$$|\ |_{\infty,\epsilon} \mapsto \epsilon$$

and the segment ℓ_p with the extended-real interval $[0,\infty]$ via

$$|\ |_{p,\epsilon} \mapsto \epsilon,$$

then the complement $\mathbf{H}_\mathbb{Z}$ in $\mathcal{M}(\mathbb{Z})$ of all points of type (Z1) becomes a metric space. We let ρ denote the corresponding metric.

REMARK 2.5.2. 1. The points of $\mathcal{M}(\mathbb{Z})$ having distance 1 from the trivial seminorm $|\ |_0$ are precisely the points corresponding to the standard absolute values $|\ |_p = |\ |_{p,1}$ and $|\ |_\infty = |\ |_{\infty,1}$.

2. If we extend ρ to a degenerate metric on all of $\mathcal{M}(\mathbb{Z})$, then a point x of type (Z1) is infinitely far away from every point $y \in \mathcal{M}(\mathbb{Z})$ distinct from x.

REMARK 2.5.3. Like $\mathbb{P}^1_{\mathrm{Berk}}$, the space $\mathcal{M}(\mathbb{Z})$ can be viewed as an inverse limit of finite graphs. Indeed, define a *finite subgraph* of $\mathcal{M}(\mathbb{Z})$ to be the "convex hull" (in the obvious sense) of finitely many points of $\mathbf{H}_\mathbb{Z}$, endowed with the usual Euclidean topology on a finite union of real segments. The collection $\mathcal{S}$ of all such finite subgraphs $\Gamma \subseteq \mathcal{M}(\mathbb{Z})$ forms an inverse system with respect to the natural retraction maps $r_{\Gamma',\Gamma} : \Gamma' \to \Gamma$ (defined whenever $\Gamma \subseteq \Gamma'$), and one can show that $\mathcal{M}(\mathbb{Z})$ is homeomorphic to the inverse limit $\varprojlim_{\Gamma\in\mathcal{S}} \Gamma$.

Equipping each finite subgraph $\Gamma \in \mathcal{S}$ with the metric induced by ρ, the space $\mathcal{M}(\mathbb{Z}) = \varprojlim_{\Gamma\in\mathcal{S}}$ becomes a profinite $\mathbb{R}$-tree, with $\mathbf{H}_\mathbb{Z} \cong \varinjlim_{\Gamma\in\mathcal{S}} \Gamma$ in the locally metric topology.

3. Harmonic functions

In this lecture, we explore the notion of a *harmonic function* in the context of the spaces $\mathcal{M}(\mathbb{Z})$ and $\mathbb{P}^1_{\mathrm{Berk}}$. We will also discuss the related notion of a *subharmonic function* on $\mathbb{P}^1_{\mathrm{Berk}}$.

By a *measure* on a space X, we will always mean a signed Borel measure on X.

3.1. Harmonic functions on $\mathcal{M}(\mathbb{Z})$. It is possible to give a natural definition of a "harmonic function" on $\mathcal{M}(\mathbb{Z})$, using the metric ρ introduced in §2.5.

We introduce the following convenient notation for points of $\mathcal{M}(\mathbb{Z})$:

$\zeta_{p,\infty}$: the point of $\mathcal{M}(\mathbb{Z})$ corresponding to $|\ |_{p,\infty}$.

ζ_0: the point of $\mathcal{M}(\mathbb{Z})$ corresponding to $|\ |_0$.

$\zeta_{p,\epsilon}$: the point of $\mathcal{M}(\mathbb{Z})$ corresponding to $|\ |_{p,\epsilon}$.

$\zeta_{\infty,\epsilon}$: the point of $\mathcal{M}(\mathbb{Z})$ corresponding to $|\ |_{\infty,\epsilon}$.

ζ_v: the point $\zeta_{p,1}$ if $v \in M_{\mathbb{Q}}$ is a non-archimedean place corresponding to the prime p, or the point $\zeta_\infty = \zeta_{\infty,1}$ if $v \in M_{\mathbb{Q}}$ is the archimedean place.

As in §1.4.3, for x in $\mathcal{M}(\mathbb{Z})$, we define the set T_x of *tangent directions* at x to be the connected components of $\mathcal{M}(\mathbb{Z})\backslash\{x\}$. When $x = \zeta_0$ is the point corresponding to the trivial seminorm $|\ |_0$ on $\mathbb{Z}$, there is a canonical bijection between T_x and the set $M_{\mathbb{Q}}$ of places of $\mathbb{Q}$; at all other points of $\mathcal{M}(\mathbb{Z})$, the space T_x has cardinality 1 or 2. For $v \in M_{\mathbb{Q}}$, we will refer to the segments ℓ_v defined above as the "branches emanating from ζ_0".

Recall also from §2.5 that $\mathbf{H}_{\mathbb{Z}}$ denotes the complement of the points of type (Z1); the points of $\mathbf{H}_{\mathbb{Z}}$ are precisely the ones at finite distance from the trivial point ζ_0 with respect to the metric ρ.

Let U be a connected open subset of $\mathcal{M}(\mathbb{Z})$ (with respect to the Berkovich topology), and let $f : U \to \mathbb{R}\cup\{\pm\infty\}$ be a continuous extended-real valued function which is finite-valued on $U\cap\mathbf{H}_{\mathbb{Z}}$. For expositional simplicity, we assume that $\zeta_0 \in U$ (which is the main case of interest, since the connected components of $\mathcal{M}(\mathbb{Z})\backslash\{\zeta_0\}$ are homeomorphic to segments in $\mathbb{R}$, and one already knows how to define the Laplacian on $\mathbb{R}$; in our terminology it is just $-f''(x)dx$).

We say that f is *continuous piecewise affine* on U, and write $f \in \mathrm{CPA}(U)$, if f is (i) continuous, (ii) piecewise-affine along each branch of $\mathcal{M}(\mathbb{Z})$ emanating from ζ_0, and (iii) constant on all but finitely many branches emanating from ζ_0. These conditions guarantee that if $f \in \mathrm{CPA}(U)$ and $x \in U \cap \mathbf{H}_{\mathbb{Z}}$, then the directional derivative $d_{\vec{v}}f(x)$ is well-defined for all $\vec{v} \in T_x$, and $d_{\vec{v}}f(x) = 0$ for all but finitely many $\vec{v} \in T_x$. Thus for all $x \in U \cap \mathbf{H}_{\mathbb{Z}}$ the quantity

$$\Delta_x(f) := -\sum_{v \in T_x} d_{\vec{v}}f(x)$$

is well-defined.

Let $x \in U$, and let $h \in \mathrm{CPA}(U)$.

DEFINITION 3.1.1. 1. If $x \in \mathbf{H}_{\mathbb{Z}}$, we say that h is *harmonic at* x if $\Delta_x(h) = 0$.

2. If x is of type (Z1), we say that h is *harmonic at* x if h is constant on an open neighborhood of x.

EXAMPLE 3.1.2. Let $n \in \mathbb{Z}$ be a nonzero integer, let $S_0 = \{\zeta_{p,\infty} : p \mid n\}$, and let $S = S_0 \cup \{\zeta_\infty\}$.

Define

$$F_n(x) = \begin{cases} +\infty & x \in S_0, \\ -\log|n|_x & x \in \mathcal{M}(\mathbb{Z})\backslash S_0. \end{cases}$$

Claim: $F_n(x)$ is continuous piecewise affine and is harmonic outside S.

To see this, first note that if Λ denotes the smallest connected subset of $\mathcal{M}(\mathbb{Z})$ containing all the points of S, then Λ is finitely branched and there is a natural retraction map $r_\Lambda : \mathcal{M}(\mathbb{Z}) \twoheadrightarrow \Lambda$. Along the branch Λ_v of $\mathcal{M}(\mathbb{Z})$ emanating from ζ_0 in the tangent direction corresponding to $v \in M_\mathbb{Q}$, the function $F_n(x)$ is linear with slope equal to $-\log|n|_v$. In particular, $F_n(x)$ is locally constant off Λ: for all $x \in \mathcal{M}(\mathbb{Z})$, we have $F_n(x) = F_n(r_\Lambda(x))$. It follows from this that $F_n(x)$ is harmonic at all points $x \notin S \cup \{\zeta_0\}$. Finally, the fact that $F_n(x)$ is harmonic at ζ_0 is equivalent to the *product formula* for $\mathbb{Q}$:

$$\Delta_{\zeta_0}(F_n) = -\sum_{\vec{v} \in T_{\zeta_0}} d_{\vec{v}} F_n(\zeta_0) = \sum_{v \in M_\mathbb{Q}} \log|n|_v = 0.$$

If we think of $n \neq \pm 1$ as an analytic function on $\mathcal{M}(\mathbb{Z})$, of S_0 as the set of "zeros" of n, and of ζ_∞ as the unique "pole"[10] of n, then this example can be rephrased, by analogy with the classical situation over $\mathbb{C}$, as saying that the function $-\log|n|$ on $\mathcal{M}(\mathbb{Z})$ is harmonic outside the zeros and poles of n.

3.2. Harmonic functions on $\mathbb{P}^1_{\mathrm{Berk}}$. In this section, we define what it means for a real-valued function on $\mathbb{P}^1_{\mathrm{Berk}}$ to be harmonic. This is somewhat more complicated than the corresponding notion for $\mathcal{M}(\mathbb{Z})$ discussed in §3.1, since the branching behavior of $\mathbb{P}^1_{\mathrm{Berk}}$ is much more complicated than that of $\mathcal{M}(\mathbb{Z})$.

We recall from §1.4.3 that if $x \in \mathbb{P}^1_{\mathrm{Berk}}$, there is a well-defined set T_x of *tangent directions* at x, and the tangent directions at x are in one-to-one correspondence with the connected components of $\mathbb{P}^1_{\mathrm{Berk}} \backslash \{x\}$.

Let U be a connected open subset of $\mathbb{P}^1_{\mathrm{Berk}}$, and let $f : U \to \mathbb{R} \cup \{\pm\infty\}$ be a continuous extended-real valued function which is finite-valued on $\mathbf{H}_{\mathrm{Berk}} = \mathbb{P}^1_{\mathrm{Berk}} \backslash \mathbb{P}^1(K)$.

We say that f is *continuous piecewise affine* on U, and write $f \in \mathrm{CPA}(U)$ if:

(CPA1) The restriction of f to $\mathbf{H}_{\mathrm{Berk}}$ is piecewise-affine with respect to the path metric ρ; concretely, this means that for each $x \in \mathbf{H}_{\mathrm{Berk}}$ and each sufficiently small path $\Lambda = \ell_{x,y}$ emanating from x, the restriction of f to Λ is affine.

(CPA2) If $f \in \mathrm{CPA}(U)$ and $x \in U \cap \mathbf{H}_{\mathrm{Berk}}$, then for each $\vec{v} \in T_x$ the *directional derivative* $d_{\vec{v}} f(x)$ is well-defined. Concretely, this means that for each $\vec{v} \in T_x$, there exists a constant $m_{\vec{v}} = d_{\vec{v}} f(x)$ such that for every $y \in \mathbf{H}_{\mathrm{Berk}}$ representing the tangent direction $\vec{v}$, there exists a point $y' \in (x, y]$ such that for every $z \in (x, y']$ we have

$$f(z) = f(x) + m_{\vec{v}} \rho(x, z).$$

(CPA3) For each $x \in U \cap \mathbf{H}_{\mathrm{Berk}}$, we have $d_{\vec{v}} f(x) = 0$ for all but finitely many $\vec{v} \in T_x$. In particular, the quantity

(3.2.1) $$\Delta_x(f) := -\sum_{v \in T_x} d_{\vec{v}} f(x)$$

is well-defined for each $x \in U \cap \mathbf{H}_{\mathrm{Berk}}$.

DEFINITION 3.2.2. Let $x \in U$, and let $h \in \mathrm{CPA}(U)$.

---

[10]Somewhat peculiarly, it seems that the point ζ_∞ should be thought of as a pole of n, despite the fact that $-\log|n|_\infty$ is finite-valued, because the function $x \mapsto -\log|n|_x$ is not locally constant near ζ_∞; see Example 4.5.8 below for another explanation.

1. If $x \in \mathbf{H}_{\mathrm{Berk}}$, we say that h is *harmonic at x* if $\Delta_x(h) = 0$. In other words, a function $h \in \mathrm{CPA}(U)$ is harmonic at a point $x \in \mathbf{H}_{\mathrm{Berk}}$ if the sum of the slopes of h in all tangent directions emanating from x is zero.

2. If $x \in \mathbb{P}^1(K)$, we say that h is *harmonic at x* if h is constant on an open neighborhood of x.

EXAMPLE 3.2.3. Consider the function $G : \mathbb{P}^1_{\mathrm{Berk}} \to \mathbb{R} \cup \{+\infty\}$ defined by

$$G(x) = \begin{cases} +\infty & x = \infty, \\ \log_v \max(|T|_x, 1) & x \in \mathbb{A}^1_{\mathrm{Berk}}, \end{cases}$$

whose restriction to K is the function $\log_v^+ |x| = \log_v \max(|x|, 1)$. Let $\Lambda = \ell_{\zeta_{\mathrm{Gauss}}, \infty}$ be the closed path from ζ_{Gauss} to ∞ in $\mathbb{P}^1_{\mathrm{Berk}}$, and let $r_\Lambda : \mathbb{P}^1_{\mathrm{Berk}} \twoheadrightarrow \Lambda$ be the natural retraction map from $\mathbb{P}^1_{\mathrm{Berk}}$ onto Λ. Recall that if $x = \zeta_{a,r} \in \mathbf{H}^{\mathbb{R}}_{\mathrm{Berk}}$, then

$$|T|_x = \sup_{z \in B(a,r)} |z|.$$

From this, one deduces easily:

- $G(x)$ is linear with slope 1 along Λ, i.e., $G(x) = \rho(\zeta_{\mathrm{Gauss}}, x)$.
- $G(x)$ is locally constant off Λ, i.e., for all $x \in \mathbb{P}^1_{\mathrm{Berk}}$, we have $G(x) = G(r_\Lambda(x))$.

It follows that $G \in \mathrm{CPA}(\mathbb{P}^1_{\mathrm{Berk}})$ and that G is harmonic on $\mathbb{P}^1_{\mathrm{Berk}} \backslash \{\zeta_{\mathrm{Gauss}}, \infty\}$, but is not harmonic at ζ_{Gauss} or ∞. For example, the sum of the slopes of G in all directions emanating from ζ_{Gauss} is 1: in the direction heading up to infinity the slope is 1, and in all other directions the slope is 0.

As an immediate consequence of the definition of harmonic functions, we have:

LEMMA 3.2.4. *If h_1, h_2 are harmonic on U and $c_1, c_2 \in \mathbb{R}$, then $c_1 h_1 + c_2 h_2$ is harmonic on U.*

As an application of Lemma 3.2.4, we discuss the following example.

EXAMPLE 3.2.5. Let $f(T) = \prod_{i=1}^n (x - a_i) \in K[T]$ be a nonconstant polynomial, and let

$$F(x) = \begin{cases} -\infty & x = \infty, \\ +\infty & x \in \{a_1, \ldots, a_n\}, \\ -\log_v |f|_x & x \in \mathbb{P}^1_{\mathrm{Berk}} \backslash \{\infty, a_1, \ldots, a_n\} \end{cases}$$

be the unique continuous function on $\mathbb{P}^1_{\mathrm{Berk}}$ extending the function $-\log_v |f(x)|$ on K.

Claim: $F(x)$ is harmonic outside $\{\infty, a_1, \ldots, a_n\}$.

Indeed, as far as type I points go, it follows from the ultrametric inequality that if $x \in K \backslash \{a_1, \ldots, a_n\}$, then $|f(x)|$ is constant on every disk around a not containing $a_1, \ldots, a_n$. Since F is continuous and K is dense in $\mathbb{A}^1_{\mathrm{Berk}}$, it follows that F is constant on a Berkovich open disk $\mathcal{B}(a, r)^-$ containing a.

It remains to see why F is harmonic on $\mathbf{H}_{\mathrm{Berk}}$. First, we consider the special case in which $f(T) = T - a$. In this case, if $\Lambda_a = \ell_{a,\infty}$ denotes the unique path in $\mathbb{P}^1_{\mathrm{Berk}}$ from a to ∞ and $F_a = -\log_v |T - a|_x$, then we have:

- $F_a(x)$ is linear with slope -1 along (a, ∞).
- $F_a(x)$ is locally constant off Λ_a, i.e., for all $x \in \mathbb{P}^1_{\mathrm{Berk}}$, we have $F_a(x) = F_a(r_{\Lambda_a}(x))$.

It follows in this special case that $F_a \in \mathrm{CPA}(\mathbb{P}^1_{\mathrm{Berk}})$, and that F_a is harmonic on $\mathbb{P}^1_{\mathrm{Berk}}\backslash\{\infty, a\}$.

In the general case we have $F(x) = \sum_{i=1}^{n} F_a(x)$, and it follows from Lemma 3.2.4 that F is harmonic outside $\{\infty, a_1, \ldots, a_n\}$, as claimed.

3.3. Properties of harmonic functions on $\mathbb{P}^1_{\mathrm{Berk}}$. By a *domain* in $\mathbb{P}^1_{\mathrm{Berk}}$, we will mean a connected open subset of $\mathbb{P}^1_{\mathrm{Berk}}$. In this section, we present a selection of results from Chapter 7 of [**BR08**] concerning harmonic functions on domains in $\mathbb{P}^1_{\mathrm{Berk}}$

3.3.1. *The maximum principle.* The following result is the Berkovich space analogue of the classical maximum principle for harmonic functions on domains in $\mathbb{C}$:

PROPOSITION 3.3.1 (Maximum Principle).
  (1) *If h is a nonconstant harmonic function on a domain $U \subset \mathbb{P}^1_{\mathrm{Berk}}$, then h does not achieve a maximum or a minimum value on U.*
  (2) *If h is a harmonic function on a domain $U \subset \mathbb{P}^1_{\mathrm{Berk}}$ which extends continuously to the closure $\bar{U}$ of U, then h achieves both its minimum and maximum values on the boundary ∂U of U.*

Recall from Lemma 2.2.3 that a *simple domain* in $\mathbb{P}^1_{\mathrm{Berk}}$ is a connected open set $U \subseteq \mathbb{P}^1_{\mathrm{Berk}}$ whose boundary is a finite subset of $\mathbf{H}^{\mathbb{R}}_{\mathrm{Berk}}$. One can show (see §3.3.2 below) that every harmonic function on a simple domain U extends continuously to $\bar{U}$. If $U = \mathbb{P}^1_{\mathrm{Berk}}$ (resp. U is a Berkovich open disk), then ∂U is empty (resp. consists of a single point). By the second part of the Maximum Principle, we therefore conclude:

COROLLARY 3.3.2. *If $U = \mathbb{P}^1_{\mathrm{Berk}}$ or U is an open Berkovich disk, then every harmonic function on U is constant.*

The conclusion of Corollary 3.3.2 can be better understood through the observation that the behavior of a harmonic function on a domain U in $\mathbb{P}^1_{\mathrm{Berk}}$ is controlled by its behavior on a certain special subset.

DEFINITION 3.3.3. If U is a domain in $\mathbb{P}^1_{\mathrm{Berk}}$, the *main dendrite* $D(U) \subset U$ is the set of all $x \in U$ belonging to paths between boundary points $y, z \in \partial U$.

The main dendrite of a domain U is empty if and only if U has at most one boundary point, which happens precisely in the following three cases:
  • $U = \mathbb{P}^1_{\mathrm{Berk}}$.
  • $U \cong \mathbb{P}^1_{\mathrm{Berk}}\backslash\{a\}$ for some point a of type I or IV.
  • U is an open Berkovich disk.

EXAMPLE 3.3.4. If $U = \mathcal{B}(a, R)^-\backslash\mathcal{B}(a, r)$ is a Berkovich open annulus (see Figure 3), then $D(U)$ is the open segment joining the two boundary points $\zeta_{a,r}$ and $\zeta_{a,R}$ of U.

EXAMPLE 3.3.5. If $K = \mathbb{C}_p$ and $U = \mathbb{P}^1_{\mathrm{Berk}}\backslash\mathbb{P}^1(\mathbb{Q}_p)$, then the main dendrite $D(U)$ is a locally finite real tree in which the set of branch points is discrete, and every branch point has degree $p + 1$. In fact, $D(U)$ can be identified with the (geometric realization of the) *Bruhat-Tits tree* associated to $\mathrm{PGL}(\mathbb{Q}_p)$ (see [**FvdP04**, Definition 4.9.3] or [**DT07**]).

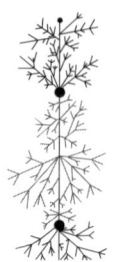

FIGURE 3. A Berkovich open annulus.

When $D(U)$ is non-empty, it is not hard to see that there is a natural retraction map $r_{U,D(U)} : U \twoheadrightarrow D(U)$. The following result is proved in Chapter 7 of [**BR08**]:

PROPOSITION 3.3.6. *Let U be a domain in $\mathbb{P}^1_{\mathrm{Berk}}$.*

  (1) *If the main dendrite $D(U)$ of a domain U is nonempty, then it is finitely branched at each point of $\mathbf{H}^{\mathbb{R}}_{\mathrm{Berk}}$.*

  (2) *Let h be harmonic in a domain U. If the main dendrite is empty, then h is constant; otherwise, h is constant on branches off the main dendrite, i.e., $h = h \circ r_{U,D(U)}$.*

3.3.2. *The Poisson formula.* In the classical theory of harmonic functions in the complex plane, if f is harmonic on an open disk V then it has a continuous extension to the closure of V, and the Poisson Formula expresses the values of f on V in terms of its values on the boundary of V.

Specifically, if $V \subseteq \mathbb{C}$ is an open disk of radius r centered at z_0, and if f is harmonic in V, then f extends continuously to $\bar{V}$ and $f(z_0) = \int_{\partial V} f \, d\mu_V$, where μ_V is the uniform probability measure $d\theta/2\pi$ on the boundary circle ∂V. More generally, for any $z \in V$ there is a measure $\mu_{z,V}$ depending only on z and V, called the *Jensen-Poisson measure*, for which

$$f(z) = \int_{\partial V} f \, d\mu_{z,V}$$

for every harmonic function f on $\bar{V}$. We seek to generalize this type of formula to the Berkovich projective line.

In $\mathbb{P}^1_{\mathrm{Berk}}$, the basic open neighborhoods are the *simple domains*. A simple domain has only a finite number of boundary points (cf. Lemma 2.2.3), and its main dendrite is the interior of a finite subgraph Γ of $\mathbb{P}^1_{\mathrm{Berk}}$. As we will see, every harmonic function f on a simple domain V has a continuous extension to its closure, and there is an analogue of the Jensen-Poisson measure which yields an explicit formula for f in terms of its values on the boundary. In other words, one can explicitly solve the Berkovich space analogue of the Dirichlet problem on any simple domain (using, as we will see, only simple linear algebra).

Recall from §1.6.2 that $\kappa_z(x,y) = -\log_v[x,y]_z$ denotes the fundamental potential kernel on $\mathbb{P}^1_{\mathrm{Berk}}$ relative to the point z.

Let V be a simple domain in $\mathbb{P}^1_{\mathrm{Berk}}$ with boundary points $x_1, \ldots, x_m \in \mathbf{H}^{\mathbb{R}}_{\mathrm{Berk}}$. For $z \in V$, let $C(z)$ be the $m \times m$ matrix whose ij^{th} entry is $\kappa_z(x_i, x_j)$. Define a *probability vector* on $\mathbb{R}^m$ to be a vector $[p_1, \ldots, p_m]^T \in \mathbb{R}^m$ such that $p_i \geq 0$ for $1 \leq i \leq m$ and $p_1 + \cdots + p_m = 1$.

PROPOSITION 3.3.7. *For each $z \in V$, there is a unique probability vector $\vec{p}(z) = [p_1(z), p_2(z), \ldots, p_m(z)]^T \in \mathbb{R}^m$ such that $C(z) \cdot \vec{p}(z)$ is a scalar multiple of $[1, 1, \ldots, 1]^T$.*

For each $1 \leq i \leq m$, define the function $h_i : V \to \mathbb{R}$, called the ith *harmonic measure* with respect to V, by setting $h_i(z) = p_i(z)$. By construction, we have $0 \leq h_i(z) \leq 1$ for all $z \in V$ and $h_1 + \cdots + h_m \equiv 1$ on V.

Explicitly, let

$$M(z) = \begin{pmatrix} 0 & 1 & \cdots & 1 \\ 1 & \kappa_z(x_1, x_1) & \cdots & \kappa_z(x_1, x_m) \\ \vdots & \vdots & \ddots & \vdots \\ 1 & \kappa_z(x_m, x_1) & \cdots & \kappa_z(x_m, x_m) \end{pmatrix}$$

and for each $i = 0, 1, \ldots, m$, let $M_i(z)$ be the matrix obtained by replacing the ith column of $M(z)$ by $[1, 0, \ldots, 0]^T$. If $C(z) \cdot \vec{p}(z) = [-\nu, \ldots, -\nu]^T$, then

$$M(z) \begin{bmatrix} \nu \\ p_1(z) \\ \vdots \\ p_m(z) \end{bmatrix} = \begin{bmatrix} 1 \\ 0 \\ \vdots \\ 0 \end{bmatrix}$$

and so by Cramer's rule, we have

$$h_i(z) = \det(M_i(z))/\det(M(z)) .$$

LEMMA 3.3.8. *For each $1 \leq i \leq m$, the function $h_i(z)$ is harmonic in V and extends continuously to $\bar{V}$ by setting $h_i(x_j) = \delta_{ij}$.*

PROPOSITION 3.3.9 (Poisson Formula). *Let V be a simple domain in $\mathbb{P}^1_{\text{Berk}}$ with boundary points $x_1, \ldots, x_m$. Then each harmonic function f on V has a continuous extension to $\bar{V}$, and there is a unique such function with a prescribed set of boundary values $A_1, \ldots, A_m$. Moreover, f can be computed from its boundary values using the formula*

$$f(z) = \sum_{i=1}^m f(x_i) \cdot h_i(z),$$

valid for all $z \in \bar{V}$, where $h_i(z)$ is the ith harmonic measure with respect to V.

A useful reformulation of Proposition 3.3.9 is as follows (compare with [**Kan89**, §4.2]). For $z \in \bar{V}$, define the *Jensen-Poisson measure* $\mu_{z,V}$ on $\bar{V}$ relative to the point z by

$$\mu_{z,V} = \sum_{i=1}^m h_i(z)\delta_{x_i}.$$

Then by Proposition 3.3.9, we have:

COROLLARY 3.3.10. *If V is a simple domain in $\mathbb{P}^1_{\text{Berk}}$, then a continuous function $f : \bar{V} \to \mathbb{R} \cup \{\pm\infty\}$ is harmonic in V if and only if*

$$f(z) = \int_{\partial V} f \, d\mu_{z,V}$$

for all $z \in V$.

Since the closures of simple domains form a fundamental system of compact neighborhoods for the topology on $\mathbb{P}^1_{\mathrm{Berk}}$, it follows that a function f is harmonic on an open set U if and only if its restriction to every simple subdomain $V \subseteq U$ is harmonic, where a *simple subdomain* of U denotes a simple domain whose closure is contained in U. With this terminology, we have:

COROLLARY 3.3.11. *If U is a domain in $\mathbb{P}^1_{\mathrm{Berk}}$ and $f : U \to \mathbb{R} \cup \{\pm\infty\}$ is a continuous function, then f is harmonic in U if and only if for every simple subdomain V of U we have*

$$f(z) = \int_{\partial V} f \, d\mu_{z,V}$$

for all $z \in V$.

Corollary 3.3.11 is the Berkovich space analogue of the mean value characterization for harmonic functions on a domain $U \subseteq \mathbb{C}$. Note that over $\mathbb{C}$, it suffices to consider small disks $V \subseteq U$ centered at z, while in the Berkovich case disks are not sufficient.

Arizona Winter School Project #1: Let $B = B_1 \cup \cdots \cup B_m$ be a finite disjoint union of closed disks in $\mathbb{C}_p$ having radii in $|\mathbb{C}_p^*| = p^{\mathbb{Q}}$. Prove that there is a polynomial $f \in \mathbb{C}_p[T]$ such that $B = \{z \in \mathbb{C}_p : |f(z)| \leq 1\}$, and find an explicit formula for $f(z)$ in terms of the Jensen-Poisson measure associated to the simple domain $V = \mathbb{P}^1_{\mathrm{Berk}} \setminus (\mathcal{B}_1 \cup \cdots \cup \mathcal{B}_m)$, where $\mathcal{B}_i$ is the closed Berkovich disk in $\mathbb{P}^1_{\mathrm{Berk}}$ associated to B_i.

3.3.3. *Uniform convergence.* The Poisson formula implies that the limit of a sequence of harmonic functions is harmonic, under a much weaker condition than is required classically (see Chapter 7 of [**BR08**]):

PROPOSITION 3.3.12. *Let U be an open subset of $\mathbb{P}^1_{\mathrm{Berk}}$. Suppose $f_1, f_2, \ldots$ are harmonic in U and converge pointwise to a function $f : U \to \mathbb{R}$. Then $f(z)$ is harmonic in U, and the $f_i(z)$ converge uniformly to $f(z)$ on compact subsets of U.*

Using the previous result, one can characterize harmonic functions as local uniform limits of logarithms of norms of rational functions (see Chapter 7 of [**BR08**]):

PROPOSITION 3.3.13. *If $U \subset \mathbb{P}^1_{\mathrm{Berk}}$ is a domain and h is harmonic in U, there are rational functions $g_1(T), g_2(T), \ldots \in K(T)$ and rational numbers $R_1, R_2, \ldots \in \mathbb{Q}$ such that*

$$h(x) = \lim_{i \to \infty} R_i \cdot \log_v(|g_i|_x)$$

uniformly on compact subsets of U.

A Berkovich space analogue of Harnack's principle holds as well (see Chapter 7 of [**BR08**]):

PROPOSITION 3.3.14 (Harnack's Principle). *Let U be a domain in $\mathbb{P}^1_{\mathrm{Berk}}$, and suppose $f_1, f_2, \ldots$ are harmonic in U, with $0 \leq f_1 \leq f_2 \leq \cdots$. Then either*
   A) $\lim_{i \to \infty} f_i(z) = \infty$ *for each $z \in U$, or*
   B) $f(z) = \lim_{i \to \infty} f_i(z)$ *is finite for all z, the $f_i(z)$ converge uniformly to $f(z)$ on compact subsets of U, and $f(z)$ is harmonic in U.*

3.4. Subharmonic functions. We give a brief introduction to the notion of a subharmonic function on $\mathbb{P}^1_{\text{Berk}}$; see [**BR08**, Chapter 8] and [**Thu05**] for further details.

DEFINITION 3.4.1. Let $U \subset \mathbb{P}^1_{\text{Berk}}$ be a domain.
A function $f : U \to [-\infty, \infty)$ with $f(x) \not\equiv -\infty$ is called *subharmonic* on U if

(SH1) f is upper semicontinuous.
(SH2) For each simple subdomain $V \subset U$ we have

$$f(z) \leq \int_{\partial V} f \, d\mu_{z,V}$$

for all $z \in V$.

A function $f : U \to (-\infty, \infty]$ with $f(x) \not\equiv +\infty$ is called *superharmonic* on U if $-f$ is subharmonic on U.

REMARK 3.4.2. By Corollary 3.3.11, f is harmonic on U if and only if it is both subharmonic and superharmonic on U.

Corollary 3.3.11 also shows that condition (SH2) can be replaced by the condition that for each simple subdomain $V \subset U$ and each harmonic function h on V, if $f(x) \leq h(x)$ on ∂V then $f(x) \leq h(x)$ on V.

EXAMPLE 3.4.3. For fixed $y, z \in \mathbb{P}^1_{\text{Berk}}$ with $y \neq z$, the function $f(x) = \kappa_z(x, y)$ is superharmonic in $\mathbb{P}^1_{\text{Berk}} \backslash \{z\}$, and is subharmonic in $\mathbb{P}^1_{\text{Berk}} \backslash \{y\}$.

EXAMPLE 3.4.4. If ν is a probability measure on $\mathbb{P}^1_{\text{Berk}}$ and $z \notin \text{Supp}(\nu)$, then the *potential function*

$$p_{\nu,z}(x) = \int_{\mathbb{P}^1_{\text{Berk}}} \kappa_z(x, y) \, d\nu(y)$$

is superharmonic in $\mathbb{P}^1_{\text{Berk}} \backslash \{z\}$ and is subharmonic in $\mathbb{P}^1_{\text{Berk}} \backslash \text{Supp}(\nu)$.

Subharmonic functions obey the following maximum principle (see Chapter 8 of [**BR08**]):

PROPOSITION 3.4.5.

(1) If f is a nonconstant subharmonic function on a domain $U \subset \mathbb{P}^1_{\text{Berk}}$, then f does not achieve a global maximum on U.
(2) If f is a subharmonic function on a domain $U \subset \mathbb{P}^1_{\text{Berk}}$ which extends continuously to $\bar{U}$, then f achieves its maximum value on ∂U.

Finally, we mention the following analogue of Proposition 3.3.6 (see Chapter 8 of [**BR08**]):

PROPOSITION 3.4.6. *Let f be subharmonic on a domain U. Then f is non-increasing on paths leading away from the main dendrite of U. If U is a disk, then f is non-increasing on paths leading away from the unique boundary point of U.*

Since the main dendrite of a domain is finitely branched, Proposition 3.4.6 implies that at any given point, there are only finitely many tangent directions in which a subharmonic function can be increasing.

4. Laplacians

In this lecture, we will define a Laplacian operator on the Berkovich projective line which is analogous in many ways to the classical Laplacian operator

$$\Delta(f) = -(\frac{\partial^2 f}{\partial x^2} + \frac{\partial^2 f}{\partial y^2})dx \wedge dy$$

on $\mathbb{C}$.

Actually, a slight abstraction of the construction from [**BR08**] of the Laplacian on $\mathbb{P}^1_{\text{Berk}}$ yields a Laplacian operator on more general one-dimensional Berkovich spaces such as $\mathcal{M}(\mathbb{Z})$ or the analytic space X_{Berk} associated to a complete nonsingular curve (see §5). The Laplacian on $\mathbb{P}^1_{\text{Berk}}$ will be constructed via a limiting process from the Laplacian on a finite $\mathbb{R}$-tree. For curves of higher genus, the associated Berkovich analytic space is no longer simply connected, so in order to construct a Laplacian in this generality, one needs to replace finite $\mathbb{R}$-trees by *metrized graphs*.

We will define a Laplacian operator in the rather abstract general setting of an *arboretum*, which is a special kind of inverse limit of metrized graphs, and then gradually specialize to the particular cases of interest to us. This involves setting up some cumbersome notation, but it has the advantage of making the entire construction more conceptually clear.

4.1. Metrized graphs.

4.1.1. *Definition of a metrized graph.* Intuitively, a metrized graph is a finite graph Γ whose edges are thought of as line segments having a well-defined length. In particular, Γ is a one-dimensional manifold except at finitely many "branch points", where it looks locally like an n-pointed star. The path-length function along each edge extends to a metric on all of Γ, making it a compact metric space. One thinks of a metrized graph as an analytic object, not just a combinatorial one.

Formally, define a *star-shaped set of valence $n_p \geq 1$* to be a set of the form

$$S(n_p, r_p) = \{z \in \mathbb{C} : z = te^{k \cdot 2\pi i/n_p} \text{ for some } 0 \leq t < r_p \text{ and some } k \in \mathbb{Z}\}.$$

Then a *metrized graph* is a compact, connected metric space Γ such that each $p \in \Gamma$ has a neighborhood U_p isometric to a star-shaped set of valence $n_p \geq 1$, endowed with the path metric.

A metrized graph with no cycles is the same thing as a compact, finite $\mathbb{R}$-tree, as defined in §2.2.

By a *vertex set* for Γ, we mean a finite set of points S such that $\Gamma \backslash S$ is a union of open intervals whose closures have distinct endpoints. (A vertex set necessarily contains all endpoints and branch points of Γ, and if Γ has loops, a vertex set also contains at least one interior point from each loop.)

There is a close connection between metrized graphs and finite weighted graphs (i.e., finite graphs whose edges are assigned positive real weights). Given a metrized graph Γ, any choice of a vertex set S for Γ gives rise to a weighted graph G which one may call a *model* for Γ; a different choice of vertex set leads to an equivalent weighted graph (with respect to a certain natural equivalence relation). Conversely, every weighted graph G determines a metrized graph in the obvious way, so there is a one-to-one correspondence between metrized graphs and equivalence classes of finite weighted graphs (see [**BR07**] or [**BF06**, Theorem 4] for further details).

Often, when given a metrized graph Γ, one chooses without explicit comment a vertex set S, together with distinguished parametrizations of the edges of the corresponding model G. The definition of the Laplacian given below in §4 is independent of these implicit choices.

By a *path* in Γ, we will mean an injective length-preserving continuous map from the real interval $[0, L]$ into Γ. We will say that a path $\gamma : [0, L] \to \Gamma$ *emanates from* p, and *terminates at* q, if $\gamma(0) = p$ and $\gamma(L) = q$. We call two paths emanating from p *equivalent* if they share a common initial segment. For each $p \in \Gamma$, we let $T_p(\Gamma)$ (the set of *tangent directions* at p) denote the set of equivalence classes of paths emanating from p. It is easy to see that $|T_p(\Gamma)| = n_p$, i.e., there is a bijection between elements of T_p and the "edges" of Γ emanating from p.

It is useful to associate to each element of $T_p(\Gamma)$ a formal "unit tangent vector" $\vec{v}$, and to write $p + t\vec{v}$ instead of $\gamma(t)$, where $\gamma : [0, L] \to \Gamma$ is a representative path. If $f : \Gamma \to \mathbb{R}$ is a function, and $\vec{v}$ is a formal unit tangent vector at p, we define the derivative of f in the direction $\vec{v}$ to be

$$d_{\vec{v}} f(p) = \lim_{t \to 0^+} \frac{f(p + t\vec{v}) - f(p)}{t} = \lim_{t \to 0^+} \frac{f(\gamma(t)) - f(p)}{t} ,$$

provided the limit exists.

4.2. The Laplacian on a metrized graph.

4.2.1. *The space* $\mathrm{CPA}(\Gamma)$. Let $\mathrm{CPA}(\Gamma)$ be the space of continuous, piecewise-affine, real-valued functions on Γ, i.e., functions which have the form $t \mapsto at + b$ on each edge of Γ (with respect to some vertex set).

If $f \in \mathrm{CPA}(\Gamma)$, then clearly the directional derivatives $d_{\vec{v}} f(p)$ are defined for all $p \in \Gamma$ and all $\vec{v} \in T_p(\Gamma)$.

Chinburg and Rumely ([**CR93**]) introduced a Laplacian operator on $\mathrm{CPA}(\Gamma)$. Their Laplacian is a map from $\mathrm{CPA}(\Gamma)$ to the space of discrete signed measures on Γ. We will take the Laplacian to be the negative of theirs, and put

$$\Delta(f) = \sum_{p \in \Gamma} \Delta_p(f)\, \delta_p(x),$$

where $\Delta_p(f) = -\sum_{\vec{v} \in T_p(\Gamma)} d_{\vec{v}} f(p)$ and $\delta_p(x)$ is the Dirac measure at p. The operator Δ coincides, in a natural sense, with the usual combinatorial Laplacian on a finite weighted graph – see [**BF06**] for details. Here are some easily verified properties of Δ:

PROPOSITION 4.2.1. *Let* $f, g \in \mathrm{CPA}(\Gamma)$. *Then*

(1) $\Delta(f) \equiv 0$ *if and only if* f *is constant on* Γ.
(2) $\Delta(f) = \Delta(g)$ *if and only if* $f = g + C$ *for some constant* C.
(3) *If* f *is nonconstant, then* $f(x)$ *achieves its maximum at a point* p *where* $\Delta(f)(p) > 0$, *and its minimum at a point* q *where* $\Delta(f)(q) < 0$.
(4) $\int_\Gamma f \, \Delta(g) = \int_\Gamma g \, \Delta(f)$.
(5) *The total mass* $\Delta(f)(\Gamma)$ *is* 0.

4.2.2. *The space* $\mathrm{BDV}(\Gamma)$. One can define a measure-valued Laplacian operator on a much larger class of functions than just $\mathrm{CPA}(\Gamma)$. The construction is motivated by the following "Mass Formula". If $S \subseteq \Gamma$, we let ∂S be the set of points belonging to the closures of both S and $\Gamma \backslash S$. (Note that under this definition, if $\Gamma = [0, 1]$ and $S = [0, \frac{1}{2}]$, for example, then the left endpoint 0 is not a boundary point of S.)

For each $p \in \partial S$, let $\mathrm{In}(p, S)$ be the set of "inward-pointing unit tangent vectors at p", i.e., the set of all $\vec{v} \in T_p(\Gamma)$ for which $p + t\vec{v}$ belongs to S for all sufficiently small $t > 0$. Similarly, let $\mathrm{Out}(p, S) = T_p(\Gamma) \setminus \mathrm{In}(p, S)$ be the collection of "outward-directed unit tangent vectors at p". For example, if p is an isolated point of S, then $\mathrm{In}(p, S) = \emptyset$ and $\mathrm{Out}(p, S) = T_p(\Gamma)$.

PROPOSITION 4.2.2 (Mass formula). *Let $f \in \mathrm{CPA}(\Gamma)$. Then the measure $\mu = \Delta(f)$ satisfies the following properties:*

(a) *If $E \subseteq \Gamma$ is a finite union of connected closed sets, then*

$$(4.2.3) \qquad \mu(E) = - \sum_{p \in \partial E} \sum_{\vec{v} \in \mathrm{Out}(p,E)} d_{\vec{v}} f(p).$$

(b) *If $V \subseteq \Gamma$ is a finite union of connected open sets, then*

$$(4.2.4) \qquad \mu(V) = \sum_{p \in \partial V} \sum_{\vec{v} \in \mathrm{In}(p,V)} d_{\vec{v}} f(p).$$

Let $\mathcal{D}(\Gamma)$ be the class of all functions on Γ whose one-sided derivatives exist everywhere, i.e.,

$$\mathcal{D}(\Gamma) = \{ f : \Gamma \to \mathbb{R} : \; d_{\vec{v}} f(p) \text{ exists for each } p \in \Gamma \text{ and } \vec{v} \in T_p(\Gamma) \} \; .$$

It is easy to see that each $f \in \mathcal{D}(\Gamma)$ is continuous.

DEFINITION 4.2.5. We will say that a continuous function $f \in \mathcal{D}(\Gamma)$ is of *bounded differential variation*, and write $f \in \mathrm{BDV}(\Gamma)$, if there is a (bounded signed Borel) measure μ on Γ satisfying properties (a) and (b) of Proposition 4.2.2. If this is the case, we define the Laplacian $\Delta(f)$ of f to be this measure. Since the connected open sets generate the topology of Γ, it is not hard to see that the measure $\Delta(f)$, if it exists, is uniquely determined by properties (a) and (b).

By Proposition 4.2.2, we have $\mathrm{CPA}(\Gamma) \subseteq \mathrm{BDV}(\Gamma)$. We now describe a larger class of functions for which $\Delta(f)$ exists and can be explicitly described.

PROPOSITION 4.2.6. *Let $\mathrm{Zh}(\Gamma)$ be the space of continuous, piecewise $\mathcal{C}^2$ functions f whose one-sided directional derivatives $d_{\vec{v}} f(p)$ exist for all $p \in \Gamma$, and for which f'' is bounded along each edge. Then $\mathrm{Zh}(\Gamma) \subset \mathrm{BDV}(\Gamma)$, and for $f \in \mathrm{Zh}(\Gamma)$ we have*

$$(4.2.7) \qquad \Delta(f) = -f''(x)\, dx + \sum_{p \in \Gamma} \Delta_p(f)\, \delta_p,$$

where f'' is computed on each segment in the complement of an appropriate vertex set X_f for Γ.

For $f \in \mathrm{Zh}(\Gamma)$, our Laplacian operator therefore coincides with the one defined by S. Zhang in [**Zha93**]. The Laplacian in (4.2.7) is a hybrid of the usual Laplacian $-f''(x)dx$ on $\mathbb{R}$ and the combinatorial Laplacian on a weighted graph. It is easy to check using integration by parts that $\int_\Gamma f\Delta(g) = \int_\Gamma g\Delta(f)$ for all $f, g \in \mathrm{Zh}(\Gamma)$. From this, it follows (taking g to be the constant function 1) that $\Delta(f)(\Gamma) = 0$.

It is clear from the definitions that the Laplacian on $\mathrm{BDV}(\Gamma)$ is a linear operator, i.e., that

$$\Delta(\alpha f + \beta g) = \alpha\Delta(f) + \beta\Delta(g)$$

for all $f, g \in \mathrm{BDV}(\Gamma)$ and all $\alpha, \beta \in \mathbb{R}$.

We note the following additional properties of the Laplacian on BDV(Γ), which extend those proved in Proposition 4.2.1 for functions in CPA(Γ). (We let μ^+, μ^- denote the positive and negative parts, respectively, of the Jordan decomposition of a measure μ.)

PROPOSITION 4.2.8. *If $f, g \in$ BDV(Γ), then*

(1) $\Delta(f) \equiv 0$ *if and only if f is constant on Γ.*
(2) $\Delta(f) = \Delta(g)$ *if and only if $f = g + C$ for some constant C.*
(3) *If f is nonconstant, then $f(x)$ achieves its maximum at a point p in the support of $\Delta(f)^+$, and its minimum at a point q in the support of $\Delta(f)^-$.*
(4) $\int_\Gamma f \, \Delta(g) = \int_\Gamma g \, \Delta(f)$.
(5) $\Delta(f)(\Gamma) = 0$.

We also have the following useful, but somewhat harder to prove, result:

THEOREM 4.2.9. *If ν is a measure of total mass zero on a metrized graph Γ, then there exists a function $h \in$ BDV(Γ), unique up to an additive constant, such that $\Delta h = \nu$.*

It follows that there is a natural bijection between measures of total mass zero on Γ and functions in BDV(Γ) modulo constant functions.

Finally, we note that the Laplacian operator on a metrized graph satisfies the following compatibility property:

LEMMA 4.2.10. *If Γ, Γ' are metrized graphs with $\Gamma \subseteq \Gamma'$ and there is a deformation retraction $r_{\Gamma',\Gamma}$ from Γ' onto Γ, and if $f \in$ BDV(Γ'), then*

$$(4.2.11) \qquad\qquad \Delta_\Gamma(f|_\Gamma) = (r_{\Gamma',\Gamma})_* \, \Delta_{\Gamma'}(f) \ .$$

Lemma 4.2.10 will be used in §4.4 to define the Laplacian on an arboretum.

4.3. Arboreta. Recall that a *directed set* is a set I together with a reflexive and transitive relation $\leq$ satisfying:

Given $\alpha, \beta \in I$, there exists $\gamma \in I$ with $\alpha \leq \gamma$ and $\beta \leq \gamma$.

DEFINITION 4.3.1. An *arboreal system of metrized graphs* is a directed set $(I, \leq)$ together with the following data:

- For each $\alpha \in I$, a metrized graph Γ_α.
- For each $\alpha \leq \beta$ in I, an isometric inclusion map $i_{\alpha,\beta} : \Gamma_\alpha \hookrightarrow \Gamma_\beta$.
- For each $\alpha \leq \beta$ in I, a deformation retraction $r_{\beta,\alpha} : \Gamma_\beta \twoheadrightarrow \Gamma_\alpha$.

REMARK 4.3.2. Concretely, the condition that $r_{\beta,\alpha} : \Gamma_\beta \to \Gamma_\alpha$ is a deformation retraction is equivalent to saying that the graph Γ_β is obtained from Γ_α by attaching finitely many finite $\mathbb{R}$-trees, and $r_{\beta,\alpha}$ is the map obtained by contracting each of these $\mathbb{R}$-trees to a point.

We will usually write $\{\Gamma_\alpha\}$ instead of $(\Gamma_\alpha, i_{\alpha,\beta}, r_{\beta,\alpha})$ to denote an arboreal system of metrized graphs.

DEFINITION 4.3.3. The *arboretum* attached to an arboreal system of metrized graphs $\{\Gamma_\alpha\}$ is the topological space $X = \varprojlim \Gamma_\alpha$, where the inverse limit is taken with respect to the maps $r_{\beta,\alpha}$.

Recall that a point of $\varprojlim \Gamma_\alpha$ is a *compatible system* $(x_\alpha) \in \prod_{\alpha \in I} \Gamma_\alpha$, where "compatible" means that if $\alpha \leq \beta$ then $r_{\beta,\alpha}(x_\beta) = x_\alpha$. The inverse limit X comes equipped with a compatible system of continuous maps $r_\alpha : X \to \Gamma_\alpha$ satisfying the following universal property:

> If Y is a topological space equipped with a compatible system of continuous maps $s_\alpha : Y \to \Gamma_\alpha$, then there is a unique continuous map $\phi : Y \to X$ such that $s_\alpha = r_\alpha \circ \phi$ for all $\alpha \in I$.

The topology on $X = \varprojlim \Gamma_\alpha$ is the weakest one for which all of the maps r_α are continuous. A fundamental system of open neighborhoods for the topology on X is given by the open sets $r_\alpha^{-1}(U_\alpha)$ for $\alpha \in I$ and $U_\alpha \subseteq \Gamma_\alpha$ a connected open set.

The following topological properties of an arboretum are easily verified:

LEMMA 4.3.4. *If X is an arboretum, then*

(1) *X is a compact, Hausdorff, and path-connected topological space.*
(2) *For each $\alpha \in I$ and each $x_0 \in \Gamma_\alpha$, the map $r_\alpha : X \to \Gamma_\alpha$ is a retraction which induces an isomorphism of fundamental groups $\pi_1(X, x_0) \cong \pi_1(\Gamma_\alpha, x_0)$.*

REMARK 4.3.5. If X is an arboretum, then for each $\alpha \in I$ and each $x \in \Gamma_\alpha$, the fiber $r_\alpha^{-1}(x)$ can be endowed in a natural way with the structure of a compact topological tree. Thus one can think of an arboretum as a family of (compact topological) trees fibered over a (finite) metrized graph (hence the name *arboretum*).

Next, we define the analogue for a general arboretum of the collection $\mathbf{H}_{\mathrm{Berk}}^{\mathbb{R}}$ of all points of type II or III in $\mathbb{P}_{\mathrm{Berk}}^1$.

DEFINITION 4.3.6. The *hyperbolic space* $\mathbf{H}^{\mathbb{R}}(X)$ associated to an arboretum X is the metric space $\varinjlim \Gamma_\alpha$, where the direct limit is taken with respect to the inclusion maps $i_{\alpha,\beta}$.

By the definition of a direct limit, there are natural continuous inclusion maps $i_\alpha : \Gamma_\alpha \hookrightarrow \mathbf{H}^{\mathbb{R}}(X)$ for each $\alpha \in I$, and $\mathbf{H}^{\mathbb{R}}(X)$ satisfies a universal property relative to these maps.

It is not hard to show that there is a natural injective map of sets $i : \mathbf{H}^{\mathbb{R}}(X) \hookrightarrow X$, and the image of $\mathbf{H}^{\mathbb{R}}(X)$ under this map is a dense, path-connected subspace of X.

REMARK 4.3.7. The metric topology on $\mathbf{H}^{\mathbb{R}}(X)$ is *not* in general the same as the subspace topology induced by the inclusion map $i : \mathbf{H}^{\mathbb{R}}(X) \hookrightarrow X$. However, i is always continuous as a map from $\mathbf{H}^{\mathbb{R}}(X)$ with its direct limit topology to X with its inverse limit topology.

DEFINITION 4.3.8. A *simple function* on an arboretum X is a function of the form $f \circ r_\alpha$ with $\alpha \in I$ and $f \in \mathrm{CPA}(\Gamma_\alpha)$.

We will make use of the following consequence of the Stone-Weierstrass theorem in §4.

PROPOSITION 4.3.9. *The simple functions are dense in the ring $\mathcal{C}(X)$ of continuous real-valued functions on X (endowed with the uniform topology).*

4.4. The Laplacian on an arboretum. Let X be an arboretum corresponding to the arboreal system $\mathcal{S} = (\Gamma_\alpha)_{\alpha \in I}$ of metrized graphs.

DEFINITION 4.4.1. A system of measures $\{\mu_\Gamma\}_{\Gamma \in \mathcal{S}}$ is called *coherent* if:

- For each $\alpha \leq \beta$, we have $(r_{\beta,\alpha})_*(\mu_\beta) = \mu_\alpha$.
- There is a constant B such that $|\mu_\Gamma|(\Gamma) \leq B$ for all $\Gamma \in \mathcal{S}$.

If μ is a measure on X and we set $\mu_\Gamma = (r_\Gamma)_*(\mu)$ for each $\Gamma \in \mathcal{S}$, then $\{\mu_\Gamma\}$ is easily seen to be a coherent system of measures. Conversely, every coherent system arises in this way:

THEOREM 4.4.2. *If $\{\mu_\Gamma\}$ is a coherent system of measures on X, there is a unique measure μ on X such that $(r_\Gamma)_*(\mu) = \mu_\Gamma$ for each $\Gamma \in \mathcal{S}$.*

The proof is based on the Riesz representation theorem, together with Proposition 4.3.9 (see Chapter 5 of [**BR08**] for a proof in the special case $X = \mathbb{P}^1_{\text{Berk}}$, which can be easily generalized to the present setting).

We define $\text{BDV}(X)$ to be the collection of all functions $f : X \to \mathbb{R} \cup \{\pm\infty\}$ such that:

(BDV1) $f|_\Gamma := f \circ i_\Gamma \in \text{BDV}(\Gamma)$ for each finite subgraph $\Gamma \in \mathcal{S}$, where $i_\Gamma : \Gamma \hookrightarrow X$ is the natural inclusion.

(BDV2) The measures $|\Delta_\Gamma(f|_\Gamma)|$ for $\Gamma \in \mathcal{S}$ have uniformly bounded total mass.

Note that belonging to $\text{BDV}(\Gamma)$ imposes no condition on the values of f at points of $X \backslash \mathbf{H}^\mathbb{R}(X)$.

Using the compatibility property (4.2.11), one shows that if $f \in \text{BDV}(X)$, then $\{\Delta_\Gamma(f|_\Gamma)\}$ is a coherent system of measures on X. By Theorem 4.4.2, there is a unique measure $\Delta_X(f)$ on the inverse limit space X for which

$$(4.4.3) \qquad\qquad (r_\Gamma)_*\Delta_X(f) = \Delta_\Gamma(f \circ i_\Gamma)$$

for all $\Gamma \in \mathcal{S}$.

DEFINITION 4.4.4. If $f \in \text{BDV}(X)$, the unique measure $\Delta_X(f)$ on X satisfying (4.4.3) is called the *Laplacian* of f on X.

REMARK 4.4.5. It follows from the definitions that $\Delta_X(f)$ is the weak limit over all $\Gamma \in \mathcal{S}$ of the measures $(i_\Gamma)_* (\Delta_\Gamma(f \circ i_\Gamma))$; this is in fact an alternate way to define $\Delta_X(f)$.

As a consequence of Theorem 4.4.2, we have:

COROLLARY 4.4.6. *Let $F = f \circ r_\Gamma$ be a simple function on X (cf. Proposition 4.3.9). Then $F \in \text{BDV}(X)$ and $\Delta_X(F) = (i_\Gamma)_*\Delta_\Gamma(f)$.*

The Laplacian on an arboretum satisfies the following properties, which can be deduced from the corresponding properties of the Laplacian on a metrized graph.

PROPOSITION 4.4.7. *Let X be an arboretum, and let $f, g \in \text{BDV}(X)$. Then:*

(1) $\Delta_X(f) \equiv 0$ *if and only if f is constant on $\mathbf{H}^\mathbb{R}(X)$.*
(2) $\int_X f \, \Delta_X(g) = \int_X g \, \Delta_X(f)$.
(3) $\Delta_X(f)$ *has total mass zero.*

We also have the following result, which is a generalization of Theorem 4.2.9.

THEOREM 4.4.8. *If ν is a measure of total mass zero on an arboretum X, then there exists a function $h \in \mathrm{BDV}(X)$, whose restriction to $\mathbf{H}^{\mathbb{R}}(X)$ is unique up to an additive constant, such that $\Delta_X(h) = \nu$.*

4.5. Examples. We now compute the Laplacian in some specific examples, using the fact that we have endowed the Berkovich spaces $\mathbb{P}^1_{\mathrm{Berk}}$ and $\mathcal{M}(\mathbb{Z})$ with a canonical arboretum structure (see Theorem 2.2.1 and Remark 2.5.3).

EXAMPLE 4.5.1. As a concrete example, fix $y \in \mathbb{A}^1_{\mathrm{Berk}}$ and let $f : \mathbb{P}^1_{\mathrm{Berk}} \to \mathbb{R} \cup \{\pm\infty\}$ be defined by

$$f(x) = \kappa_\infty(x, y) = -\log_v[x, y]_\infty$$

(cf. §1.6.2). Then $f \in \mathrm{BDV}(\mathbb{P}^1_{\mathrm{Berk}})$, and

$$(4.5.2) \qquad \Delta_{\mathbb{P}^1_{\mathrm{Berk}}}(f) = \delta_y - \delta_\infty$$

is a discrete measure on $\mathbb{P}^1_{\mathrm{Berk}}$ supported on $\{y, \infty\}$. The explanation for the formula (4.5.2) is as follows. The function f is locally constant off the path $\Lambda = \ell_{y,\infty}$ from y to ∞; more precisely, we have $f(x) = f(r_\Lambda(x))$ for all $x \in \mathbb{P}^1_{\mathrm{Berk}}$. Moreover, the restriction of f to the open segment (y, ∞) is linear (with respect to the metric ρ) with slope -1. It follows that for every finite subgraph Γ of $\mathbb{P}^1_{\mathrm{Berk}}$, we have

$$(4.5.3) \qquad \Delta_\Gamma(f) = \delta_{r_\Gamma(y)} - \delta_{r_\Gamma(\infty)}.$$

The weak limit over all $\Gamma \in \mathcal{S}$ of the right-hand side of (4.5.3) is $\delta_y - \delta_\infty$, which establishes (4.5.2).

Equation (4.5.2) shows that $\kappa_\infty(x, y) = -\log_v[x, y]_\infty$, like its classical counterpart $-\log|x - y|$ over $\mathbb{C}$, is a "fundamental solution of the Laplace equation".

EXAMPLE 4.5.4. Generalizing the previous example, fix $y, z \in \mathbb{P}^1_{\mathrm{Berk}}$ and let $f : \mathbb{P}^1_{\mathrm{Berk}} \to \mathbb{R} \cup \{\pm\infty\}$ be defined by

$$f(x) = \kappa_z(x, y).$$

Then one can show that $f(x)$ is the unique solution in $\mathrm{BDV}(\mathbb{P}^1_{\mathrm{Berk}})$ to the differential equation

$$\Delta_{\mathbb{P}^1_{\mathrm{Berk}}}(f) = \delta_y - \delta_z$$

and satisfying the initial condition

$$f(\zeta_{\mathrm{Gauss}}) = \log_v \|y, z\|$$

(cf. §1.6.2).

EXAMPLE 4.5.5. Consider the function $G : \mathbb{P}^1_{\mathrm{Berk}} \to \mathbb{R} \cup \{+\infty\}$ from Example 3.2.3, defined as

$$G(x) = \begin{cases} +\infty & x = \infty \\ \log_v \max(|T|_x, 1) & x \in \mathbb{A}^1_{\mathrm{Berk}}. \end{cases}$$

If Λ denotes the closed path from ζ_{Gauss} to ∞ in $\mathbb{P}^1_{\mathrm{Berk}}$, and $r_\Lambda : \mathbb{P}^1_{\mathrm{Berk}} \to \Lambda$ is the natural retraction map from $\mathbb{P}^1_{\mathrm{Berk}}$ onto Λ, then we saw in Example 3.2.3 that $G = g \circ r_\Lambda$, where $g : \Lambda \to \mathbb{R} \cup \{+\infty\}$ is the linear map of slope 1 along Λ for which $g(\zeta_{\mathrm{Gauss}}) = 0$.

It follows as in Example 4.5.1 that

$$\Delta_{\mathbb{P}^1_{\mathrm{Berk}}}(G) = \delta_\infty - \delta_{\zeta_{\mathrm{Gauss}}}.$$

Note that $G(x) = -\log_v \|x, \infty\|$, so that this example is in fact a special case of Example 4.5.4.

EXAMPLE 4.5.6. Let $f(T) = \prod_{i=1}^{k}(x - a_i)^{m_i} \in K[T]$ be a nonconstant polynomial of degree n, with $a_1, \ldots, a_k$ distinct, and let $F(x)$ be the unique continuous function on $\mathbb{P}^1_{\mathrm{Berk}}$ extending the function $-\log_v |f(x)|$ on K.

From Example 3.2.5 and the linearity of the Laplacian, we deduce easily that

$$\Delta_{\mathbb{P}^1_{\mathrm{Berk}}}(F) = \left(\sum_{i=1}^{k} m_i \delta_{a_i}\right) - n\delta_\infty.$$

More generally, let $\varphi \in K(T)$ be a nonzero rational function with zeros and poles given by the divisor $\mathrm{Div}(\varphi)$ on $\mathbb{P}^1(K)$. As discussed in §2.1, the usual action of φ on $\mathbb{P}^1(K)$ extends naturally to an action of φ on $\mathbb{P}^1_{\mathrm{Berk}}$, and there is a unique continuous function $-\log_v |\varphi| : \mathbb{P}^1_{\mathrm{Berk}} \to \mathbb{R} \cup \{\pm\infty\}$ in $\mathrm{CPA}(\mathbb{P}^1_{\mathrm{Berk}})$ extending the usual map $x \mapsto -\log_v |\varphi(x)|$ on $\mathbb{P}^1(K)$. One can derive from (4.5.2) the following analogue for $\mathbb{P}^1_{\mathrm{Berk}}$ of the classical *Poincaré-Lelong formula*:

$$(4.5.7) \qquad \Delta_{\mathbb{P}^1_{\mathrm{Berk}}}(-\log_v |\varphi|) = \delta_{\mathrm{Div}(\varphi)} .$$

EXAMPLE 4.5.8. Following Example 3.1.2, let $n \in \mathbb{Z}$ be a nonzero integer. As we saw previously, along the branch Λ_v of $\mathcal{M}(\mathbb{Z})$ emanating from ζ_0 in the tangent direction corresponding to $v \in M_{\mathbb{Q}}$, the function $F_n(x) = -\log |n|_x$ is linear with slope equal to $-\log |n|_v$. Taking a limit over all finite subgraphs of $\mathcal{M}(\mathbb{Z})$ yields the formula

$$\Delta_{\mathcal{M}(\mathbb{Z})}(-\log |n|_x) = \sum_{v \in M_{\mathbb{Q}} \text{ finite}} -\log(|n|_v) \cdot \delta_{\zeta_{v,\infty}} - \log(|n|_\infty) \cdot \delta_{\zeta_\infty}.$$

For example, if $n = 12$ then

$$\Delta_{\mathcal{M}(\mathbb{Z})}(-\log |12|) = \log(4) \cdot \delta_{\zeta_{2,\infty}} + \log(3) \cdot \delta_{\zeta_{3,\infty}} - \log(12) \cdot \delta_{\zeta_\infty}.$$

Thus we see that the Laplacian operator on $\mathcal{M}(\mathbb{Z})$ is basically *factoring* the integer n.

EXAMPLE 4.5.9. More generally, it is not hard to see that if $\alpha = m/n$ is a nonzero rational number written in lowest terms, then $\Delta_{\mathcal{M}(\mathbb{Z})}(-\log |\alpha|)$ has total mass zero, and in the Jordan decomposition

$$\Delta_{\mathcal{M}(\mathbb{Z})}(-\log |\alpha|) = \mu^+ - \mu^-,$$

each of μ^+ and μ^- has total mass equal to the standard logarithmic Weil height $h(\alpha) = \log \max\{|m|, |n|\}$.

For example, if $\alpha = 4/3$ then

$$\Delta_{\mathcal{M}(\mathbb{Z})}(-\log |4/3|) = \log(4) \cdot \delta_{\zeta_{2,\infty}} - \log(3) \cdot \delta_{\zeta_{3,\infty}} - \log(4/3) \cdot \delta_{\zeta_\infty}$$

and

$$\Delta_{\mathcal{M}(\mathbb{Z})}(-\log |4/3|)^+(\mathcal{M}(\mathbb{Z})) = \Delta_{\mathcal{M}(\mathbb{Z})}(-\log |4/3|)^-(\mathcal{M}(\mathbb{Z})) = \log(4) = h(4/3).$$

Thus we see that the Laplacian on $\mathcal{M}(\mathbb{Z})$ provides an intriguing way of interpreting the Weil height on $\mathbb{Q}^*$.

Arizona Winter School Project #2: Extend the results of Examples 4.5.8 and 4.5.9 to the Berkovich analytic space associated to the ring of integers in an arbitrary number field k, and interpret the Arakelov class group of k (and in particular the usual class group and unit group) in terms of functions on this space.

4.6. Harmonic functions and the Laplacian. Let U be a domain in $\mathbb{P}^1_{\text{Berk}}$. In this section, we will define the Laplacian $\Delta_{\bar{U}}(f)$ of a function f on U as a measure supported on $\bar{U}$, and explain the connection between harmonic functions on U and the Laplacian.

The key point is that $\bar{U}$ is itself an arboretum. Indeed, define $\mathcal{S}_U$ to be the collection of all finite subgraphs Γ of $\mathbb{P}^1_{\text{Berk}}$ which are contained in U. With respect to the natural inclusion and retraction maps, we have:

LEMMA 4.6.1. *There are homeomorphisms*

$$U \cap \mathbf{H}_{\text{Berk}} \cong \varinjlim_{\Gamma \in \mathcal{S}_U} \Gamma$$

(with respect to the locally metric topology on $\mathbf{H}_{\text{Berk}}$) and

$$\bar{U} \cong \varprojlim_{\Gamma \in \mathcal{S}_U} \Gamma$$

(with respect to the Berkovich subspace topology on $\bar{U}$).

It follows that $\bar{U}$ is naturally endowed with the structure of an arboretum, and therefore we have defined a Laplacian operator on $\bar{U}$. Since the Laplacian of a function f on $\bar{U}$ depends only on the restriction of f to $\mathbf{H}(U) = U \cap \mathbf{H}_{\text{Berk}}$, we see that it makes sense to speak of the Laplacian of a (suitably nice) function f on U as a measure on $\bar{U}$.

REMARK 4.6.2. The definition of $f \in \text{BDV}(\bar{U})$ only concerns the restrictions $f|_\Gamma$ to finite subgraphs $\Gamma \subset U$. By definition, such subgraphs only contain points of type II or III; in fact, as the subgraphs vary they exhaust $U \cap \mathbf{H}^{\mathbb{R}}_{\text{Berk}}$. In particular, requiring that $f \in \text{BDV}(\bar{U})$ imposes no conditions on the behavior of f on $\mathbb{P}^1(K)$. In practice, that behavior must be deduced from auxiliary hypotheses, such as continuity or upper-semicontinuity.

Using Lemma 4.2.10, we have:

LEMMA 4.6.3. *If $V \subseteq U$ are domains in $\mathbb{P}^1_{\text{Berk}}$, and if $r = r_{\bar{U}, \bar{V}} : \bar{U} \to \bar{V}$ denotes the natural retraction map, then for all $f \in \text{BDV}(\bar{U})$, we have $f|_V \in \text{BDV}(\bar{V})$ and*

$$r_*(\Delta_{\bar{U}}(f)) = \Delta_{\bar{V}}(f|_V).$$

The relation between harmonic functions and the Laplacian is the following:

THEOREM 4.6.4. *Let U be a domain in $\mathbb{P}^1_{\text{Berk}}$, and let $f \in \text{BDV}(\bar{U})$ be a continuous real-valued function. Then f is harmonic on U if and only if $\Delta_{\bar{V}}(f)$ is supported on ∂V for every simple subdomain V of U.*

Similarly:

THEOREM 4.6.5. *Let U be a domain in $\mathbb{P}^1_{\text{Berk}}$, and let $f \in \text{BDV}(\bar{U})$ be a continuous extended-real valued function. Then f is subharmonic on U if and only if $\Delta_{\bar{V}}(f)|_V \leq 0$ for every simple subdomain V of U.*

4.7. Valuation polygons. Let $\varphi \in K(T)$ be a nonzero rational function. We saw in §2.1 how φ induces in a natural way a map

$$\varphi : \mathbb{P}^1_{\text{Berk}} \to \mathbb{P}^1_{\text{Berk}}$$

which, when restricted to $\mathbb{P}^1(K)$, coincides with the usual induced map.

There is a unique simple function $f = -\log_v |\varphi| : \mathbb{P}^1_{\text{Berk}} \to \mathbb{R} \cup \{\pm\infty\}$ whose restriction to $\mathbb{P}^1(K)$ is the map $x \mapsto -\log_v |\varphi(x)|$ on $\mathbb{P}^1(K)$. As we have seen in (4.5.7), one has the following Berkovich space analogue of the *Poincaré-Lelong formula*:

$$\Delta_{\mathbb{P}^1_{\text{Berk}}}(f) = \delta_{\text{Div}(\varphi)} .$$

As a consequence, we obtain:

COROLLARY 4.7.1. *For any Borel subset A of $\mathbb{P}^1_{\text{Berk}}$, let $N_0(\varphi, A)$ be the number of zeros of φ in A, and let $N_\infty(\varphi, A)$ be the number of poles of φ in A (counting multiplicities). Then:*

$$(4.7.2) \qquad \Delta_{\mathbb{P}^1_{\text{Berk}}}(-\log_v |\varphi|)(A) = N_0(\varphi, A) - N_\infty(\varphi, A).$$

We wish to interpret the formula (4.7.2) in terms of the classical theory of *valuation polygons* (see [**Rob00**, Chapter 6, §1,3]) using the *Mass Formula* (Proposition 4.2.2).

If D is a Berkovich open disk in $\mathbb{P}^1_{\text{Berk}}$, then by Lemma 1.4.1, we have $D = U(x; \vec{v})$ for some $x \in \mathbf{H}^{\mathbb{R}}_{\text{Berk}}$ and $\vec{v} \in T_x$ (and conversely, each pair $(x, \vec{v})$ with $x \in \mathbf{H}^{\mathbb{R}}_{\text{Berk}}$ and $\vec{v} \in T_x$ determines a unique Berkovich open disk). Following Rivera-Letelier [**RL03a, RL03b**] we call x the *boundary* of D and $\vec{v}$ the corresponding *end* (French: "bout").

More generally, let V be a simple domain in $\mathbb{P}^1_{\text{Berk}}$. By Lemma 2.2.3, we can write $V = r_\Gamma^{-1}(U)$ for some finite subgraph Γ of $\mathbb{P}^1_{\text{Berk}}$ and some connected open subset U of Γ. Let $\partial U = \{x_1, \ldots, x_n\} \subseteq \Gamma$ be the boundary of U in Γ (which is the same as the boundary of U in $\mathbb{P}^1_{\text{Berk}}$). For each $1 \le i \le n$, let $\vec{v}_i \in T_{x_i}(\Gamma)$ denote the unique inward-pointing tangent vector at x_i (with respect to U). We can identify $\vec{v}_i$ with an element of T_{x_i} (the set of tangent directions in $\mathbb{P}^1_{\text{Berk}}$ at x_i) in a natural way, since there is a canonical inclusion $T_{x_i}(\Gamma) \subset T_{x_i}$. We call $x_1, \ldots, x_n$ the *boundary points* of V and $\vec{v}_1, \ldots, \vec{v}_n$ the *ends* of V. It is easy to see that the set of ends of V is *canonical*, and does not depend on our choice of the pair (Γ, U).

REMARK 4.7.3. If we write $D_i = U(x_i; \vec{v}_i)$, then each D_i is a Berkovich open disk and

$$V = \bigcap_{i=1}^{n} D_i.$$

Using the *Mass Formula* (Proposition 4.2.2) and Corollary 4.4.6, we find:

PROPOSITION 4.7.4. *Let V be a simple domain with boundary $\{x_1, \ldots, x_n\}$ and ends $\{\vec{v}_1, \ldots, \vec{v}_n\}$, and let f be a simple function on $\mathbb{P}^1_{\text{Berk}}$. Then:*

$$\Delta_{\mathbb{P}^1_{\text{Berk}}}(f)(V) = \sum_{i=1}^{n} d_{\vec{v}_i} f(x_i).$$

Combining Proposition 4.7.4 and Corollary 4.7.1, we obtain the following result:

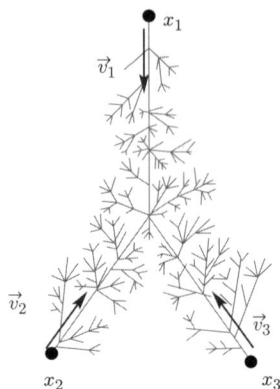

FIGURE 4. A simple domain with three boundary points and three ends.

COROLLARY 4.7.5. *Let V be a simple domain with boundary $\{x_1, \ldots, x_n\}$ and ends $\{\vec{v}_1, \ldots, \vec{v}_n\}$, let $\varphi \in K(T)$ be a nonzero rational function, and let $f = -\log_v |\varphi|$. Then:*

$$N_0(\varphi, V) - N_\infty(\varphi, V) = \sum_{i=1}^{n} d_{\vec{v}_i} f(x_i).$$

In other words, one can compute the difference between the number of zeros and the number of poles of a rational function in a simple domain V in terms of the inward-pointing derivatives of $-\log_v |\varphi|$ at each boundary point of V. This is one of the main consequences of the theory of valuation polygons.

REMARK 4.7.6. For simplicity, we have stated Corollary 4.7.5 just for rational functions, but using the Weierstrass preparation theorem, one can extend Corollary 4.7.5 more generally to meromorphic functions.

EXAMPLE 4.7.7. Let $\varphi(T) = T^2 - T$, and let $f(x) = -\log_v |T^2 - T|_x$ for $x \in \mathbb{A}^1_{\mathrm{Berk}}$. Let $D = \mathcal{B}(0, 1)^-$ be the open unit disk in $\mathbb{A}^1_{\mathrm{Berk}}$, and let $\vec{v} \in T_{\zeta_{\mathrm{Gauss}}}$ be the corresponding end (i.e., the unique tangent vector at ζ_{Gauss} pointing in the direction of 0). Since $f(x)$ is linear with slope 1 along the path from ζ_{Gauss} to 0 (with respect to the path-distance ρ), we have $d_{\vec{v}} f(\zeta_{\mathrm{Gauss}}) = 1$. This agrees with the prediction of Corollary 4.7.5, since $T^2 - T$ has precisely one zero (namely 0) in the open unit disk $B(0, 1)^-$. In other words:

$$\text{\# of zeros of } T^2 - T \text{ on } B(0, 1)^- = d_{\vec{v}} f(\zeta_{\mathrm{Gauss}}) = 1.$$

EXAMPLE 4.7.8. There is a variant of Corollary 4.7.5 for intersections of *closed* Berkovich disks (which are precisely the connected affinoid subdomains of $\mathbb{P}^1_{\mathrm{Berk}}$). For example, to compute the number of zeros of $T^2 - T$ in the *closed* unit disk $B(0, 1)$, we look at the tangent vector $\vec{w} \in T_{\zeta_{\mathrm{Gauss}}}$ pointing in the direction of ∞ (which is the unique end of the open Berkovich disk $\mathbb{P}^1_{\mathrm{Berk}} \setminus \mathcal{B}(0, 1)$), and we find that the number of zeros of $T^2 - T$ in $B(0, 1)$ is

$$\text{\# of zeros of } T^2 - T \text{ on } B(0, 1) = -d_{\vec{w}} f(\zeta_{\mathrm{Gauss}}) = 2.$$

We leave the details to the reader.

We conclude this section by explaining more precisely the relationship between Corollary 4.7.5 and the theory of valuation polygons. For this, we shift our attention to analytic functions on a closed disk.

Let $B(a, R)$ be a closed disk in K, and let $B = \mathcal{B}(a, R)$ denote the corresponding Berkovich closed disk in $\mathbb{A}^1_{\mathrm{Berk}}$. We let $x = \zeta_{a,R} \in \mathbf{H}^{\mathbb{R}}_{\mathrm{Berk}}$ denote the unique boundary point of B, and let $\vec{w}$ be the corresponding "outward-pointing" tangent direction at x. More precisely, we define $\vec{w}$ to be the end of the open disk $\mathbb{P}^1_{\mathrm{Berk}} \setminus B$, and call $\vec{w}$ the end of B. Let F be an analytic function on a strictly larger disk $B(a, R')$ containing $B(a, R)$, i.e., F is an element of the affinoid algebra

$$\{\sum_{n=0}^{\infty} a_n(T - a)^n \; : \; a_n \in K, \; \lim_{n \to \infty} |a_n|(R')^n = 0\}$$

for some $R' > R$.

We define the *growth modulus* of F to be the function $M_r(F) : [0, R'] \to \mathbb{R}$ defined by

$$M_r(F) = \max_{n \geq 0} |a_n| r^n = \sup_{z \in B(a,r)} |F(z)| = |F|_{\mathcal{B}(a,r)},$$

so that $M_r(F)$ is a positive, increasing real-valued function on $[0, R')$.

Replacing open disks by closed disks in the Mass Formula, and using the Weierstrass Preparation Theorem to pass from polynomials to analytic functions, one obtains the following variant of Proposition 4.7.4:

PROPOSITION 4.7.9. *Let B be a closed Berkovich disk with boundary point x and end $\vec{w}$, and let F be analytic on an open neighborhood U of B in $\mathbb{P}^1_{\mathrm{Berk}}$. Then*

$$\Delta_{\mathbb{P}^1_{\mathrm{Berk}}}(\log_v |F|)(B) = d_{\vec{w}}(\log_v |F|)(x).$$

It follows from the definitions that if $B = \mathcal{B}(a, R)$ is a closed Berkovich disk with unique boundary point $x = \zeta_{a,r}$ and end $\vec{w}$, then

$$d_{\vec{w}} \log_v |F|(\zeta_{a,R}) = \frac{d}{dr} \log_v M_r(F)|_{r=R}.$$

We therefore obtain a new proof using basic properties of the Laplacian on $\mathbb{P}^1_{\mathrm{Berk}}$ of the following classical result from the theory of valuation polygons:

COROLLARY 4.7.10. *Let $B(a, R)$ be a closed disk in K, and let F be analytic on $B(a, R')$ for some $R' > R$. Then*

$$\# \text{ of zeros of } F \text{ on } B(a, R) \; = \; \frac{d}{dr} \log_v M_r(F) |_{r=R}.$$

4.8. The modulus of an open annulus. In this section, we explore some more "classical" results from p-adic analysis from the point of view of Berkovich's theory. Our treatment of this material draws inspiration from [**RL03a, RL03b**].

DEFINITION 4.8.1. An *open annulus in K* is a subset of K of the form $A = B(a, R)^- \setminus B(a, r)$ with $0 < r < R$, i.e., an open disk in K with a closed disk removed from it. A is called a *rational open annulus* if $r, R \in |K^*|$.

The *standard open annulus A_r of height r* for $0 < r < 1$, is the subset of K given by $A_r = B(0, 1)^- \setminus B(0, r)$.

The *modulus* of an open annulus A is defined to be

$$\mathrm{Mod}(A) = \log_v(R/r).$$

REMARK 4.8.2. If $\mathcal{A} = \mathcal{B}(a, R)^- \backslash \mathcal{B}(a, r)$ is the Berkovich open annulus in $\mathbb{A}^1_{\text{Berk}}$ corresponding to the open annulus A in K, then by definition, the modulus $\text{Mod}(A)$ coincides with the distance $\rho(\zeta_{a,r}, \zeta_{a,R})$ between the two boundary points of $\mathcal{A}$ in $\mathbf{H}_{\text{Berk}}$. The invariance of ρ under Möbius transformations is therefore equivalent to the fact that the modulus of an annulus is invariant under Möbius transformations.

DEFINITION 4.8.3. An *analytic function* on an open annulus $A \subset K$ is a map $\varphi : A \to K$ given by a power series $\sum_{n=-\infty}^{\infty} a_n(T-a)^n$ with $a_n \in K$ which converges at every point of A.

An *analytic isomorphism* between open annuli is an analytic function $\varphi : A \to A'$ with an analytic inverse.

The following proposition shows that, just as in the complex case, the only "conformal invariant" of an open annulus is its modulus.

PROPOSITION 4.8.4. *Let A, A' be rational open annuli in K. Then there exists an analytic isomorphism between A and A' if and only if $\text{Mod}(A) = \text{Mod}(A')$.*

SKETCH. By applying suitable Möbius transformations, one is reduced to showing that the standard open annuli A_r and A_s are analytically isomorphic if and only if $r = s$. Suppose that $\phi : A_r \to A_s$ is an analytic isomorphism. Since ϕ is a nonzero analytic function on A_r, one can show that ϕ extends to a function $\mathcal{A}_r \to \mathcal{A}_s$, where $\mathcal{A}_r = \mathcal{B}(0, 1)^- \backslash \mathcal{B}(0, r)$ denotes the corresponding simple domain in $\mathbb{P}^1_{\text{Berk}}$, and that the function $f = -\log_v |\phi| : \mathcal{A}_r \to \mathbb{R}$ is harmonic and extends continuously to $\partial \mathcal{A}_r = \mathcal{A}_r \cup \{\zeta_{0,1}, \zeta_{0,r}\}$. Consequently, $\Delta_{\overline{\mathcal{A}_r}}(f) = d\delta_{\zeta_{0,1}} - d\delta_{\zeta_{0,r}}$ for some integer $d \in \mathbb{Z}$. As $g(x) = -d\log_v |T|$ has the same property, it follows that $f(x) = g(x) + C'$ for some constant C', and therefore that $\varphi(x) = C \cdot |x|^d$ for all $x \in A_r$. Letting $|x| \to 1$ shows that $C = 1$, and then letting $|x| \to r$ shows that $r = s$. $\square$

PROPOSITION 4.8.5. *Let $B = B(a, R)^-$ be an open disk in K, and let $\phi : B \to K$ be an analytic function on B defined by a power series $\sum_{n=0}^{\infty} a_n(T-a)^n$ which converges for all $x \in B$. Then $B' = \phi(B)$ is also an open disk, and there exists an integer $d \geq 1$ such that $\phi : B \to B'$ has degree d (i.e., such that every point of B' has exactly d preimages in B, counting multiplicities).*

SKETCH. By applying a suitable translation, we may assume that $\phi(0) = 0$. Let $\vec{v}$ be the end corresponding to the unique boundary point $\zeta_{a,R}$ of the open Berkovich disk $\mathcal{B} = \mathcal{B}(a, R)^-$. Then

$$R' := |\phi|_{\zeta_{a,R}} = \sup_{z \in B(a,R)} |\phi(z)|.$$

We claim that $\phi(B) = B(0, R')^-$. Indeed, $\phi(B) \subseteq B(0, R')^-$ by the maximum modulus principle, since $\log_v |\phi|$ is harmonic on $\mathcal{B}$ outside the zeros of ϕ (where it takes the value $-\infty$). And if $w \in B(0, R')^-$ is any point, then $|\phi|_{B(0,r)} = |\phi - w|_{B(0,r)}$ for all r with $|w| < r < R'$ by the ultrametric inequality. By considerations analogous to those in §4.7, the number of zeros of $\phi - w$ in B coincides with the directional derivative $d_{\vec{v}}(-\log_v |\phi|)(\zeta_{a,R})$, which is equal to d for some integer $d \geq 1$ that is independent of w. $\square$

To study the effect of ϕ on open annuli instead of open disks, one uses the following lemma, which can be proved using the theory of valuation polygons (see [**RL03b**, Lemma 4.3]:

LEMMA 4.8.6. *Let $\phi \in K(T)$ be a non-constant rational function. Then there is an $\varepsilon > 0$ (depending only on ϕ) such that if A is a rational open annulus in K with $\mathrm{Mod}(A) < \varepsilon$, then $\phi(A)$ is also a rational open annulus.*

PROPOSITION 4.8.7. *Let A be a rational open annulus in K, and let $\phi : A \to K$ be an analytic function on A for which $\phi(A)$ is also an open annulus. Then $\mathrm{Mod}(\phi(A)) = d\,\mathrm{Mod}(A)$ for some integer $d \geq 1$, and the map $\phi : A \to \phi(A)$ has degree d.*

SKETCH. By applying suitable Möbius transformations, we may assume that $0 \notin \phi(A)$, so that ϕ has no zeros or poles on A, and that $A = A_r$ is a standard open annulus. Write $\mathcal{A} = \mathcal{B}(0,1)^- \backslash \mathcal{B}(0,r)$, so that the two boundary points of $\mathcal{A}$ are $\zeta = \zeta_{0,1}$ and $\zeta' = \zeta_{0,r}$, with corresponding ends $\vec{v}$ and $\vec{v}'$. Let $f = -\log_v |\phi| : \mathcal{A} \to \mathbb{R}$. As in the proof of Proposition 4.8.4, we find that there is a nonzero integer d such that $|\phi(z)| = \alpha |z|^d$ for all $z \in A$. Applying suitable Möbius transformations, we may assume (since $\phi(A)$ is an open annulus by hypothesis) that $d \geq 1$ and that $\alpha = 1$. Thus if $A' = B(0,1)^- \backslash B(0, r^d)$, then $\phi(A) \subseteq A'$.

We claim that $\phi(A) = A'$, and more generally that every point of A' has exactly d preimages in A, counting multiplicities. To see this, choose a point $w \in A'$. Then the number $m_w(\phi)$ of zeros of $\phi - w$ in A is given by

$$m_w(\phi) = d_{\vec{v}}(-\log_v |\phi - w|)(\zeta) + d_{\vec{v}'}(-\log_v |\phi - w|)(\zeta').$$

Since $|\phi(z) - w| = |z|^d$ for $|w| < |z|^d < 1$ and $|\phi(z) - w| = |w|$ for $r < |z|^d < |w|$, we find that

$$m_w(\phi) = d + 0 = d$$

as desired. $\qquad\square$

Using Remark 4.8.2, Lemma 4.8.6, and Proposition 4.8.7, one deduces the following result, due originally to Rivera-Letelier [**RL03a**].

THEOREM 4.8.8. *Let $\varphi \in K(T)$ be a nonzero rational function of degree $d \geq 1$, let $x \in \mathbf{H}^{\mathbb{R}}_{\mathrm{Berk}}$, and let $\vec{v} \in T_x$. Then there is an integer $\deg_{\vec{v}}(\varphi)$ with $1 \leq \deg_{\vec{v}}(\varphi) \leq d$ such that for all $y \in \mathbf{H}^{\mathbb{R}}_{\mathrm{Berk}} \cap U(x; \vec{v})$ sufficiently close to x (with respect to the locally metric topology on $\mathbf{H}_{\mathrm{Berk}}$), we have*

$$\rho(\varphi(x), \varphi(y)) = \deg_{\vec{v}}(\varphi) \cdot \rho(x, y).$$

This result shows that, locally in the direction of a tangent vector $\vec{v}$, a rational function φ stretches distances in $\mathbf{H}_{\mathrm{Berk}}$ by an integer factor between 1 and d, which Rivera-Letelier calls the *local degree* of φ with respect to $\vec{v}$.

As a consequence of Theorem 4.8.8, one deduces from a simple compactness argument the following result:

COROLLARY 4.8.9. *Let $\varphi \in K(T)$ be a nonzero rational function of degree $d \geq 1$. Then for all $x, y \in \mathbf{H}_{\mathrm{Berk}}$, we have*

$$\rho(\varphi(x), \varphi(y)) \leq d \cdot \rho(x, y).$$

5. Introduction to potential theory on Berkovich curves

Following Berkovich [**Ber90**] and Thuillier [**Thu05**], we describe in this lecture how to generalize some of our constructions and "visualization techniques" from $\mathbb{P}^1_{\mathrm{Berk}}$ to more general Berkovich curves. The basic idea, explained using different terminology in [**Ber90**], is that the Berkovich analytic space X_{Berk} attached to a smooth, proper, geometrically integral curve X/K admits a deformation retraction onto a finite topological graph called the *skeleton* of X_{Berk}. A very special case of the results of [**Ber99**] and [**Ber04**] is that the skeleton of X_{Berk} can be naturally endowed with the structure of a *metrized graph*. More generally, the entire space X_{Berk} can be viewed in a canonical way as an *arboretum*, in the sense of §4.3. This observation will allow us to define a Laplacian operator on X_{Berk} which generalizes the one we have already defined on the Berkovich projective line.

Out of necessity, this lecture assumes significantly more background in rigid analysis, algebraic geometry, and Berkovich's global theory of K-analytic spaces than the previous lectures did. (See [**Con07**] for an overview of much of the background material that we will be assuming, and also [**BL85**].) Nevertheless, we have tried to keep the exposition as elementary as possible.

5.1. Visualizing Berkovich curves via the semistable reduction theorem. As before, K will denote a complete, algebraically closed non-archimedean field. Let X be a smooth, proper, and geometrically integral algebraic curve over K, and let X^{rig} denote the corresponding rigid-analytic space in the sense of Tate. In this section, we will explain how to visualize the associated Berkovich analytic space X_{Berk} (see §2.4 above) in a manner similar to the way we visualized $\mathbb{P}^1_{\mathrm{Berk}}$ and $\mathcal{M}(\mathbb{Z})$ in §1.3 and §2.5, respectively. To do this, we will use a deep result from rigid analysis, namely the existence of semistable models. We will also utilize the rigid-analytic description (due to Bosch and Lütkebohmert) of the formal fibers of a semistable model.

Let $R = \{x \in K \ : \ |x| \le 1\}$ denote the valuation ring of K, and let $\tilde{K}$ be its (algebraically closed) residue field. By a *formal model* for X, we will mean an admissible formal model in the sense of Raynaud (see [**Con07**] for further discussion). Recall that a projective curve Z over k is called *semistable* if it is reduced, and if the only singularities of Z are ordinary double points. The semistable reduction theorem implies that every curve X as above has a semistable formal model.

From now on, unless otherwise specified, we choose without comment a particular semistable formal model $\mathfrak{X}$, and let Z denote its special fiber. We denote by π the natural reduction map $\pi : X(K) \to Z(\tilde{K})$. If $z \in Z(\tilde{K})$, we call $X(z) = \pi^{-1}(z)$ the *formal fiber* of X over z. It carries the structure of a rigid analytic space in a natural way. Let $B(0, r)$ denote the closed disk of radius r in K (i.e., the affinoid space corresponding to $K\langle r^{-1}T\rangle$), and let $B(0, r)^-$ be the corresponding open disk, which is the inverse image of 0 under the canonical reduction map from $\mathrm{Sp}\, K\langle r^{-1}T\rangle$ to the affine line over $\tilde{K}$. Also, for $0 < \alpha < 1$, define

$$A(\alpha) = \mathrm{Sp}\, K\langle T, \alpha T^{-1}\rangle \leftrightarrow \{z \in B(0, 1) \ : \ \alpha \le |z| \le 1\}$$

to be the *standard closed annulus of height α* in $B(0, 1)$. The canonical reduction of $A(\alpha)$ is the scheme $\mathrm{Spec}\, \tilde{K}[S, T]/(ST)$ consisting of two affine lines intersecting at an ordinary double point. We also let

$$A(\alpha)^- = \{z \in B(0,1) \ : \ \alpha < |z| < 1\}$$

be the corresponding *standard open annulus of height* α, which is the inverse image in $A(\alpha)$ of the singular point of $\mathrm{Spec}\, \tilde{K}[S,T]/(ST)$ under the canonical reduction map.

We recall the following result due to Bosch and Lütkebohmert:

PROPOSITION 5.1.1. *Let $z \in Z(\tilde{K})$. Then:*

(i) *If z is a nonsingular point of Z, then the formal fiber $X(z)$ is analytically isomorphic to the open unit disk $B(0,1)^-$.*

(ii) *If z is a singular point of Z (which by hypothesis is an ordinary double point), then the formal fiber $X(z)$ is analytically isomorphic to a standard open annulus $A(\alpha)^-$ for some unique $\alpha \in |K^*|$ with $0 < \alpha < 1$.*

REMARK 5.1.2. In fact, the proof of (i) shows that there exists an affinoid neighborhood V of $X(z)$ and an analytic map $V \to B(0,1)$ which restricts to an isomorphism $X(z) \xrightarrow{\sim} B(0,1)^-$. Similarly, the proof of (ii) shows that if z lies on two different components of Z, then there exists an affinoid neighborhood V of $X(z)$ and an analytic map $V \to A(\alpha)$ which restricts to an isomorphism $X(z) \xrightarrow{\sim} A(\alpha)^-$.

For our purposes, it is important to note that, just as over the complex numbers, the height α is a "conformal invariant" of an annulus. More precisely, define a *closed annulus* to be an affinoid of the form

$$A(\alpha, \beta) = \mathrm{Sp}\, K\langle \beta^{-1}T, \alpha T^{-1}\rangle = \{z \in \mathbb{A}^1 \ : \ \alpha \le |z| \le \beta\},$$

and define an *open annulus* to be a rigid space of the form $A(\alpha, \beta)^-$ (i.e., the formal fiber over the singular point of the canonical reduction of a closed annulus). As in §4.8, we define the *modulus* of a closed (resp. open) annulus $A = A(\alpha, \beta)$ (resp. $A = A(\alpha, \beta)^-$) to be $m(A) = \log_v \beta - \log_v \alpha$. Generalizing Proposition 4.8.4, we have:

LEMMA 5.1.3.   (i) *Any two closed (resp. open) rational (i.e., with radii in $|K^*|$) disks are isomorphic as rigid spaces.*

(ii) *Two closed (resp. open) rational annuli A, A' are isomorphic as rigid spaces if and only if they have the same modulus.*

For simplicity, we assume for the moment that the special fiber Z of the formal model $\mathfrak{X}$ satisfies $Z = \bigcup_{i=1}^t Z_i$, where the Z_i are smooth and irreducible projective curves, each containing at least two double points of Z. Let Z_i^* be the nonsingular affine curve obtained from Z_i by removing all double points of Z lying on Z_i. The rigid analytic space $X_i^* = \pi^{-1}(Z_i^*)$ is an affinoid with nonsingular canonical reduction Z_i^*. For each ordinary double point $p \in Z(\tilde{K})$, lying on the components Z_i and Z_j, we also have an affinoid $X_p = \pi^{-1}(Z_i^* \cup Z_j^* \cup \{p\})$ whose canonical reduction is $(Z_i^* \cup Z_j^* \cup \{p\})$. As a rigid space, X is obtained by gluing the affinoids X_p along the affinoid subsets X_i^*.

Berkovich's construction of global K-analytic spaces in [**Ber93**] (cf. §2.4) yields a space X_{Berk} (the "Berkovich analytification" of X^{rig}) obtained by gluing the affinoid spaces $(X_p)_{\mathrm{Berk}}$ along the affinoid subsets $(X_i^*)_{\mathrm{Berk}}$. The reduction map $\pi : X(K) \to Z(\tilde{K})$ extends naturally to a map $\mathrm{red} : X_{\mathrm{Berk}} \to Z$, where the target space is the set of *scheme-theoretic* points of Z. According to [**Ber90**, Proposition 2.4.4], for each irreducible component Z_i of Z, there is a *unique* point ζ_i of X_{Berk} reducing to the generic point of Z_i. We now describe, following [**FvdP04**, §7.2], how the underlying topological space of X_{Berk} can be understood in terms of the *dual graph* of Z.

Recall that the spaces $(X_i^*)_{\mathrm{Berk}}$ correspond to the irreducible components of Z, and the spaces $(X_p)_{\mathrm{Berk}}$ to the intersections between components. This can be conveniently encoded via the dual graph $\Sigma_{\mathfrak{X}}$ of Z, which is the topological graph having a vertex for each component of Z, and an edge for each point of intersection between two components. The gluing data for the space X_{Berk} corresponds in a natural way to $\Sigma_{\mathfrak{X}}$. According to [**Ber90**, Chapter 4], since each X_i^* is a 1-dimensional affinoid with nonsingular canonical reduction, the spaces $(X_i^*)_{\mathrm{Berk}}$ are all contractible. More concretely, there is a deformation retraction $r_i : (X_i^*)_{\mathrm{Berk}} \twoheadrightarrow \{\zeta_i\}$.

One can visualize the retraction r_i as follows. By Proposition 5.1.1, the rigid-analytic formal fiber $\pi^{-1}(z)$ over any closed point $z \in Z_i^*(\tilde{K})$ is isomorphic to the open disk $B(0,1)^-$. Applying the Berkovich analytification functor, one sees that the inverse image $\mathrm{red}^{-1}(z)$ is isomorphic to the Berkovich open disk $\mathcal{B}(0,1)^-$. The closure of $\mathrm{red}^{-1}(z)$ in $(X_i^*)_{\mathrm{Berk}}$ (or equivalently, in X_{Berk}) is precisely $\mathrm{red}^{-1}(z) \cup \{\zeta_i\}$, which is isomorphic to $\mathcal{B}(0,1)^- \cup \{\zeta_{\mathrm{Gauss}}\}$, the closure of $\mathcal{B}(0,1)^-$ in $\mathcal{B}(0,1)$. Using the tree structure of $\mathcal{B}(0,1)^-$, one therefore sees that each Berkovich formal fiber $\mathrm{red}^{-1}(z)$ admits a deformation retraction onto $\{\zeta_i\}$, and the deformation retraction $r : (X_i^*)_{\mathrm{Berk}} \twoheadrightarrow \{\zeta_i\}$ is obtained by simultaneously retracting all of these formal fibers to this point.

Similarly, for each singular point p of Z, corresponding to an intersection of Z_i and Z_j, the space $(X_p)_{\mathrm{Berk}}$ deformation retracts onto a line segment e_p with endpoints ζ_i and ζ_j. One can again visualize this retraction using Proposition 5.1.1, as there is an isomorphism $\psi : \pi^{-1}(p) \to A(\alpha_p)^-$ from the rigid-analytic formal fiber $\pi^{-1}(p)$ to an open annulus $A(\alpha_p)^-$ which extends to an isomorphism on Berkovich spaces. By considering the reduction map $\mathrm{red} : X_p \to Z_i^* \cup Z_j^* \cup \{p\}$, one sees that $(X_p)_{\mathrm{Berk}} = \mathrm{red}^{-1}(p) \cup (X_i)_{\mathrm{Berk}} \cup (X_j)_{\mathrm{Berk}}$ as sets. The closure of $\mathrm{red}^{-1}(p)$ in $(X_p)_{\mathrm{Berk}}$ is $\mathrm{red}^{-1}(p) \cup \{\zeta_i, \zeta_j\}$, and the open annulus $A(\alpha_p)^- \subset \mathcal{B}(0,1)$ deformation retracts onto the open segment connecting its two boundary points $\psi(\zeta_i)$ and $\psi(\zeta_j)$ in $\mathbb{A}^1_{\mathrm{Berk}}$, which are the type II points corresponding to the disks $B(0,1)$ and $B(0,\alpha_p)$. (Of course, $A(\alpha_p)^-$ is in fact contractible, but to get a retraction which is compatible with the gluing maps, we need to keep the points $\psi(\zeta_i)$ and $\psi(\zeta_j)$ fixed.) The segment e_p is the unique path in $(X_p)_{\mathrm{Berk}}$ from ζ_i to ζ_j, and can be thought of as a "line of embedded disks" $\{\psi^{-1}(B(0,r)) : \alpha_p \leq r \leq 1\}$ linking the "Gauss point" of X_i^* to the "Gauss point" of X_j^*.

Globalizing, we now see that the entire Berkovich space X_{Berk} admits a deformation retraction onto the topological space obtained by gluing the segments e_p along the points ζ_i, which is precisely the topological dual graph $\Sigma_{\mathfrak{X}}$. Let $r : X_{\mathrm{Berk}} \to \Sigma_{\mathfrak{X}}$ be this retraction map. For $x \in \Sigma_{\mathfrak{X}}$, each connected component of $r^{-1}(x) \backslash \{x\}$ is isomorphic to $\mathcal{B}(0,1)^-$, and in particular is a topological tree.

FIGURE 5. The Berkovich analytic space associated to an elliptic curve with multiplicative reduction.

We now define a path metric $\rho(x, y)$ on $\mathbf{H}(X) = X_{\text{Berk}} \backslash X(K)$. It is enough to define ρ locally, so we may assume that either x, y lie in the closure of the same connected component V of $r^{-1}(z)$ for some $z \in \Sigma_{\mathfrak{X}}$, or that $x, y \in e_p$ for some singular point $p \in Z$. Since both $\overline{V}$ and e_p are isomorphic to subsets of $\mathcal{B}(0, 1)$, we can use the path metric ρ on $\mathcal{B}(0, 1)$ to define distances locally on $\mathbf{H}(X)$. Lemma 5.1.3 shows that this is well-defined. It is not hard to show that the metric on $\mathbf{H}(X)$ is canonical, i.e., does not depend on our choice of a particular semistable formal model $\mathfrak{X}$ of X.

REMARK 5.1.4. The set $S(X_{\text{Berk}})$ of points of X_{Berk} corresponding to generic points of irreducible components of Z having arithmetic genus at least 1 is independent of the formal model $\mathfrak{X}$. We will refer to $S(X_{\text{Berk}})$ as the set of *marked points* of X_{Berk}. The set of marked points of X_{Berk} is empty if and only if X is a *Mumford curve* in the sense of [**GvdP80**].

REMARK 5.1.5. If the genus of X is at least one, there is a maximal subgraph Σ of $\Sigma_{\mathfrak{X}}$ containing the set $S(X_{\text{Berk}})$ of marked points and having no unmarked vertices of degree 1. Following Berkovich, the metrized graph Σ is called the *skeleton* of X_{Berk}. The skeleton is canonical, and in particular does not depend on the choice of a formal model $\mathfrak{X}$ for X. We refer to the metrized graph $\Sigma_{\mathfrak{X}}$, which admits a deformation retraction onto Σ, as the skeleton of X_{Berk} with respect to the formal model $\mathfrak{X}$.

The existence of the skeleton of a curve of genus at least one, and the non-existence of such a canonical skeleton for curves of genus zero, is closely related to the theory of minimal models for arithmetic surfaces (see, e.g., [**Liu02**]). For example, if X has genus at least 2 and is defined over a discretely valued subfield K' of K with valuation ring R', then the skeleton of the formal covering associated to the (unique) minimal regular model of X over R' coincides with the skeleton of X_{Berk}.

5.2. Tate elliptic curves. The semistable reduction theorem is not the only way to study Berkovich curves. For example, when studying curves with totally degenerate reduction, one can also profit considerably from the point of view offered by p-adic uniformization theory. To illustrate the utility of this point of view, in this section we give a detailed topological description of the Berkovich analytic space associated to a Tate elliptic curve (i.e., an elliptic curve with multiplicative reduction) from the point of view of Tate's non-archimedean uniformization theory.

Let E/K be an elliptic curve with multiplicative reduction, let E^{rig} be the corresponding rigid analytic space, and let E_{Berk} be the corresponding Berkovich analytic space. According to Tate's theory, $E^{\text{rig}} \cong (\mathbf{G}_m^{\text{rig}})/q^{\mathbb{Z}}$, where $\mathbf{G}_m^{\text{rig}}$ is the rigid analytic space associate to the multiplicative group over K and $q \in K^*$ satisfies $|q| < 1$. In particular, $E(K) \cong K^*/q^{\mathbb{Z}}$. By choosing coordinates on $\mathbb{P}^1$, we can identify $\mathbf{G}_m^{\text{rig}}$ with $(\mathbb{P}^1)^{\text{rig}} \backslash \{0, \infty\}$. We can then view the Berkovich analytic space E_{Berk} somewhat informally as the space obtained from the closed annulus $V = \mathcal{B}(0,1)\backslash\mathcal{B}(0,|q|)^-$ by identifying the affinoid subspaces $V_1 = \mathcal{B}(0,1)\backslash\mathcal{B}(0,1)^-$ and $V_2 = \mathcal{B}(0,|q|)\backslash\mathcal{B}(0,|q|)^-$ in the "obvious" way. Let $\pi : V \to E_{\text{Berk}}$ denote the corresponding quotient map.

We may visualize the resulting graph structure on E_{Berk}, and the path distance ρ on $\mathbf{H}(E_{\text{Berk}}) := E_{\text{Berk}}\backslash E(K)$, as follows. Let ζ, ζ' be the points of $\mathbb{P}^1_{\text{Berk}}$ corresponding to $B(0,1)$ and $B(0,|q|)$, respectively, and let I denote the unique path in $\mathbb{P}^1_{\text{Berk}}$ from ζ to ζ', so that I is isometric to a segment of length $\ell = -\log_v |q|$ with $0 < \ell < \infty$. Then the image $\pi(I)$ of I in E_{Berk} is isometric to a circle Σ of length ℓ, which we call the *skeleton* of E_{Berk}. There is also a retraction map $r : E_{\text{Berk}} \to \Sigma$ induced by the retraction (which we also denote by r) of V onto I. Note that if $x \in I$ corresponds to the disk $B(0,R)$ with $R \in |K^*|$ (i.e., x is a point of type II), then the space $\{z \in V : r(z) = x\}$ is naturally identified with $\mathcal{B}(0,R)\backslash\mathcal{B}(0,R)^-$, which itself is isomorphic (and isometric) to $\mathcal{B}(0,1)\backslash\mathcal{B}(0,1)^-$. If $R \notin |K^*|$ (i.e., x is a point of type III), then $r^{-1}(x) = \{x\}$. In this way, we can view E_{Berk} as fibered over the circle Σ, and each fiber $r^{-1}(x)$ for $x \in \Sigma$ is isomorphic either to a point or to the closed annulus $W = \mathcal{B}(0,1)\backslash\mathcal{B}(0,1)^-$, which has the structure of an infinite metrized tree. In particular, E_{Berk} can be thought of as a family of infinite metrized trees fibered over the rational points of the circle Σ. For any two points $x, y \in E_{\text{Berk}}$, we can define the path-distance $\rho(x,y)$ as the length of the shortest path from x to y, where path lengths are given on each annulus W by the usual path distance on $\mathbb{P}^1_{\text{Berk}}$, and on Σ by identifying it with a circle of length ℓ. If $x \neq y$, we have $\rho(x,y) = \infty$ if and only if one of x and y is in $E(K)$. In particular, $\rho(x,y) < \infty$ for all $x, y \in \mathbf{H}(E_{\text{Berk}})$.

The Berkovich topology on E_{Berk} is the quotient topology induced by the map π; a fundamental system of open neighborhoods for the topology on E_{Berk} is given by $\{U_{\Gamma,V}\}$, where Γ is a finitely branched connected subgraph of $\mathbf{H}(E_{\text{Berk}})$ containing Σ, $V \subseteq \Gamma$ is open in the metric topology on Γ, and $U_{\Gamma,V} = r^{-1}(V)$.

REMARK 5.2.1. The space $\mathbb{P}^1_{\text{Berk}} \backslash \{0, \infty\}$ is simply connected, and is in fact the universal covering space of E_{Berk} via the natural map from $\mathbb{P}^1_{\text{Berk}} \backslash \{0, \infty\}$ to E_{Berk}. The fundamental group of E_{Berk} is $\mathbb{Z}$, since it admits a deformation retraction onto the circle Σ. This is yet another illustration of the power and utility of Berkovich's theory: it allows one to study non-archimedean analytic spaces using standard tools of algebraic topology!

5.3. Harmonic functions on Berkovich curves. In this section, we define what it means for a real-valued function on a Berkovich curve X_{Berk} to be harmonic. This is analogous to the discussion in §3.2.

5.3.1. *Definition of a harmonic function.* As in §1.4.3, if $x \in X_{\mathrm{Berk}}$ then there is a well-defined set T_x of *tangent directions* at x, defined as the set of equivalence classes of paths $\ell_{x,y}$ emanating from x, where y is any point of X_{Berk} not equal to x and two paths $\ell_{x,y_1}, \ell_{x,y_2}$ are *equivalent* if they share a common initial segment.

Define $\mathbf{H}^{\mathbb{R}}(X_{\mathrm{Berk}})$ to be the set of all points $x \in X_{\mathrm{Berk}}$ for which T_x has more than one element. (This coincides with the definition of $\mathbf{H}^{\mathbb{R}}_{\mathrm{Berk}}$ when $X = \mathbb{P}^1_{\mathrm{Berk}}$.) As with $\mathbf{H}_{\mathrm{Berk}}$, we view $\mathbf{H}^{\mathbb{R}}(X_{\mathrm{Berk}})$ as equipped with the metric ρ defined in §5.1.

Let U be a connected open subset of X_{Berk}, and let $f : U \to \mathbb{R} \cup \{\pm\infty\}$ be a continuous extended-real valued function which is finite-valued on $\mathbf{H}^{\mathbb{R}}(X_{\mathrm{Berk}})$.

As in §3.2, we say that f is *continuous piecewise affine* on U, and write $f \in \mathrm{CPA}(U)$ if:

(CPA1) The restriction of f to $\mathbf{H}^{\mathbb{R}}(X_{\mathrm{Berk}})$ is piecewise-affine with respect to the path metric ρ.

(CPA2) If $f \in \mathrm{CPA}(U)$ and $x \in U \cap \mathbf{H}^{\mathbb{R}}(X_{\mathrm{Berk}})$, then for each $\vec{v} \in T_x$ the *directional derivative* $d_{\vec{v}} f(x)$ is well-defined.

(CPA3) For each $x \in U \cap \mathbf{H}^{\mathbb{R}}(X_{\mathrm{Berk}})$, we have $d_{\vec{v}} f(x) = 0$ for all but finitely many $\vec{v} \in T_x$. In particular, the quantity

$$\Delta_x(f) := - \sum_{v \in T_x} d_{\vec{v}} f(x)$$

is well-defined for each $x \in U \cap \mathbf{H}^{\mathbb{R}}(X_{\mathrm{Berk}})$.

Let $x \in U$, and let $h \in \mathrm{CPA}(U)$. As in §3.2, we make the following definition.

DEFINITION 5.3.1. 1. If $x \in \mathbf{H}^{\mathbb{R}}(X_{\mathrm{Berk}})$, we say that h is *harmonic at x* if $\Delta_x(h) = 0$.

2. If $x \notin \mathbf{H}^{\mathbb{R}}(X_{\mathrm{Berk}})$, we say that h is *harmonic at x* if h is constant on an open neighborhood of x.

5.3.2. *Example: Logarithms of absolute values of meromorphic functions.* As before, we let X_{Berk} be a Berkovich curve, and we let f be a nonzero meromorphic function on X_{Berk}. We also let $\mathfrak{X}$ be a formal model for the underlying rigid-analytic space X^{rig}, and we let $r : X_{\mathrm{Berk}} \twoheadrightarrow \Sigma_{\mathfrak{X}}$ be the corresponding retraction onto the skeleton of X_{Berk} with respect to $\mathfrak{X}$.

PROPOSITION 5.3.2. *Let φ be the unique continuous function from X_{Berk} to $\mathbb{R} \cup \{\pm\infty\}$ for which $\varphi(x) = -\log_v |f(x)|$ outside the zeros and poles of f. Then $\varphi \in \mathrm{CPA}(X_{\mathrm{Berk}})$.*

SKETCH. Since the property of belonging to $\mathrm{CPA}(X_{\mathrm{Berk}})$ is local, and since the zeros and poles of a meromorphic function are finite in number, it suffices to check the result for an affinoid V on which either f or $1/f$ is analytic. In this case, φ is continuous by definition. Moreover, the linearity of φ can be checked locally on residue classes U isomorphic to open disks or open annuli in $\mathbb{P}^1_{\mathrm{Berk}}$. So we can reduce the assertion that φ satisfies (CPA1) to the corresponding fact on $\mathbb{P}^1_{\mathrm{Berk}}$, which we already know to be true. Furthermore, note that $X_{\mathrm{Berk}} \setminus \Sigma_{\mathfrak{X}}$ is a disjoint union of branches emanating from $\Sigma_{\mathfrak{X}}$, each of which is isomorphic to an open disk U in $\mathbb{P}^1_{\mathrm{Berk}}$. By the corresponding fact for open subsets of $\mathbb{P}^1_{\mathrm{Berk}}$, we obtain that φ satisfies (CPA2). Finally, if f has no zeros or poles on such a branch, then the corresponding function $\tilde{\phi} = -\log_v |\tilde{f}| : U \to \mathbb{R}$ is harmonic on an open disk, so by

Corollary 3.3.2 it is constant. It follows that φ is constant in all but finitely many tangent directions emanating from $\Sigma_{\mathfrak{x}}$, proving (CPA3). $\qquad\square$

We will now show that φ is harmonic outside the zeros and poles of f. In order to do this, we describe the translation into the language of Berkovich spaces of some results of Bosch and Lütkebohmert from rigid analysis (see [**BL85**, Section 3]). As before, let $Z_1, \ldots, Z_t$ be the irreducible components of Z, and for each $i = 1, \ldots, t$, choose an open affine subset $W_i \subset Z_i$. Then the preimage $V_i = \pi^{-1}(W_i)$ is an affinoid subdomain of X^{rig}.

Let $\zeta_i \in X_{\mathrm{Berk}}$ be the "Gauss point" corresponding to an irreducible component Z_i of Z. The tangent space T_{ζ_i}, which is the set of equivalence classes of paths emanating from ζ_i, is naturally in bijection with the set $\tilde{Z}_i(\tilde{K})$ of closed points of $\tilde{Z}_i$, where $\tilde{Z}_i$ is the normalization of Z_i. We make this bijection explicit by writing $T_{\zeta_i} = \{\vec{v}_{\tilde{z}}\}_{\tilde{z} \in \tilde{Z}_i(\tilde{K})}$, where $\vec{v}_{\tilde{z}}$ is a formal unit vector emanating from ζ_i in the direction corresponding to $\tilde{z} \in \tilde{Z}_i(\tilde{K})$.

REMARK 5.3.3. Our explicit "visualization" of X_{Berk} using the semistable reduction theorem shows that every point $x \in X_{\mathrm{Berk}} \backslash \{\zeta_1, \ldots, \zeta_t\}$ has an open neighborhood isomorphic to an open disk or open annulus in $\mathbb{P}^1_{\mathrm{Berk}}$, and by the corresponding result on $\mathbb{P}^1_{\mathrm{Berk}}$, in order to show that $\varphi(x)$ is harmonic outside the zeros and poles of f, it suffices to check that $\Delta_{\zeta_i}(\varphi) = 0$ at the finitely many points $\zeta_i \in X_{\mathrm{Berk}}$ corresponding to the generic points of the irreducible components of Z.

For each component Z_i, there exists a nonzero scalar $c_i = c_i(f) \in K^*$, depending on Z_i and on f, such that $c_i^{-1} f$ reduces to a nonzero rational function on Z_i. The absolute value $|c_i|$ of c_i is uniquely determined, and is in fact equal to $|f|_{\zeta_i}$. The divisor of $\overline{c_i^{-1} f}$ on the normalization $\tilde{Z}_i$ of Z_i depends only on f, not on the choice of a particular constant c_i. We can thus define the *order of f at a point* $\tilde{z} \in \tilde{Z}_i(\tilde{K})$ to be

$$\mathrm{ord}_{\tilde{z}, \tilde{Z}_i}(f) := \mathrm{ord}_{\tilde{z}}(\overline{c_i^{-1} f}).$$

Part (2) of the following proposition is Proposition 3.1 of [**BL85**], and part (1) is established as the key step in the proof of this proposition.

PROPOSITION 5.3.4. *Let f be a nonzero meromorphic function on X_{Berk}.*

(1) *If $\tilde{z} \in \tilde{Z}_i$, then the directional derivative $d_{\vec{v}_{\tilde{z}}}(-\log_v |f|)(\zeta_i)$ is equal to $-\mathrm{ord}_{\tilde{z}, \tilde{Z}_i}(f)$.*

(2) *Let $z \in Z(\tilde{K})$, and denote by $\tilde{z}_1, \ldots, \tilde{z}_r$ the points in the normalization of Z lying over z (so that r equals 1 or 2, depending on whether z is nonsingular or singular). Then the divisor of f has degree*

$$\deg(\mathrm{Div}(f)|_{\pi^{-1}(z)}) = \sum_{i=1}^{r} \mathrm{ord}_{\tilde{z}_i, \tilde{Z}_i}(f)$$

on the formal fiber $\pi^{-1}(z)$.

As an immediate consequence of part (1) of the proposition, we obtain:

COROLLARY 5.3.5. *Let f be a nonzero meromorphic function on X_{Berk}, let Z_i be an irreducible component of Z, and let $\zeta_i \in X_{\mathrm{Berk}}$ be the corresponding "Gauss point". Then*

$$\Delta_{\zeta_i}(-\log_v|f|) = \sum_{\tilde{z} \in \tilde{Z}_i(\tilde{K})} d_{\vec{v}_{\tilde{z}}}(\log_v|f|)(\zeta_i) = 0.$$

PROOF. The sum in question is equal to

$$\sum_{\tilde{z} \in \tilde{Z}_i(\tilde{K})} \mathrm{ord}_{\tilde{z},\tilde{Z}}(\overline{c_i^{-1}f}) = 0,$$

since a rational function on a complete nonsingular curve has the same number of zeros as poles (counting multiplicities). □

Combining Corollary 5.3.5 and Remark 5.3.3, we find:

COROLLARY 5.3.6. *Let f be a nonzero meromorphic function on X_{Berk}, and let $\varphi = -\log_v|f|$. Then φ is harmonic outside the zeros and poles of f.*

5.4. The Laplacian on a Berkovich curve.

5.4.1. *A higher genus analogue of Berkovich's classification theorem.* Let X_{Berk} be a Berkovich curve over K.

Define a *finite subgraph* of X_{Berk} to be a (finite and connected) metrized graph contained in $\mathbf{H}^{\mathbb{R}}(X_{\mathrm{Berk}})$. It follows from our description in §5.1 that the collection $\mathcal{S}$ of all finite subgraphs Γ_α of X_{Berk} containing $\Sigma_{\mathfrak{X}}$ forms an arboreal system of metrized graphs.

Moreover, we have the following generalization of Theorem 2.2.1, which can be thought of as an extension of Berkovich's classification theorem from $\mathbb{P}^1_{\mathrm{Berk}}$ to arbitrary Berkovich curves.

THEOREM 5.4.1. *X_{Berk} is homeomorphic to $\varprojlim_{\Gamma \in \mathcal{S}} \Gamma$.*

There is a useful and more natural version of Theorem 5.4.1 which can be formulated in terms of formal models. If $\mathfrak{X}$ is a semistable formal model of X, the dual graph $\Gamma_{\mathfrak{X}}$ of the special fiber of $\mathfrak{X}$ can be naturally endowed with the structure of a metrized graph. Moreover, given any two such formal models $\mathfrak{X}_1$ and $\mathfrak{X}_2$, there is a third semistable formal model $\mathfrak{X}_3$ of X which dominates both $\mathfrak{X}_1$ and $\mathfrak{X}_2$. Consequently, one finds that the collection

$$\{\Gamma_{\mathfrak{X}} \ : \ \mathfrak{X} \text{ is a semistable formal model of } X\}$$

forms an arboreal system of metrized graphs. The reformulation of Theorem 5.4.1 in terms of formal models is that X_{Berk} is homeomorphic to $\varprojlim_{\mathfrak{X}} \Gamma_{\mathfrak{X}}$.

5.4.2. *The Laplacian on X_{Berk}.* The discussion in the previous section shows that X_{Berk} is naturally endowed with the structure of an arboretum, and in particular there is a well-defined Laplacian operator on X_{Berk}.

Here is an example which generalizes both Example 4.5.6 and Corollary 5.3.6.

EXAMPLE 5.4.2. Let φ be a nonzero meromorphic function on the algebraic curve X/K. Then there is a continuous function $-\log_v|\varphi| : X_{\mathrm{Berk}} \to \mathbb{R} \cup \{\pm\infty\}$ in $\mathrm{CPA}(X_{\mathrm{Berk}})$ extending the usual map $x \mapsto -\log_v|\varphi(x)|$ on $X(K)$. One can deduce from the discussion in §5.3.2 the following *Poincaré-Lelong formula for Berkovich curves*:

$$(5.4.3) \qquad\qquad \Delta_{X_{\mathrm{Berk}}}(-\log_v|\varphi|) = \delta_{\mathrm{Div}(\varphi)}.$$

(See Thuillier [**Thu05**] for a generalization of this result.)

Arizona Winter School Project #3: The left-hand side of formula (5.4.3) depends on the metric structure on $\mathbf{H}^{\mathbb{R}}(X_{\mathrm{Berk}})$. Show that ρ is the *unique* metric on $\mathbf{H}^{\mathbb{R}}(X_{\mathrm{Berk}})$ for which the formula (5.4.3) holds.

5.4.3. *Examples.* We give two examples of Laplacians of functions on Berkovich curves.

EXAMPLE 5.4.4. If X_{Berk} is a Berkovich curve and $y, z \in X_{\mathrm{Berk}}$, there is, up to an additive constant, a unique function $f(x) = \kappa_z(x, y) : X_{\mathrm{Berk}} \to \mathbb{R} \cup \{\pm\infty\}$ in BDV(X_{Berk}) for which $\Delta_{X_{\mathrm{Berk}}}(f) = \delta_y - \delta_z$. We call any such function a *fundamental potential kernel* (or *generalized Hsia kernel*), and the corresponding function $[x, y]_z = q_v^{-\kappa_z(x,y)}$ a *canonical distance function* on X_{Berk} relative to z; cf. (4.5.4).

The restriction of the canonical distance $[x, y]_z$ to $x, y, z \in X(K)$ agrees with the canonical distance constructed by Rumely in [**Rum89**] (which is also defined only up to a constant).

As in Proposition 1.6.10(5), the canonical distance can be used to factorize absolute values of meromorphic functions on X: if f is a nonzero meromorphic function with divisor $\mathrm{Div}(f) = \sum m_i(a_i)$, then for any $z \in X_{\mathrm{Berk}}$ there is a constant C (depending on z and f) such that

$$|f(x)| = C \cdot \prod [x, a_i]_z^{m_i}$$

for all $x \in X_{\mathrm{Berk}}$.

EXAMPLE 5.4.5. Let E_{Berk} be the Berkovich analytic space associated to a Tate elliptic curve E/K, and let Σ be the skeleton of E_{Berk}, which is isometric to a circle of length $\ell > 0$. Let μ be the normalized Haar measure supported on $\Sigma \subset E_{\mathrm{Berk}}$. Let $\lambda : E(K) \backslash \{O\} \to \mathbb{R}$ be the Néron canonical local height relative to the origin, as defined in [**Sil94**, §VI.1]. Then λ extends in a canonical way to a function $\lambda : E_{\mathrm{Berk}} \to \mathbb{R} \cup \{+\infty\}$ which is singular only at the origin O. Moreover, $\lambda \in \mathrm{BDV}(E_{\mathrm{Berk}})$ and

(5.4.6) $\Delta_{E_{\mathrm{Berk}}}(\lambda) = \delta_O - \mu,$

so that λ can be considered as a kind of *Green's function* on E_{Berk}. Moreover, the normalization of λ given in [**Sil94**, §VI.1] ensures that

(5.4.7) $\displaystyle\int_{E_{\mathrm{Berk}}} \lambda \, d\mu = 0.$

The canonical local height λ can in fact be completely characterized as the unique function satisfying (5.4.6) and (5.4.7).

5.4.4. *The Laplacian on a simple subdomain of X_{Berk}.* For domains in $\mathbb{P}^1_{\mathrm{Berk}}$, we defined a Laplacian operator $\Delta_{\bar{U}}$ on the closure $\bar{U}$ of U by viewing $\bar{U}$ as the inverse limit of all finite subgraphs of $\mathbb{P}^1_{\mathrm{Berk}}$ contained in U. If $U \subseteq X_{\mathrm{Berk}}$ is an arbitrary domain, it is no longer true in general that $\bar{U}$ is the inverse limit of all finite subgraphs of X_{Berk} contained in U. For one thing, such graphs need not belong to $\mathcal{S}$, since they might not contain the skeleton $\Sigma_{\mathfrak{X}}$. Another problem is illustrated by the following observation: if X is a Tate elliptic curve, $r : X_{\mathrm{Berk}} \to \Sigma$ is the corresponding retraction, and $V = \Sigma \backslash \{p\}$ for some point $p \in \Sigma$, then $U = r^{-1}(V)$

has closure equal to $U \cup \{p\}$, which is not the same as the inverse limit of all finite subgraphs of X_{Berk} contained in U.

There are various possible ways to remedy the situation. The most natural is probably the development of potential theory on Berkovich curves as in [**Thu05**] in terms of formal and rigid geometry.

Here is another possible approach:

DEFINITION 5.4.8. A *simple compact set* in X_{Berk} is a subset of the form $r_\Gamma^{-1}(W)$ with $\Gamma \in \mathcal{S}$ and W a connected union of closed intervals in Γ.

If V is a simple compact set in X_{Berk}, we define

$$\mathcal{S}_V = \{\Gamma \cap V \ : \ \Gamma \in \mathcal{S} \text{ and } \Gamma \cap V \neq \emptyset\}.$$

With a bit of thought, one can show that $\mathcal{S}_V$ is an arboreal system of metrized graphs, and that $(V, \mathcal{S}_V)$ is an arboretum. In particular, V is homeomorphic to $\varprojlim_{\Gamma' \in \mathcal{S}_V} \Gamma'$, and one can construct a corresponding Laplacian operator Δ_V.

Bibliography

[Bak07] M.Baker. A finiteness theorem for canonical heights attached to rational maps over function fields, to appear in *J. Reine Angew. Math.*

[Ber90] V. G. Berkovich. *Spectral theory and analytic geometry over non-Archimedean fields*, volume 33 of *Mathematical Surveys and Monographs*. American Mathematical Society, Providence, RI, 1990.

[Ber93] V. G. Berkovich. Étale cohomology for non-Archimedean analytic spaces. *Inst. Hautes Études Sci. Publ. Math.*, 78:5–161, 1993.

[Ber95] V. G. Berkovich. The automorphism group of the Drinfeld half-plane. *C. R. Acad. Sci. Paris Ser. I Math.*, 321:1127–1132, 1995.

[Ber99] V. G. Berkovich. Smooth p-adic analytic spaces are locally contractible. *Invent. Math.*, 137(1):1–84, 1999.

[Ber04] V. G. Berkovich. Smooth p-adic analytic spaces are locally contractible. II. In *Geometric aspects of Dwork theory*, pages 293–370. Walter de Gruyter and Co. KG, Berlin, 2004.

[BF06] M. Baker and X. Faber. Metrized graphs, Laplacian operators, and electrical networks. *Contemp. Math.*, 415, 2006.

[BL85] S. Bosch and W. Lütkebohmert. Stable reduction and uniformization of abelian varieties. I. *Math. Ann.*, 270(3):349–379, 1985.

[BR06] M. Baker and R. Rumely. Equidistribution of small points, rational dynamics, and potential theory. *Ann. Inst. Fourier (Grenoble)*, 56(3):625–688, 2006.

[BR07] M. Baker and R. Rumely. Harmonic analysis on metrized graphs. *Canad. J. Math.*, 59(2):225–275, 2007.

[BR08] M. Baker and R. Rumely. Potential theory on the Berkovich projective line. In preparation, 270 pages, 2008. Available at
http://www.math.gatech.edu/~mbaker/papers.html.

[CL06] A. Chambert-Loir. Mesures et équidistribution sur les espaces de Berkovich. *J. Reine Angew. Math.*, 595:215–235, 2006.

[CR93] T. Chinburg and R. Rumely. The capacity pairing. *J. Reine angew. Math.*, 434:1–44, 1993.

[CHL07] T. Coulbois, A. Hilion, and M. Lustig. Non-unique ergodicity, observers' topology and the dual algebraic lamination for $\mathbb{R}$-trees. arXiv:0706.1313, to appear in the *Illinois J. Math.*

[Con07] B. Conrad. Several approaches to non-archimedean geometry. This volume.

[DT07] S. Dasgupta and J. Teitelbaum. The p-adic upper half plane. This volume.

[Duc06] A. Ducros. Espaces analytiques p-adiques au sens de Berkovich. preprint. Available at
http://math.unice.fr/~ducros/asterisque.pdf, 38 pages, 2006.

[FJ04] C. Favre and M. Jonsson. The valuative tree, volume 1853 of *Lecture Notes in Mathematics*. Springer, Berlin, 2004.

[FJ05]   C. Favre and M. Jonsson. Valuative analysis of planar plurisubharmonic functions. *Invent. Math.*, 162(2): 271–311, 2005.

[FJ07]   C. Favre and M. Jonsson. Eigenvaluations. *Ann. Sci. École Norm. Sup.*, 40(2): 309–349, 2007.

[FRL04]   C. Favre and J. Rivera-Letelier. Théorème d'équidistribution de Brolin en dynamique *p*-adique. *C. R. Math. Acad. Sci. Paris*, 339(4):271–276, 2004.

[FRL06]   C. Favre and J. Rivera-Letelier. Équidistribution quantitative des points de petite hauteur sur la droite projective. *Math. Ann.*, 335(2):311–361, 2006.

[FvdP04]   J. Fresnel and M. van der Put. *Rigid analytic geometry and its applications*, volume 218 of *Progress in Mathematics*. Birkhäuser Boston Inc., Boston, MA, 2004.

[GvdP80]   L. Gerritzen and M. van der Put. *Schottky groups and Mumford curves*, volume 817 of *Lecture Notes in Mathematics*. Springer, Berlin, 1980.

[Kan89]   E. Kani. Potential theory on curves. In *Théorie des nombres (Quebec, PQ, 1987)*, pages 475–543. de Gruyter, Berlin, 1989.

[Liu02]   Q. Liu. *Algebraic geometry and arithmetic curves*, volume 6 of *Oxford Graduate Texts in Mathematics*. Oxford University Press, Oxford, 2002. Translated from the French by Reinie Erné, Oxford Science Publications.

[RL03a]   J. Rivera-Letelier. Dynamique des fonctions rationnelles sur des corps locaux. *Astérisque*, (287):xv, 147–230, 2003. Geometric methods in dynamics. II.

[RL03b]   J. Rivera-Letelier. Espace hyperbolique *p*-adique et dynamique des fonctions rationnelles. *Compositio Math.*, 138(2):199–231, 2003.

[Rob00]   A. M. Robert. *A course in p-adic analysis*, volume 198 of *Graduate Texts in Mathematics*. Springer-Verlag, New York, 2000.

[Rum89]   R. Rumely. *Capacity theory on algebraic curves*, volume 1378 of *Lecture Notes in Mathematics*. Springer-Verlag, Berlin, 1989.

[Sil94]   J. H. Silverman. *Advanced topics in the arithmetic of elliptic curves*, volume 151 of *Graduate Texts in Mathematics*. Springer-Verlag, New York, 1994.

[Thu05]   A. Thuillier. *Théorie du potentiel sur les courbes en géométrie analytique non archimédienne. Applications à la théorie d'Arakelov*. PhD thesis, University of Rennes, 2005. preprint. Available at `http://tel.ccsd.cnrs.fr/documents/archives0/00/01/09/90/index.html`.

[Zha93]   S. Zhang. Admissible pairing on a curve. *Invent. Math.*, 112(1):171–193, 1993.

p-adic cohomology: from theory to practice

KIRAN S. KEDLAYA

Introduction

These notes (somewhat revised from the version presented at the 2007 AWS) present a few facets of the relationship between p-adic analysis, algebraic de Rham cohomology, and zeta functions of algebraic varieties. A key theme is the explicit, computable nature of these constructions, which makes them suitable for numerical calculations. For instance, if you ask the computer algebra system Magma for the order of the Jacobian of a hyperelliptic curve over a field of small characteristic, this order is computed using p-adic cohomology. The same is true if you ask the system Sage for the p-adic regulator of an elliptic curve over $\mathbb{Q}$, for p a good ordinary prime.

1. Algebraic de Rham cohomology

In this section, we introduce the notion of algebraic de Rham cohomology for smooth varieties, as originally introduced by Grothendieck [**19**] based on ideas of Atiyah and Hodge.

NOTATION 1.0.1. Throughout this section, let K be a field of characteristic zero. By a "variety over K", we will mean a K-scheme of finite type which is reduced and separated, but not necessarily irreducible.

1.1. de Rham cohomology of smooth affine varieties. To deal with de Rham cohomology for general varieties, we will need some machinery of sheaf cohomology and hypercohomology. Before doing so, however, let us consider the case of smooth affine varieties, for which no such machinery is needed.

DEFINITION 1.1.1. Let R be a finitely generated, reduced K-algebra, and let $X = \operatorname{Spec} R$ be the corresponding affine variety over K. Let $\Omega_{R/K}$ denote the module of Kähler differentials; that is, $\Omega_{R/K}$ is the R-module generated by symbols dr for $r \in R$, modulo the relations dr for $r \in K$, and $d(ab) - a\,db - b\,da$ for $a, b \in R$. The module $\Omega_{R/K}$ is finitely generated over R, and is equipped with a derivation $d : R \to \Omega_{R/K}$ carrying r to dr; it has the universal property that for any K-linear derivation $D : R \to M$ into an R-module, there is a unique R-linear map $\psi : \Omega_{R/K} \to M$ such that $D = \psi \circ d$.

The author thanks Ralf Gerkmann, Alan Lauder, Doug Ulmer, and the participants of STAGE (the Seminar on Topics in Algebra, Geometry, Etc.) for feedback on preliminary versions of these notes. The author was supported by the Southwest Center for Arithmetic Geometry (NSF grant DMS-0602287), NSF CAREER grant DMS-0545904, and a Sloan Research Fellowship.

We assume hereafter that X/K is smooth, which forces $\Omega_{R/K}$ to be a locally free R-module of rank $\dim(X)$. Let

$$\Omega^i_{R/K} = \wedge^i_R \Omega_{R/K}$$

be the i-th *alternating power* (or *wedge power*) of $\Omega_{R/K}$ over R. That is, $\Omega^i_{R/K}$ is the free R-module generated by symbols $\omega_1 \wedge \cdots \wedge \omega_i$ with $\omega_1, \ldots, \omega_i \in \Omega_{R/K}$, modulo the submodule generated by

$$(r\omega_1 + r'\omega'_1) \wedge \omega_2 \wedge \cdots \wedge \omega_i - r\omega_1 \wedge \omega_2 \wedge \cdots \wedge \omega_i - r'\omega'_1 \wedge \omega_2 \wedge \cdots \wedge \omega_i$$

for $r, r' \in R$, and by $\omega_1 \wedge \cdots \wedge \omega_i$ whenever two of the factors are equal. Beware that the elements of the form $\omega_1 \wedge \cdots \wedge \omega_i$, the so-called *decomposable elements*, do not comprise all of $\Omega^i_{R/K}$; e.g., if $R = k[x_1, x_2, x_3, x_4]$, then $dx_1 \wedge dx_2 + dx_3 \wedge dx_4$ is a nondecomposable element of $\Omega^2_{R/K}$.

The map d induces maps $d : \Omega^i_{R/K} \to \Omega^{i+1}_{R/K}$. Moreover, the composition $d \circ d$ is always zero. We thus have a complex $\Omega^{\cdot}_{R/K}$, called the *de Rham complex* of X. The cohomology of this complex is called the *(algebraic) de Rham cohomology* of X, denoted $H^i_{\mathrm{dR}}(X)$; it is contravariantly functorial in X. Note that for $i > \dim(X)$, $\Omega^i_{R/K} = 0$ and so $H^i_{\mathrm{dR}}(X) = 0$.

EXERCISE 1.1.2. Put $R = K[x_1, \ldots, x_n]$, so that $X = \mathbb{A}^n_K$ is the affine n-space over K. Check that

$$H^0(X) = K, \qquad H^i(X) = 0 \quad (i > 0).$$

TERMINOLOGY 1.1.3. The elements of $\Omega^i_{R/K}$ are referred to as *i-forms*. An i-form is *closed* if it is in the kernel of $d : \Omega^i_{R/K} \to \Omega^{i+1}_{R/K}$, and *exact* if it is in the image of $d : \Omega^{i-1}_{R/K} \to \Omega^i_{R/K}$. In this terminology, $H^i_{\mathrm{dR}}(X)$ is the quotient of the space of closed i-forms by the subspace of exact i-forms.

REMARK 1.1.4. There is a construction of algebraic de Rham cohomology that allows affines which are not smooth. Roughly speaking, given a closed immersion of the given affine scheme into a smooth affine variety (e.g., an affine space), one may use the cohomology of the de Rham complex on the formal neighborhood of the image; this does not depend on the choice of the immersion. This constructed is developed by Hartshorne in [**22**].

1.2. Example: an incomplete elliptic curve.

EXAMPLE 1.2.1. Assume that $P(x) = x^3 + ax + b \in K[x]$ has no repeated roots, and put

$$R = K[x, y]/(y^2 - P(x)),$$

so that $X = \mathrm{Spec}\, R$ is the affine part of an elliptic curve over K (i.e., the complete elliptic curve minus the one point at infinity). Then $H^0_{\mathrm{dR}}(X) = K$, and $H^i_{\mathrm{dR}}(X)$ vanishes for $i > 1$.

The interesting space $H^1_{\mathrm{dR}}(X)$ is simply the cokernel of $d : R \to \Omega_{R/K}$. To describe it, we use the relation

$$0 = d(y^2 - P(x)) = 2y\, dy - P'\, dx$$

in $\Omega_{R/K}$. (Throughout this example, primes denote differentiation with respect to x.) Since P has no repeated roots, we can choose polynomials $A, B \in K[x]$ such that $AP + BP' = 1$. Now put

$$\omega = Ay\,dx + 2B\,dy$$

so that

$$dx = y\omega, \qquad dy = \frac{1}{2}P'\omega.$$

Consequently, every element of $\Omega_{R/K}$ has a unique representation as $(C+Dy)\omega$ for some $C, D \in K[x]$. For this form to be exact, there must exist $E, F \in K[x]$ such that

$$(C + Dy)\omega = d(E + Fy)$$
$$= E'\,dx + F'y\,dx + F\,dy$$
$$= \left(\frac{1}{2}P'F + F'P\right)\omega + E'y\omega.$$

In particular, $Dy\omega$ is always exact. As for $C\omega$, if F has leading term cx^d, then $\frac{1}{2}P'F + F'P$ has leading term $\left(\frac{3}{2} + d\right)cx^{d+2}$. Since $\frac{3}{2} + d$ is never an integer, we can choose c so that subtracting $\left(\frac{3}{2} + d\right)cx^{d+2}$ removes the leading term of C.

Repeating this process (of clearing leading terms from C) allows us to write $C\omega$ as an exact differential plus a K-linear combination of

$$\omega, \quad x\omega.$$

These two thus form a basis of $H^1_{\mathrm{dR}}(X)$.

REMARK 1.2.2. Note that by writing

$$\omega = \frac{dx}{y} = \frac{2\,dy}{P'(x)},$$

we can see that ω actually extends to a 1-form on the complete elliptic curve, whereas $x\omega$ has a double pole at infinity.

REMARK 1.2.3. From the point of view of making machine computations, what is crucial here is not simply that we were able to compute the dimension of $H^1_{\mathrm{dR}}(X)$, or write down a basis. Rather, it is crucial that given any (closed) 1-form, we have a simple algorithm for presenting it as an exact 1-form plus a linear combination of basis elements.

Later, we will need a slight variation of the above example.

EXAMPLE 1.2.4. Define P as in Example 1.2.1, but this time put

$$R = K[x, y, z]/(y^2 - P(x), yz - 1),$$

so that $X = \operatorname{Spec} R$ is an elliptic curve over K minus the point at infinity and the three points of order 2. Again, $H^0_{\mathrm{dR}}(X) = K$, and $H^i_{\mathrm{dR}}(X)$ vanishes for $i > 1$.

In this case, to calculate $H^1_{\mathrm{dR}}(X)$, we will work not with ω but directly with dx. Any given element of $\Omega_{R/K}$ can be written as $(C + Dy)y^{-2i}\,dx$ for some nonnegative integer i and some $C, D \in K[x]$. Now the relevant calculation of an exact differential (for $E, F \in K[x]$) is

$$d((E + Fy)y^{-2j}) = E'y^{-2j}\,dx + F'y^{-2j+1}\,dx - 2jEy^{-2j-1}\,dy - (2j-1)Fy^{-2j}\,dy$$
$$= (E'P - jP'E)\,y^{-2j-2}\,dx + \left(F'P + \frac{1-2j}{2}P'F\right)y^{-2j-1}\,dx.$$

We interpret this as saying that in $H^1_{\mathrm{dR}}(X)$,

$$(jP'E)\, y^{-2j-2}\, dx + \left(\frac{2j-1}{2} P'F\right) y^{-2j-1}\, dx$$

is cohomologous to something of the form $(G + Hy)y^{-2j}\, dx$.

At this point we may treat the even and odd powers of y completely independently. (This is explained by the fact that the curve admits a hyperelliptic involution $x \mapsto x, y \mapsto -y$; the even and odd powers of y correspond to the plus and minus eigenspaces under this involution.) In both cases, we use the fact that P and P' generate the unit ideal of $K[x]$.

In the odd powers of y, we can reduce all the way down to $Dy^{-1}\, dx$, then eliminate multiples of $P'y^{-1}\, dx$. We are thus left with generators $dx/y, x\, dx/y$.

In the even powers of y, we can reduce all the way down to $Cy^{-2}\, dx$, then eliminate multiples of $Py^{-2}\, dx$. We are thus left with generators $dx/y^2, x\, dx/y^2, x^2\, dx/y^2$.

To conclude, we have the following basis for $H^1_{\mathrm{dR}}(X)$:

$$\frac{dx}{y}, \quad \frac{x\, dx}{y}, \quad \frac{dx}{y^2}, \quad \frac{x\, dx}{y^2}, \quad \frac{x^2\, dx}{y^2},$$

and we can explicitly rewrite any 1-form as a K-linear combination of these plus an exact 1-form.

EXERCISE 1.2.5. Let $P(x) \in K[x]$ be a squarefree polynomial. Compute $H^i_{\mathrm{dR}}(X)$ for the punctured affine line $X = \operatorname{Spec} K[x,y]/(yP(x) - 1)$; again, this means that you should have an explicit recipe for presenting any 1-form as an exact 1-form plus a linear combination of basis elements.

EXERCISE 1.2.6. Repeat the derivations of Example 1.2.1 and 1.2.4 for a hyperelliptic curve $y^2 = P(x)$. Note that the net result depends on whether $\deg(P)$ is odd or even; for an explanation of this, see Exercise 1.6.4.

1.3. Sheaf cohomology. In order to move past affines, we must work with sheaf cohomology and hypercohomology. We give here a rapid summary of the key points; we presume that the reader has encountered sheaf cohomology previously, e.g., in [**23**, Chapter III].

DEFINITION 1.3.1. Let X be a scheme, and let $\mathbf{Ab}_X$ denote the category of sheaves of abelian groups on X. Given two complexes $C^\cdot = (0 \to C^0 \to C^1 \to \cdots)$ and $D^\cdot$ in $\mathbf{Ab}_X$, a *morphism* $C^\cdot \to D^\cdot$ is a commuting diagram

$$
\begin{array}{ccccccccc}
0 & \longrightarrow & C^0 & \longrightarrow & C^1 & \longrightarrow & C^2 & \longrightarrow & \cdots \\
& & \downarrow & & \downarrow & & \downarrow & & \\
0 & \longrightarrow & D^0 & \longrightarrow & D^1 & \longrightarrow & D^2 & \longrightarrow & \cdots.
\end{array}
$$

This morphism is a *quasi-isomorphism* if the induced maps on cohomology:

$$\frac{\ker(C^i \to C^{i+1})}{\operatorname{image}(C^{i-1} \to C^i)} \to \frac{\ker(D^i \to D^{i+1})}{\operatorname{image}(D^{i-1} \to D^i)}$$

are isomorphisms.

DEFINITION 1.3.2. Let $\mathbf{C}$ be a full subcategory of $\mathbf{Ab}_X$ (i.e., retain some of the objects but keep *all* morphisms between such objects). For $C^\cdot$ a complex, a *resolution by* $\mathbf{C}$ of $C^\cdot$ is a second complex $D^\cdot$ in $\mathbf{C}$ equipped with a quasi-isomorphism

$C^{\cdot} \to D^{\cdot}$. As a special case, if $C \in \mathbf{Ab}_X$ is a single object, we may identify C with the complex $C^{\cdot}$ defined by $C^0 = C$, $C^i = 0$ for $i > 0$ and all morphisms zero; this gives the notion of a resolution by $\mathbf{C}$ of the object C, which is just an exact sequence $0 \to C \to D^0 \to D^1 \to \cdots$ with each $D^i \in \mathbf{C}$. (If $\mathbf{C}$ is defined by an adjective describing certain objects of $\mathbf{Ab}_X$, e.g., "injective", we will refer to an "injective resolution" instead of a "resolution by $\mathbf{C}$".)

DEFINITION 1.3.3. We say $\mathscr{F} \in \mathbf{Ab}_X$ is *injective* if the functor $\mathrm{Hom}(\cdot, \mathscr{F})$ is exact. For any $\mathscr{F} \in \mathbf{Ab}_X$, there exists a monomorphism $\mathscr{F} \to \mathscr{G}$ with $\mathscr{G}$ injective [**23**, Proposition III.2.2]; this is commonly described by saying that $\mathbf{Ab}_X$ is an abelian category which *has enough injectives*.

DEFINITION 1.3.4. Since $\mathbf{Ab}_X$ has enough injectives, any $\mathscr{F}$ admits an injective resolution $\mathscr{F}^{\cdot}$; moreover, given two injective resolutions, there is a third one to which each admits a quasi-isomorphism. We define the *sheaf cohomology* $H^i(X, \mathscr{F})$ as the i-th cohomology of the complex $\Gamma(X, \mathscr{F}^{\cdot})$; it turns out to be canonically independent of the choice of an injective resolution. The construction $\mathscr{F} \mapsto H^i(X, \mathscr{F})$ is covariantly functorial in $\mathscr{F}$, with $H^0(X, \cdot)$ being the global sections functor $\Gamma(X, \cdot)$. Also, given a short exact sequence $0 \to \mathscr{F} \to \mathscr{G} \to \mathscr{H} \to 0$, we obtain a long exact sequence

$$\cdots \to H^i(X, \mathscr{F}) \to H^i(X, \mathscr{G}) \to H^i(X, \mathscr{H}) \xrightarrow{\delta_i} H^{i+1}(X, \mathscr{F}) \to \cdots;$$

the maps $\delta_i : H^i(X, \mathscr{H}) \to H^{i+1}(X, \mathscr{F})$ (which are themselves functorial in the short exact sequence) are the *connecting homomorphisms*. (Here $H^{-1}(X, \mathscr{H}) = 0$, so we start $0 \to H^0(X, \mathscr{F}) \to H^0(X, \mathscr{G}) \to \cdots$.)

DEFINITION 1.3.5. We say $\mathscr{F}$ is *acyclic* if $H^i(X, \mathscr{F}) = 0$ for $i > 0$. One can then show that sheaf cohomology can be computed using any acyclic resolution, not just any injective resolution. However, we must use injectives in order to define sheaf cohomology, because the notion of acyclicity is not available until after cohomology has been defined. In other words, the notion of injectivity depends only on the category $\mathbf{Ab}_X$, whereas acyclicity depends on the choice of the functor $\Gamma(X, \cdot)$ to serve as H^0.

REMARK 1.3.6. What makes the previous observation helpful is that it is quite easy to construct acyclic resolutions in many cases, using Čech complexes. The key input in the next definition is the fact that if X is affine, then *any* quasicoherent sheaf of $\mathscr{O}_X$-modules on X is acyclic, because the global sections functor on quasicoherent sheaves is exact.

DEFINITION 1.3.7. Let X be a separated scheme, let $\mathscr{F}$ be a quasicoherent sheaf of $\mathscr{O}_X$-modules on X, let I be a finite totally ordered set, and let $\{U_i\}_{i \in I}$ be a cover of X by open affine subschemes. Since X is separated, any nonempty finite intersection of U_i's is again affine. (This is the algebro-geometric analogue of a "good cover" in the parlance of [**6**], i.e., a cover of a manifold by open subsets such that each nonempty finite intersection is contractible.) For each finite subset J of I, put $U_J = \cap_{j \in J} U_j$ and let $j_J : U_J \hookrightarrow X$ be the implied open immersion. Then j_J is an affine morphism, so $(j_J)_*$ is exact. Since every quasicoherent sheaf on the affine U_J is acyclic, $(j_J)_* j_J^* \mathscr{F}$ is acyclic on X.

We thus obtain an acyclic resolution $\mathscr{F}^{\cdot}$ of $\mathscr{F}$ as follows.

- The term $\mathscr{F}^i$ consists of the direct sum of $(j_J)_* j_J^* \mathscr{F}$ over all $(i+1)$-element subsets J of I.

- The map $\mathscr{F}^i \to \mathscr{F}^{i+1}$, applied to an element of $\mathscr{F}^i$ with component x_J in $(j_J)_* j_J^* \mathscr{F}$, produces an element of $\mathscr{F}^{i+1}$ whose component in $(j_J)_* j_J^* \mathscr{F}$ with $J = \{j_0 < \cdots < j_{i+1}\}$ is

$$\sum_{h=0}^{i+1} (-1)^h x_{J \setminus \{j_h\}}.$$

In particular, the sheaf cohomology of $\mathscr{F}$ is given the cohomology of the corresponding complex of global sections. In the i-th position, this consists of the direct sum of $\Gamma(U_J, j_J^* \mathscr{F})$ over all $(i+1)$-element subsets J of I. (Compare [23, Theorem III.4.5], [20, Proposition 1.4.1].)

1.4. Hypercohomology and de Rham cohomology. We now pass from sheaf cohomology to sheaf hypercohomology as in [20, §0.11.4], then give the definition of algebraic de Rham cohomology for an arbitrary smooth variety.

DEFINITION 1.4.1. Let $C^{\cdot}$ be a complex in $\mathbf{Ab}_X$. The *sheaf hypercohomology* of $C^{\cdot}$, denoted $\mathbb{H}^i(C^{\cdot})$, is defined as the cohomology of $\Gamma(X, D^{\cdot})$ for any acyclic resolution $D^{\cdot}$ of $C^{\cdot}$; again, this is independent of the choice of the resolution (one can always compare to an injective resolution).

To make this definition useful, one needs a good way to manufacture acyclic resolutions of complexes, rather than of individual sheaves.

REMARK 1.4.2. Suppose we can construct a double complex $D^{\cdot,\cdot}$ such that the diagram

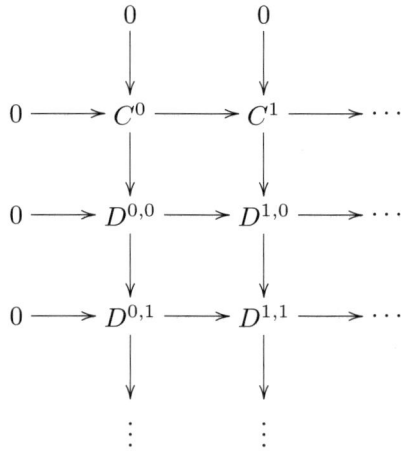

commutes, and each column gives an acyclic resolution of C^i. (In a double complex, the composition of two consecutive horizontal arrows, or two consecutive vertical arrows, must vanish.) Let $D^{\cdot}$ be the associated *total complex* of $D^{\cdot,\cdot}$, constructed as follows.

- We take $D^i = \oplus_{j+k=i} D^{j,k}$.
- The map $D^i \to D^{i+1}$, applied to an element of D^i with component $x_{j,k}$ in $D^{j,k}$, produces an element of D^{i+1} whose component in $D^{j,k}$ is

$$d_{\text{horizontal}}(x_{j-1,k}) + (-1)^j d_{\text{vertical}}(x_{j,k-1}).$$

(The factor of $(-1)^j$ is needed to ensure that the composition of two of these derivations is zero; it forces the "cross terms" involving one horizontal and one vertical derivation to cancel each other out.)

Then $D^\cdot$ forms an acyclic resolution of $C^\cdot$.

DEFINITION 1.4.3. Since the construction of the Čech resolution in Definition 1.3.7 is functorial in the sheaf, applying it to a complex immediately gives a diagram of the sort discussed in Remark 1.4.2.

We are now ready to give Grothendieck's definition of the algebraic de Rham cohomology of a smooth variety. (Although [19] is the source of the original definition, it is very cursory; see [23] for a fuller treatment.)

DEFINITION 1.4.4. Let X be a variety over K. We now obtain a sheaf $\Omega_{X/K}$ of Kähler differentials, which is coherent. Let us assume further that X is smooth of relative dimension n; then $\Omega_{X/K}$ is locally free of rank n. The exterior derivative is now a map of sheaves (but not of $\mathscr{O}_X$-modules!) $d : \mathscr{O}_X \to \Omega_{X/K}$, and using it we construct the *de Rham complex* of X:

$$0 \to \Omega^0_{X/K} \to \Omega^1_{X/K} \to \cdots \to \Omega^n_{X/K} \to 0.$$

We now define the algebraic de Rham cohomology of X, denoted $H^i_{\mathrm{dR}}(X)$, as the hypercohomology $\mathbb{H}^i(\Omega^\cdot_{X/K})$ of the de Rham complex; again, this is contravariantly functorial in X. (Formation of $H^i_{\mathrm{dR}}(X)$ commutes with extension of the base field, because formation of sheaf hypercohomology commutes with flat base change.)

REMARK 1.4.5. The fact that we use hypercohomology to define algebraic de Rham cohomology means that unlike in the affine case, we no longer automatically have $H^i_{\mathrm{dR}}(X) = 0$ whenever $i > \dim(X)$. This will be apparent in Example 1.5.1 below.

1.5. Example: a complete elliptic curve. Let us watch the definition of algebraic de Rham cohomology in action, for a complete elliptic curve.

EXAMPLE 1.5.1. Assume again that $P(x) = x^3 + ax + b \in K[x]$ has no repeated roots. However, now let X be the *complete* elliptic curve $y^2 = P(x)$, i.e.,

$$X = \operatorname{Proj} K[X, Y, W]/(Y^2W - X^3 - aXW^2 - bW^3).$$

Since X admits a one-dimensional space of everywhere holomorphic 1-forms, generated by $\omega = dx/y$, you might be tempted to think that $H^1_{\mathrm{dR}}(X)$ is one-dimensional. However, this is not what happens!

To compute $H^i_{\mathrm{dR}}(X)$, we use a simple Čech resolution. Let U be the affine curve, i.e., X minus the point at infinity $[0 : 1 : 0]$. Let V be X minus the three geometric points of the form $(x, 0)$, where x is a root of P. We now find that the de Rham cohomology of X is computed by the cohomology of the complex $D^\cdot$, where

$$D^0 = \Gamma(U, \mathscr{O}_U) \oplus \Gamma(V, \mathscr{O}_V),$$
$$D^1 = \Gamma(U, \Omega_U) \oplus \Gamma(V, \Omega_V) \oplus \Gamma(U \cap V, \mathscr{O}_{U \cap V}),$$
$$D^2 = \Gamma(U \cap V, \Omega_{U \cap V}).$$

Let us start with $H^0_{\mathrm{dR}}(X)$; this consists of pairs (f, g) with $f \in \Gamma(U, \mathscr{O}_U)$, $g \in \Gamma(V, \mathscr{O}_V)$, and $\operatorname{Res}_{U, U \cap V}(f) - \operatorname{Res}_{V, U \cap V}(g) = 0$. In other words, these are just elements of $\Gamma(X, \mathscr{O}_X)$, and the only such elements are the constant functions.

We next consider $H^1_{\mathrm{dR}}(X)$. First, note that the 1-cochains are triples (ω_U, ω_V, f) with $\omega_U \in \Gamma(U, \Omega_U)$, $\omega_V \in \Gamma(V, \Omega_V)$, $f \in \Gamma(U \cap V, \mathscr{O}_{U \cap V})$. The differential takes such a triple to $\mathrm{Res}_{V,U \cap V}(\omega_V) - \mathrm{Res}_{U,U \cap V}(\omega_U) - df$.

The 1-coboundaries are expressions of the form $(df, dg, g - f)$ with $f \in \Gamma(U, \mathscr{O}_U)$ and $g \in \Gamma(V, \mathscr{O}_V)$. Thus the projection $(\omega_U, \omega_V, f) \mapsto \omega_U$ induces a map

$$H^1_{\mathrm{dR}}(X) \to H^1_{\mathrm{dR}}(U);$$

we wish to show that this map is a bijection. We first check injectivity. If (ω_U, ω_V, f) is a 1-cocycle on X, and ω_U is a 1-coboundary on U, then $(0, \omega_V, f)$ is also a 1-cocycle on X, that is, $df = \mathrm{Res}_{V,U \cap V}(\omega_V)$. This means that f cannot have a pole at any point of V, as otherwise df would have at least a double pole at that point. Consequently, ω_V is a 1-coboundary on V, so $(0, 0, f)$ is a 1-cocycle on X. This is only possible if f is constant, in which case $(0, 0, f)$ is a 1-coboundary on X.

We now check surjectivity. By the computation we made of $H^1_{\mathrm{dR}}(U)$ in Example 1.2.1, it suffices to check that for any $c_1, c_2 \in K$, we can find a 1-cocycle (ω_U, ω_V, f) with $\omega_U = c_1 \, dx/y + c_2 x \, dx/y$. Namely, we take

$$\omega_V = c_1 \frac{dx}{y} + c_2 \left(\frac{x \, dx}{y} - 2d\left(\frac{y}{x}\right) \right), \qquad f = -2c_2 \frac{y}{x}.$$

(To see that $x \, dx/y - 2d(y/x)$ is holomorphic at ∞, we compute in terms of the local coordinate $t = y/x$ at ∞, using the local expansions $x = t^{-2} + O(t^0)$ and $y = t^{-3} + O(t^{-1})$.)

Finally, we compute $H^2_{\mathrm{dR}}(X)$. This consists of elements of $\Gamma(U \cap V, \Omega_{U \cap V})$ modulo expressions of the form $\mathrm{Res}_{V,U \cap V}(\omega_V) - \mathrm{Res}_{U,U \cap V}(\omega_U) - df$. By the computation we made of $H^1_{\mathrm{dR}}(U \cap V)$ in Example 1.2.4, we know that every element of $H^1_{\mathrm{dR}}(U \cap V)$ can be expressed as a K-linear combination of dx/y, $x \, dx/y$, dx/y^2, $x \, dx/y^2$, $x^2 \, dx/y^2$. The same is true of $H^2_{\mathrm{dR}}(X)$, but with some redundancy: the terms dx/y, $x \, dx/y$ are holomorphic on U, so they can be absorbed by ω_U, while the terms dx/y^2, $x \, dx/y^2$ are holomorphic on V, so they can be absorbed by ω_V. This proves that $H^2_{\mathrm{dR}}(X)$ is at most one-dimensional.

To show that $H^2_{\mathrm{dR}}(X)$ is in fact one-dimensional, we must use some properties of residues (see Definition 1.5.2 below). Any K-linear combination of dx/y, $x \, dx/y$, dx/y^2, $x \, dx/y^2$, $x^2 \, dx/y^2$ with a nonzero coefficient of $x^2 \, dx/y^2$ has a simple pole at ∞ with nonzero residue. On the other hand, the residues at ∞ of the three quantities $\mathrm{Res}_{V,U \cap V}(\omega_V), \mathrm{Res}_{U,U \cap V}(\omega_U), df$ must all vanish: the first is holomorphic at ∞, the third has residue zero by a local computation, and the second has residue zero because it has no other poles, and the sum of the residues at all poles must vanish.

To conclude, we find that

$$\dim_K H^0_{\mathrm{dR}}(X) = 1, \qquad \dim_K H^1_{\mathrm{dR}}(X) = 2, \qquad \dim_K H^2_{\mathrm{dR}}(X) = 1,$$

in agreement with the topological Betti numbers. (This agreement is explained by a comparison theorem with topological cohomology; see Theorem 1.7.2.)

Here are the properties of residues used in the previous example. See [**41**] for proofs and further discussion; also see [**23**, Remark 7.14].

DEFINITION 1.5.2. For z a smooth geometric point on a curve X over K, let t be a local parameter of X at z. Then any meromorphic 1-form ω on X can be expanded around z as an expression $\sum_{i=m}^{\infty} c_i t^i \, dt$, with $c_i \in \kappa(z)$. The *residue of* ω

at z is the value c_{-1}; it is independent of the choice of the parameter t. Moreover, if X is smooth proper, then the sum of the residues of any given ω at all z is zero.

REMARK 1.5.3. In Example 1.5.1, we can form an injection $H^0(X, \Omega^1_{X/K}) \hookrightarrow H^1_{\mathrm{dR}}(X)$ by converting $\omega \in \Gamma(X, \Omega^1_{X/K})$ into the triple $(\mathrm{Res}_{X,U}(\omega), \mathrm{Res}_{X,V}(\omega), 0)$. This is a special case of the Hodge filtration on de Rham cohomology; see Remark 1.8.12.

1.6. Excision in de Rham cohomology.

REMARK 1.6.1. There is an excision exact sequence in algebraic de Rham cohomology. If X is a smooth K-variety, Z is a smooth subvariety of pure codimension d, and $U = X \setminus Z$, then

$$(1.6.2) \qquad \cdots \to H^{i-2d}_{\mathrm{dR}}(Z) \to H^i_{\mathrm{dR}}(X) \to H^i_{\mathrm{dR}}(U) \to H^{i-2d+1}_{\mathrm{dR}}(Z) \to \cdots.$$

Note the shift by the codimension of Z.

REMARK 1.6.3. There is a more general version of Remark 1.6.1 with Z not required to be smooth, but one must replace the terms corresponding to Z by some sort of local cohomology which will not be introduced here. However, (1.6.2) is enough to deduce the following additivity property of the Euler characteristic

$$\chi_{\mathrm{dR}}(X) = \sum_i (-1)^i \dim_K H^i_{\mathrm{dR}}(X).$$

Namely, if Z is any subvariety of X, then

$$\chi_{\mathrm{dR}}(X) = \chi_{\mathrm{dR}}(Z) + \chi_{\mathrm{dR}}(X \setminus Z).$$

EXERCISE 1.6.4. Use (1.6.2) to show that, if C is a smooth, geometrically connected, projective curve over K, Z is a nonempty zero-dimensional subscheme of length d, and $U = C \setminus Z$, then

$$\dim_K H^0_{\mathrm{dR}}(U) = 1,$$
$$\dim_K H^1_{\mathrm{dR}}(U) = \dim_K H^1_{\mathrm{dR}}(C) + d - 1,$$
$$\dim_K H^2_{\mathrm{dR}}(U) = 0.$$

The quantity $\dim_K H^1_{\mathrm{dR}}(C)$ will turn out to be twice the genus of C, by the topological comparison (Theorem 1.7.2).

1.7. Comparison with topological cohomology.

DEFINITION 1.7.1. Let X be a scheme locally of finite type over $\mathbb{C}$. Then there is a functorial way to associate to X a complex analytic variety X^{an}, called the *analytification* of X. For instance, if $X = \mathrm{Spec}\, R$ with $R = \mathbb{C}[x_1, \ldots, x_n]/(f_1, \ldots, f_n)$ reduced, then X^{an} is the common zero locus of $f_1, \ldots, f_n$ as an analytic subvariety of $\mathbb{C}^n$. See [21, Exposé XII, Théorème et definition 1.1] for a detailed construction.

The definition of algebraic de Rham cohomology is ultimately justified by the following result.

THEOREM 1.7.2 (Grothendieck). *Let X be a smooth variety over $\mathbb{C}$. Then there is a functorial isomorphism $H^i_{\mathrm{dR}}(X) \to H^i_{\mathrm{Betti}}(X^{\mathrm{an}}, \mathbb{C})$.*

This is really two results bundled together. The first of these is that the Betti cohomology of X^{an} is functorially isomorphic to the *analytic* de Rham cohomology of X^{an}, i.e., what you get by using the definition of algebraic de Rham cohomology but replacing the scheme with a complex analytic variety. This is a consequence of de Rham's theorem (which relates Betti cohomology to C^∞ de Rham cohomology) and Dolbeaut's theorem (which relates C^∞ de Rham cohomology to analytic de Rham cohomology). See [**17**, §0.3] for further discussion of these components.

The second result is that the natural map $H^i_{\mathrm{dR}}(X) \to H^i_{\mathrm{dR}}(X^{\mathrm{an}})$ (i.e., view an algebraic cocycle as an analytic one) is an bijection. Since both sides of the comparison satisfy excision (Remark 1.6.1), by induction on dimension (and the fact that X can always be embedded into a smooth projective variety, by Hironaka's resolution of singularities) we may reduce Theorem 1.7.2 to the case of X smooth and projective. In this case, the claim becomes an instance of Serre's *GAGA principle*. (GAGA is an acronym for "Géométrie algébrique et géométrie analytique", the title of Serre's paper [**40**] introducing this principle.)

THEOREM 1.7.3 (Complex-analytic GAGA). *Let X be a proper variety over $\mathbb{C}$.*
  (a) *Any coherent sheaf on X^{an} is the analytification of a coherent sheaf on X.*
  (b) *For any coherent sheaves $\mathscr{E}, \mathscr{F}$ on X, any morphism $\mathscr{E}^{\mathrm{an}} \to \mathscr{F}^{\mathrm{an}}$ is induced by a morphism $\mathscr{E} \to \mathscr{F}$.*
  (c) *For any coherent sheaf $\mathscr{E}$ on X, with analytification $\mathscr{E}^{\mathrm{an}}$, the natural maps $H^i(X, \mathscr{E}) \to H^i(X^{\mathrm{an}}, \mathscr{E}^{\mathrm{an}})$ are bijections.*

PROOF. For X projective, proceed directly to Serre's original paper [**40**]. For the general case (which reduces to the projective case using Chow's lemma), see SGA1 [**21**, Exposé XII]. □

1.8. Spectral sequences and the Hodge filtration. There is a useful extra structure on de Rham cohomology given by the Hodge filtration. To explain how it arises algebraically, we must introduce spectral sequences; these will appear again later in the construction of Gauss-Manin connections.

REMARK 1.8.1. The notion of a spectral sequence is a generalization of the long exact sequence

$$\cdots \to H^i(C_1^{\cdot}) \to H^i(C^{\cdot}) \to H^i(C_2^{\cdot}) \xrightarrow{\delta} H^{i+1}(C_1^{\cdot}) \to \cdots$$

associated to a short exact sequence of complexes

$$0 \to C_1^{\cdot} \to C^{\cdot} \to C_2^{\cdot} \to 0.$$

The relevance of spectral sequences is described aptly by [**17**, §3.5] (which see for more details): "[to] someone who works with cohomology, they are essential in the same way that the various integration techniques are essential to a student of calculus."

DEFINITION 1.8.2. A *filtered complex* is a decreasing sequence of complexes

$$C^{\cdot} = F^0 C^{\cdot} \supseteq F^1 C^{\cdot} \supseteq \cdots \supseteq F^n C^{\cdot} \supseteq F^{n+1} C^{\cdot} = 0;$$

given a filtered complex, the *associated graded complex* is

$$\mathrm{Gr}\, C^{\cdot} = \oplus_{p \geq 0} \mathrm{Gr}^p C^{\cdot}, \qquad \mathrm{Gr}^p C^{\cdot} = \frac{F^p C^{\cdot}}{F^{p+1} C^{\cdot}}.$$

For instance, a short exact sequence as above gives a filtration with $F^2 C^{\cdot} = 0$, $\mathrm{Gr}^1 C^{\cdot} \cong F^1 C^{\cdot} = C_1^{\cdot}$, and $\mathrm{Gr}^0 C^{\cdot} = C_2^{\cdot}$. (Note that one could also talk about infinite filtrations, but a few statements below would have to be modified.)

EXAMPLE 1.8.3. For instance, if we start with a double complex $D^{p,q}$ and flatten it into a single complex $C^i = \oplus_{p+q=i} D^{p,q}$ as in Remark 1.4.2, we can filter this by taking

$$F^p C^i = \bigoplus_{p'+q=i, p' \geq p} D^{p',q},$$

yielding

$$\mathrm{Gr}^p C^i = D^{p,i-p}.$$

We can also flip p and q to get a second filtration; these two filtrations will end up giving two distinct spectral sequences related to the cohomology of $C^{\cdot}$.

As payback for being easy to use, spectral sequences sacrifice the computation of the entire cohomology of filtered complexes. Instead, they only compute the graded pieces of a certain filtration on the cohomology.

DEFINITION 1.8.4. Let Z^q and B^q be the cocycles and coboundaries in a filtered complex C^q. The *filtered cohomology* is

$$F^p H^q(C^{\cdot}) = \frac{F^p Z^q}{F^p B^q} = \mathrm{image}(H^q(F^p C^{\cdot}) \to H^q(C^{\cdot}));$$

note that the map $H^q(F^p C^{\cdot}) \to H^q(C^{\cdot})$ need not be injective. The *associated graded cohomology* is

$$\mathrm{Gr}\, H^{\cdot}(C^{\cdot}) = \oplus_{p,q} \mathrm{Gr}^p H^q(C^{\cdot}), \qquad \mathrm{Gr}^p H^q(C^{\cdot}) = \frac{F^p H^q(C^{\cdot})}{F^{p+1} H^q(C^{\cdot})}.$$

DEFINITION 1.8.5. A *spectral sequence* is a sequence $\{E_r, d_r\}_{r=r_0}^{\infty}$, where each E_r is a bigraded group

$$E_r = \oplus_{p,q \geq 0} E_r^{p,q}$$

and

$$d_r : E_r^{p,q} \to E_r^{p+r,q-r+1}, \qquad d_r^2 = 0$$

is a map (usually called a *differential*) such that

$$E_{r+1}^{p,q} = H^{p,q}(E_r) = \frac{\ker(d_r : E_r^{p,q} \to E_r^{p+r,q-r+1})}{\mathrm{image}(d_r : E_r^{p-r,q+r-1} \to E_r^{p,q})}.$$

If at some point $E_r = E_{r+1} = \cdots$, we call this stable value the *limit* of the spectral sequence, denoted E_∞. One also says that the sequence *degenerates at* E_r, and that the sequence *converges to* E_∞.

REMARK 1.8.6. Pictures of spectral sequences speak louder than words. Here are the first three differentials of a spectral sequence represented diagramatically:

$$
\begin{array}{ccc}
E_0^{0,1} & E_0^{1,1} & E_0^{2,1} \\
d_0 \uparrow & d_0 \uparrow & d_0 \uparrow \\
E_0^{0,0} & E_0^{1,0} & E_0^{2,0}
\end{array}
\qquad
\begin{array}{ccc}
E_1^{0,1} \xrightarrow{d_1} E_1^{1,1} \xrightarrow{d_1} E_1^{2,1} \\
\\
E_1^{0,0} \xrightarrow{d_1} E_1^{1,0} \xrightarrow{d_1} E_1^{2,0}
\end{array}
\qquad
\begin{array}{ccc}
E_2^{0,1} & E_2^{1,1} & E_2^{2,1} \\
& \searrow^{d_2} & \\
E_2^{0,0} & E_2^{1,0} & E_2^{2,0}
\end{array}
$$

THEOREM 1.8.7. *Let $F^p C^{\cdot}$ be a filtered complex. Then there is a spectral sequence $\{E_r\}_{r=0}^{\infty}$ with*

$$E_0^{p,q} = \frac{F^p C^{p+q}}{F^{p+1} C^{p+q}},$$
$$E_1^{p,q} = H^{p+q}(\mathrm{Gr}^p C^{\cdot}),$$
$$E_{\infty}^{p,q} = \mathrm{Gr}^p H^{p+q}(C^{\cdot}).$$

PROOF. See [**17**, §3.5] or [**6**, §14].   □

TERMINOLOGY 1.8.8. In the previous theorem, one writes

$$E_r \Rightarrow H^{\cdot}(C^{\cdot})$$

and says that the spectral sequence *abuts to* $H^{\cdot}(C^{\cdot})$.

EXAMPLE 1.8.9. For instance, in the example of a short exact sequence, we have

$$E_1^{0q} = H^q(C_2^{\cdot}), \quad E_1^{1q} = H^{q+1}(C_1^{\cdot}),$$

$d_1 : H^q(C_2^{\cdot}) \to H^{q+1}(C_1^{\cdot})$ is the connecting homomorphism δ, and $d_2 = d_3 = \cdots = 0$ because the arrows always have a zero at one or both endpoints. For a filtered complex with $F^{n+1} = 0$, we similarly have $E_{\infty} = E_{n+1}$.

REMARK 1.8.10. A map between filtered complexes is a quasi-isomorphism if the same is true at any single stage of the spectral sequence; the converse is also true here because our filtrations are finite, so the spectral sequence must degenerate at some stage. For instance, in the example of a short exact sequence, this is an instance of the five lemma.

EXAMPLE 1.8.11. We now apply the observation of Example 1.8.3 to the Čech complex that computes the de Rham cohomology of X. In the notation of Definition 1.3.7, we start with the double complex

$$D^{p,q} = \bigoplus_J \Gamma(U_J, \Omega_{U_J/K}^p)$$

where J runs over all $(q+1)$-element subsets of I.

Suppose we filter this complex as in Example 1.8.3 with the indices as written. Using Theorem 1.8.7 gives a spectral sequence in which we first compute the sheaf cohomology of $\Omega_{X/K}^p$ for each p, i.e.,

$$E_1^{pq} = H^q(X, \Omega_{X/K}^p);$$

this is the *Hodge-de Rham spectral sequence*, and the filtration it determines on $H_{\mathrm{dR}}^i(X)$ is called the *Hodge filtration* (see Remark 1.8.12 for further discussion). It is a deep theorem that for X smooth proper, the Hodge-de Rham spectral sequence degenerates already at E_1; this is usually established for compact Kähler manifolds using analytic techniques [**17**, §3.5, p. 466], but can also be proved for varieties algebraically [**11**].

Now filter as in Example 1.8.3 but with the indices reversed. Using Theorem 1.8.7 now gives a spectral sequence with

$$E_1^{pq} = \prod_J H_{\mathrm{dR}}^q(U_J),$$

where J runs over $(p+1)$-element subsets of I. Keep in mind that the E_∞ term of the spectral sequence only computes the graded cohomology, whereas one must use the full Čech complex to compute the cohomology itself. Nonetheless, the information provided is often quite useful (e.g., as in Remark 1.8.10).

REMARK 1.8.12. For X smooth proper, the degeneration at E_1 of the Hodge-de Rham spectral sequence implies that

$$\mathrm{Gr}^p H^i_{\mathrm{dR}}(X) \cong H^{i-p}(X, \Omega^p_{X/K}).$$

The Hodge numbers $h^{p,q} = \dim_K H^q(X, \Omega^p_{X/K})$ satisfy some additional nice properties, such as the Serre duality symmetry satisfy the symmetry relation

$$h^{q,p} = h^{n-q,n-p} \qquad (n = \dim(X))$$

and the complex conjugation symmetry

$$h^{q,p} = h^{p,q};$$

the latter is usually proved analytically for compact Kähler manifolds, by constructing the Hodge decomposition [**17**, §0.7], but can also be proved algebraically [**15**].

As an example of the above, we point out the short exact sequence

$$0 \to \mathrm{Gr}^1 H^1_{\mathrm{dR}}(X) \cong H^0(X, \Omega^1_{X/K}) \to H^1_{\mathrm{dR}}(X) \to \mathrm{Gr}^0 H^1_{\mathrm{dR}}(X) \cong H^1(X, \mathscr{O}_X) \to 0$$

where the two factors have equal dimension. In particular, if X is a geometrically connected curve, then $\dim_K H^1_{\mathrm{dR}}(X) = 2g$ for $g = \dim_K H^0(X, \Omega^1_{X/K})$ the genus of the curve.

1.9. Cohomology with logarithmic singularities. For various reasons (primarily the finiteness of sheaf cohomology for coherent sheaves), one prefers to manipulate cohomology on proper schemes whenever possible. In order to detect the cohomology of a nonproper scheme while doing calculations on a compactification, we may use logarithmic differentials.

DEFINITION 1.9.1. By a *smooth (proper) pair* over a base S, we will mean a pair (X, Z) in which X is a smooth (proper) scheme over S and Z is a relative (to S) strict normal crossings divisor. Over a field, this means each component of Z is smooth (no self-intersections allowed; that's the "strict" part), and the components of Z always meet transversely.

REMARK 1.9.2. If (X, Z) is a smooth pair, then étale locally, X should look like an affine space over S and Z should look like an intersection of coordinate hyperplanes. The converse is not quite true, but only because we chose to require Z not to have self-intersections.

EXAMPLE 1.9.3. For instance, if X is a smooth proper curve over $\mathbb{Z}_p$, then you can form a smooth proper pair (X, Z) by taking Z to be the Zariski closure in X of a set of closed points of $X_{\mathbb{Q}_p}$ which have *distinct* images in $X_{\mathbb{F}_p}$.

DEFINITION 1.9.4. Let (X, Z) be a smooth pair over K. Put $U = X \setminus Z$ and let $j : U \hookrightarrow X$ be the implied open immersion. The *sheaf of logarithmic differentials* on X, denoted $\Omega_{(X,Z)/K}$, is the subsheaf of $j_*\Omega_{U/K}$ generated by $\Omega_{X/K}$ and by sections of the form df/f, where f is a regular function on some open subset V of X which only vanishes along components of Z. Again, we write $\Omega^i_{(X,Z)/K}$ for the i-th exterior power of $\Omega^1_{(X,Z)/K}$ over $\mathscr{O}_X$; this yields a *logarithmic de Rham complex* $\Omega^{\cdot}_{(X,Z)/K}$. We define the *logarithmic de Rham cohomology* $H^i_{\mathrm{dR}}(X, Z)$ as the hypercohomology $\mathbb{H}^i(\Omega^{\cdot}_{(X,Z)/K})$.

THEOREM 1.9.5. *The evident map of complexes*

$$\Omega^{\cdot}_{(X,Z)/K} \to j_* \Omega^{\cdot}_{U/K}$$

is a quasi-isomorphism. Hence we obtain an isomorphism $H^i_{\mathrm{dR}}(X, Z) \cong H^i_{\mathrm{dR}}(U)$.

PROOF. The first assertion was originally proved by Deligne [10] using GAGA (Theorem 1.7.3). It is possible, and important for applications to p-adic cohomology, to give a completely algebraic proof, e.g., [1, Theorem 2.2.5].

The second assertion follows by considering the spectral sequence that goes from cohomology sheaves to hypercohomology (see §1.8). □

EXERCISE 1.9.6. Prove Theorem 1.9.5 directly for the example in Exercise 1.2.5.

REMARK 1.9.7. One might be tempted to deduce from Theorem 1.9.5 that for a smooth projective curve X and a point $x \in X(K)$, for $U = X \setminus \{x\}$, every class in $H^1_{\mathrm{dR}}(X) \cong H^1_{\mathrm{dR}}(U)$ is represented by a 1-form on U with a logarithmic singularity at x. This is false; for instance, in Example 1.5.1, you need to allow either a double pole at one point (to pick up $x\omega$), or poles at two different points. (This is because the sum of the residues of the poles of a 1-form is always zero; see Definition 1.5.2.) A related observation is that $\Omega^{\cdot}_{(X,Z)/K}$ is a sheaf on X, not on U, so that even if U is affine one must use hypercohomology to compute $\Omega^{\cdot}_{(X,Z)/K}$.

1.10. Example: a smooth hypersurface in projective space. We close this section by mentioning one higher-dimensional example due to Griffiths. We first need the following fact.

EXERCISE 1.10.1. Prove that the space $H^i_{\mathrm{dR}}(\mathbb{P}^n_K)$ is one-dimensional if $i = 0, 2, \ldots, 2n$ and zero otherwise. This uses the Hodge-de Rham spectral sequence plus the calculation of the cohomology $H^q(\mathbb{P}^n_K, \Omega^p_{\mathbb{P}^n_K/K})$, as partially given by [23, Theorem III.5.1] (see also [20, §1.2.1]).

EXAMPLE 1.10.2. Let X be a smooth hypersurface in $\mathbb{P}^n_K$, defined by the homogeneous polynomial $P(x_0, x_1, \ldots, x_n)$. Then there are natural maps $H^i_{\mathrm{dR}}(\mathbb{P}^n_K) \to H^i_{\mathrm{dR}}(X)$ which by the Lefschetz hyperplane theorem [17, §1.2] are isomorphisms for $i < n - 1$ and injective for $i = n - 1$. (That's actually a fact about complex manifolds, but by GAGA it transfers to the algebraic setting.) Since the cohomology of projective space is simple (Exercise 1.10.1), the only interesting cohomology group of X is $H^{n-1}_{\mathrm{dR}}(X)$.

There is a short exact sequence

$$0 \to \Omega^{\cdot}_{\mathbb{P}^n/K} \to \Omega^{\cdot}_{(\mathbb{P}^n, X)/K} \xrightarrow{\mathrm{Res}} j_* \Omega^{\cdot+1}_{X/K} \to 0,$$

where $j : X \to \mathbb{P}^n$ is the implied closed immersion. The map Res is a residue map, which can be described as follows: locally on $\mathbb{P}^n$, a section of $\Omega^{\cdot}_{(\mathbb{P}^n, X)/K}$ can be written as $df/f \wedge \omega$, where f is a dehomogenized form of P; Res takes this section to the restriction of ω to X. Taking cohomology and using Theorem 1.9.5 gives a long exact sequence

$$\cdots \to H^i_{\mathrm{dR}}(\mathbb{P}^n) \to H^i_{\mathrm{dR}}(U) \to H^{i-1}_{\mathrm{dR}}(X) \to H^{i+1}_{\mathrm{dR}}(\mathbb{P}^n) \to \cdots,$$

where $U = \mathbb{P}^n \setminus X$.

The upshot of this is that if n is even, then $H_{\mathrm{dR}}^n(U)$ is isomorphic to $H_{\mathrm{dR}}^{n-1}(X)$. If n is odd, then (using Poincaré duality) $H_{\mathrm{dR}}^n(U)$ is isomorphic to the quotient of $H_{\mathrm{dR}}^{n-1}(X)$ by the (one-dimensional) image of the map $H_{\mathrm{dR}}^{n-1}(\mathbb{P}^n) \to H_{\mathrm{dR}}^{n-1}(X)$, the so-called *primitive middle cohomology* of X.

The point is that U is affine, so you can compute its de Rham cohomology on global sections. For the recipe for doing this easily, see Griffiths [16, §4, 5].

REMARK 1.10.3. All of the above extends easily to smooth hypersurfaces in toric varieties, providing a rich source of examples for the study of mirror symmetry. The analogue of the Griffiths recipe (attributed to Dwork-Griffiths-Katz) is described in [9, §5.3].

2. Frobenius actions on de Rham cohomology

In this section, we explain how to define and compute a Frobenius action on the algebraic de Rham cohomology of a smooth proper variety over a p-adic field with good reduction.

NOTATION 2.0.1. Throughout this section, let q be a power of the prime p, let $\mathbb{Q}_q$ be the unramified extension of $\mathbb{Q}_p$ with residue field $\mathbb{F}_q$, and let $\mathbb{Z}_q$ be the integral closure of $\mathbb{Z}_p$ in $\mathbb{Q}_q$. We now use the marker "an" to denote rigid analytification for algebraic varieties over $\mathbb{Q}_q$, rather than complex analytification.

REMARK 2.0.2. One could in principle deal with ramified extensions of $\mathbb{Q}_q$ also. Our choice not to do so skirts a couple of complicating issues, including the choice of a Frobenius lift, and failure of the integral comparison theorem between de Rham and crystalline cohomology in case the absolute ramification index is greater than $p - 1$.

2.1. de Rham and crystalline cohomology.

DEFINITION 2.1.1. Let (X, Z) be a smooth pair over $\mathbb{Z}_q$. Put

$$X_{\mathbb{Q}} = X \times_{\operatorname{Spec}\mathbb{Z}_q} \operatorname{Spec}\mathbb{Q}_q,$$
$$Z_{\mathbb{Q}} = Z \times_{\operatorname{Spec}\mathbb{Z}_q} \operatorname{Spec}\mathbb{Q}_q,$$
$$\overline{X} = X \times_{\operatorname{Spec}\mathbb{Z}_q} \operatorname{Spec}\mathbb{F}_q,$$
$$\overline{Z} = Z \times_{\operatorname{Spec}\mathbb{Z}_q} \operatorname{Spec}\mathbb{F}_q,$$
$$U = X \setminus Z,$$
$$U_{\mathbb{Q}} = X_{\mathbb{Q}} \setminus Z_{\mathbb{Q}},$$
$$\overline{U} = \overline{X} \setminus \overline{Z},$$
$$\widehat{U} = \text{formal completion of } U \text{ along } \overline{U}.$$

We define the relative logarithmic de Rham cohomology $H_{\mathrm{dR}}^i(X, Z)$ as the hypercohomology $\mathbb{H}^i(\Omega_{(X,Z)/\mathbb{Z}_q}^{\cdot})$ (where the logarithmic de Rham complex is defined as in Definition 1.9.4 replacing K with $\mathbb{Z}_q$); by flat base change, the map $H_{\mathrm{dR}}^i(X, Z) \otimes_{\mathbb{Z}_q} \mathbb{Q}_q \to H_{\mathrm{dR}}^i(X_{\mathbb{Q}}, Z_{\mathbb{Q}})$ is an isomorphism.

One then has the following comparison theorem of Berthelot [4] in the nonlogarithmic case; the logarithmic generalization is similar. (We will not define crystalline cohomology here; see [5] for the construction in the nonlogarithmic case $Z = \emptyset$, and [27] for the logarithmic case.)

THEOREM 2.1.2. *There is a canonical isomorphism between $H^i_{\mathrm{dR}}(X, Z)$ and the logarithmic crystalline cohomology $H^i_{\mathrm{crys}}(\overline{X}, \overline{Z})$.*

REMARK 2.1.3. Even without the definition of crystalline cohomology, there is still an essential piece of content that one should carry away from Theorem 2.1.2. It is that the de Rham cohomology $H^i_{\mathrm{dR}}(X, Z)$ is functorial in the mod p reduction $(\overline{X}, \overline{Z})$. In particular, if (X', Z') is a second smooth pair and $\overline{f} : \overline{X} \to \overline{X}'$ is a morphism carrying $\overline{Z}$ into $\overline{Z}'$ (i.e., $\overline{f}$ induces a morphism $(\overline{X}, \overline{Z}) \to (\overline{X}', \overline{Z}')$ of smooth pairs), then $\overline{f}$ functorially induces a morphism $H^i_{\mathrm{dR}}(X', Z') \to H^i_{\mathrm{dR}}(X, Z)$. What is most surprising about this is that in order to obtain this functoriality, it is not necessary for $\overline{f}$ to lift to a morphism $f : X \to X'$.

One crucial instance of Remark 2.1.3 is that the q-power Frobenius map $F_q : \overline{X} \to \overline{X}$ induces maps $H^i_{\mathrm{dR}}(X, Z) \to H^i_{\mathrm{dR}}(X, Z)$. These satisfy the following Lefschetz trace formula.

THEOREM 2.1.4. *We have*
$$\#\overline{U}(\mathbb{F}_q) = \sum_i (-1)^i \operatorname{Trace}(F_q, H^i_{\mathrm{dR}}(X_{\mathbb{Q}}, Z_{\mathbb{Q}})).$$

PROOF. By Theorem 2.1.2, we may replace $H^i_{\mathrm{dR}}(X_{\mathbb{Q}}, Z_{\mathbb{Q}})$ with $H^i_{\mathrm{crys}}(\overline{X}, \overline{Z})$. By excision, we may reduce to the case where $Z = \emptyset$. In this case, this is a result of Katz and Messing [**28**]. □

REMARK 2.1.5. In the usual manner (e.g., as in [**23**, Appendix C]), Theorem 2.1.4 leads to a product formula for the zeta function
$$\zeta(\overline{U}, T) = \exp\left(\sum_{i=1}^{\infty} \#\overline{U}(\mathbb{F}_{q^i}) \frac{T^i}{i} \right);$$
namely,
$$(2.1.6) \qquad \zeta(\overline{U}, T) = \prod_i \det(1 - F_q T, H^i_{\mathrm{dR}}(X_{\mathbb{Q}}, Z_{\mathbb{Q}}))^{(-1)^{i+1}}.$$

However (as in the analogous ℓ-adic situation), there is a key difference between the case $Z = \emptyset$ and the general case. In case $Z = \emptyset$, the Riemann hypothesis component of the Weil conjectures (now a theorem of Deligne) asserts that the polynomial $\det(1 - F_q T, H^i_{\mathrm{dR}}(X_{\mathbb{Q}}, Z_{\mathbb{Q}})) \in \mathbb{Z}[T]$ has roots in $\mathbb{C}$ of norm $q^{-i/2}$. In particular, since the roots for different i lie in disjoint subsets of $\mathbb{C}$, (2.1.6) uniquely determines the factors on the right side. This is no longer true for Z general, though, because the roots can get mixed during the excision process.

2.2. Rigid cohomology. The definition of crystalline cohomology is not suitable for explicit calculations. Fortunately, there is a related construction that, at the expense of involving some auxiliary choices, is much more computable; it is Berthelot's rigid cohomology. The new book [**35**] provides a comprehensive development; here are the salient points for our purposes. (The strategy described below is essentially that of [**31**].)

DEFINITION 2.2.1. Suppose that Z is the inverse image of the infinity section under the morphism $f : X \to \mathbb{P}^1_{\mathbb{Z}_q}$. For $\eta \in (1, \infty]$, put
$$U_\eta = \{x \in X_{\mathbb{Q}}^{\mathrm{an}} : |f(x)| < \eta\};$$

note that $U_\infty = U_{\mathbb{Q}}^{\mathrm{an}}$, whereas

$$\bigcap_{\eta > 1} U_\eta = \{x \in X_{\mathbb{Q}}^{\mathrm{an}} : |f(x)| \leq 1\}$$

is the generic fibre of the formal scheme $\widehat{U}$. Using the formalism of algebraic de Rham cohomology, we can define the de Rham cohomology $H_{\mathrm{dR}}^i(U_\eta)$ as the hypercohomology $\mathbb{H}^i(U_\eta, \Omega_{U_\eta/\mathbb{Q}_q}^{\cdot})$.

EXAMPLE 2.2.2. For instance, if $X = \mathbb{P}_{\mathbb{Z}_q}^1$ and f is the identity map, then U_∞ is the whole affine line over $\mathbb{Q}_q$, whereas $\cap_{\eta > 1} U_\eta$ is only the closed unit disc.

By Berthelot's fibration theorem, we have the following.

THEOREM 2.2.3. *For* $1 < \eta \leq \eta' \leq \infty$, *the inclusion* $U_{\eta'} \hookrightarrow U_\eta$ *induces an isomorphism* $H_{\mathrm{dR}}^i(U_{\eta'}) \cong H_{\mathrm{dR}}^i(U_\eta)$. *In particular, we have* $H_{\mathrm{dR}}^i(U_\eta) \cong H_{\mathrm{dR}}^i(U_\infty) \cong H_{\mathrm{dR}}^i(X_{\mathbb{Q}}, Z_{\mathbb{Q}})$.

REMARK 2.2.4. The isomorphism $H_{\mathrm{dR}}^i(U_\infty) \cong H_{\mathrm{dR}}^i(X_{\mathbb{Q}}, Z_{\mathbb{Q}})$ in Theorem 2.2.3 amounts to a rigid analytic version of the corresponding complex analytic result, which is the combination of Theorems 1.7.2 and 1.9.5. In particular, it relies on an analogue of GAGA for formal schemes [20, §5] or rigid analytic spaces [33].

REMARK 2.2.5. Without the extra hypothesis on the existence of f, we would have to replace the U_η in Theorem 2.2.3 with a cofinal system of strict neighborhoods of the generic fibre of $\widehat{U}$. Moreover, for best results, we would have to work not with the individual $H_{\mathrm{dR}}^i(U_\eta)$ but with their direct limit as $\eta \to 1^+$.

More to the point, one has a comparison theorem with crystalline cohomology, which manifests as follows.

THEOREM 2.2.6. *Let* (X', Z') *be a second smooth pair over* $\mathbb{Z}_q$ *such that* Z' *is the inverse image of the infinity section under* $f' : X' \to \mathbb{P}_{\mathbb{Z}_q}^1$. *Suppose that* $g : \widehat{U} \to \widehat{U'}$ *is a morphism of formal schemes which induces a map* $U_\eta \to U_{\eta'}'$ *for some* $\eta, \eta' \in (1, \infty)$, *and which induces a map* $(\overline{X}, \overline{Z}) \to (\overline{X}', \overline{Z}')$. *Using Theorem 2.2.3, obtain from this a map* $H_{\mathrm{dR}}^i(X_{\mathbb{Q}}', Z_{\mathbb{Q}}') \to H_{\mathrm{dR}}^i(X_{\mathbb{Q}}, Z_{\mathbb{Q}})$. *Then this map corresponds, via Theorem 2.1.2, to the map* $H_{\mathrm{crys}}^i(\overline{X}', \overline{Z}') \otimes_{\mathbb{Z}_q} \mathbb{Q}_q \to H_{\mathrm{crys}}^i(\overline{X}, \overline{Z}) \otimes_{\mathbb{Z}_q} \mathbb{Q}_q$.

REMARK 2.2.7. The reader familiar with Dwork's proof of the rationality of zeta functions of varieties over finite fields [13], which also uses p-adic analytic methods, may wonder how those methods relate to rigid cohomology. The short answer is that there is a sort of duality between the two approaches; for more precise answers, see [2].

REMARK 2.2.8. The cohomology $H_{\mathrm{dR}}^i(U_\eta)$ in Theorem 2.2.3, or more properly its direct limit as $\eta \to 1^+$, may be interpreted as an instance of the "formal cohomology" of Monsky and Washnitzer [39, 37, 38]. See [42] for a useful overview of that construction.

2.3. Example: an elliptic curve. It is time to make the previous construction explicit in an example, and once again we opt to consider an elliptic curve. This amounts to a paraphrase of [30]. (See also Edixhoven's course notes [14]. For $p = 2$, one needs a different approach, given by Denef and Vercauteren [12].)

EXAMPLE 2.3.1. Assume $p \neq 2$. Let $P(x) = x^3 + ax + b \in \mathbb{Z}_q[x]$ be such that the reduction $\overline{P}(\overline{x}) = \overline{x}^3 + \overline{ax} + \overline{b} \in \mathbb{F}_q[x]$ has no repeated roots. Put

$$X = \operatorname{Proj} \mathbb{Z}_q[W, X, Y]/(Y^2W - X^3 - aXW^2 - bW^3),$$
$$Z = \operatorname{Proj} \mathbb{Z}_q[W, X, Y]/(Y^2W - X^3 - aXW^2 - bW^3, WY),$$

so that

$$U_{\mathbb{Q}} = \operatorname{Spec} R, \qquad R = \mathbb{Q}_q[x, y, z]/(y^2 - x^3 - ax - b, yz - 1).$$

(Note that (X, Z) form a smooth pair over $\mathbb{Z}_q$ as in Example 1.9.3.) We may obtain a map f as in Definition 2.2.1 by taking $f(x, y) = y + y^{-1}$.

In Example 1.2.4, we computed that $H^1_{\mathrm{dR}}(U_{\mathbb{Q}})$ admits a basis over $\mathbb{Q}_q$ given by

$$\frac{dx}{y}, \frac{x\,dx}{y}, \frac{dx}{y^2}, \frac{x\,dx}{y^2}, \frac{x^2\,dx}{y^2}.$$

We now know that the q-power Frobenius on $\overline{U}$ induces a $\mathbb{Q}_q$-linear action on $H^1_{\mathrm{dR}}(U_{\mathbb{Q}})$, and we wish to compute the matrix of this action.

To this end, we define a $\mathbb{Q}_q$-linear map $F_q : \widehat{U} \to \widehat{U}$ as follows:

$$F_q(x) = x^q,$$
$$F_q(y) = y^q(1 + (x^{3q} + ax^q + b - (x^3 + ax + b)^q)z^{2q})^{1/2},$$
$$F_q(z) = z^q(1 + (x^{3q} + ax^q + b - (x^3 + ax + b)^q)z^{2q})^{-1/2}.$$

This then extends to a map $U_\eta \to U_{\eta'}$ for some $\eta, \eta' > 1$, so we may apply Theorem 2.2.6 to deduce that this indeed induces the desired action on $H^1_{\mathrm{dR}}(U_{\mathbb{Q}})$. Better yet, F_q commutes with the hyperelliptic involution $y \mapsto -y$, so we may compute its action separately on the plus and minus eigenspaces of $H^1_{\mathrm{dR}}(U_{\mathbb{Q}})$.

In short, we may use F_q to compute the Frobenius action on the minus eigenspace of $H^1_{\mathrm{dR}}(U_{\mathbb{Q}})$, which is equal to $H^1_{\mathrm{dR}}(X_{\mathbb{Q}})$. To do this, we must rewrite the pullback $F_q^*(dx/y) = x^{q-1}F_q(z)\,dx$ as a $\mathbb{Q}_q$-linear combination of $dx/y, x\,dx/y$ and an exact 1-form (and likewise for $F_q^*(x\,dx/y)$). This calculation in general cannot be made exactly; for purposes of machine computation, one can only expect to work to a prescribed level of p-adic accuracy. So modulo some power of p, we write for instance

$$F_q^*(dx/y) \equiv \sum_{i=0}^{N} \frac{Q_i(x)\,dx}{y^{2i+1}}$$

for some $Q_i(x) \in \mathbb{Q}_q[x]$, then uses the rules from Example 1.2.4 to rewrite this as a $\mathbb{Q}_q$-linear combination of $dx/y, x\,dx/y$ plus an exact 1-form. If one has some information about the p-adic size of the reductions of the neglected terms in the expansion of $F_q^*(dx/y)$ (e.g., as in [**30**, Lemma 2]), one can then compute the matrix of action of F_q on the basis $dx/y, x\,dx/y$ to any desired p-adic accuracy.

REMARK 2.3.2. For the purposes of computing the zeta function of $\overline{X}$, one ultimately wants to compute the characteristic polynomial of the matrix of action of F_q to a certain amount of p-adic accuracy, which one can specify in advance. Namely, one is trying to compute some integer whose archimedean norm can be limited, so one needs only enough p-adic accuracy to pin down the integer uniquely in the archimedean range. (Put briefly, if one is trying to compute some $n \in \mathbb{Z}$ with $|n| \leq N$, it suffices to compute $n \pmod{p^k}$ for any $k > \log_p(2N)$.)

REMARK 2.3.3. If $q \neq p$, one does not typically compute the action of F_q directly. Instead, one starts with a map $F_p : \widehat{U} \to \widehat{U}$ which is $\mathbb{Q}_q$-semilinear for the Frobenius automorphism σ of $\mathbb{Q}_q$, and acts on x, y, z as follows:

$$F_p(x) = x^p,$$
$$F_p(y) = y^p(1 + (x^{3p} + \sigma(a)x^p + \sigma(b) - (x^3 + ax + b)^p)z^{2p})^{1/2},$$
$$F_p(z) = z^p(1 + (x^{3p} + \sigma(a)x^p + \sigma(b) - (x^3 + ax + b)^p)z^{2p})^{-1/2}.$$

One then computes the matrix of action of F_p on dx/y, $x\,dx/y$, and takes an appropriate norm of this matrix to recover the action of F_q. The fact that one can do this makes the use of p-adic cohomology particularly advantageous for computing zeta functions over fields of small characteristic; for example, the computer algebra system Magma uses p-adic cohomology to compute zeta functions of hyperelliptic curves in small characteristic. (In the elliptic case, there exist faster alternatives.)

REMARK 2.3.4. By contrast, if one wants to use p-adic cohomology to compute zeta functions over $\mathbb{F}_p$, one is forced to take p not too large. The complexity estimates for the above procedure include a factor of p; however, it has been shown recently by David Harvey [24] that one can restructure the algorithm to improve the dependence on p to a factor of $p^{1/2}$ times a power of $\log p$. This makes it conceivable to work with p as big as 2^{64}.

REMARK 2.3.5. The matrix of action of Frobenius has some uses beyond simply determining the zeta function; here is an interesting example due to Mazur, Stein, and Tate [36]. Following a suggestion of Katz, they give a formula for the p-adic (cyclotomic) canonical height of an elliptic curve over $\mathbb{Q}$ in terms of the Frobenius action on the de Rham cohomology over $\mathbb{Q}_p$. (This height is not the Néron local height; it is a global height with p-adic values, which computes the regulator term in Mazur-Tate-Teitelbaum's p-adic analogue of the Birch-Swinnerton-Dyer conjecture.) Using this formula, one can then compute p-adic canonical heights much more rapidly than any other known method; this was carried out in 2006 by a group of students (Jennifer Balakrishnan, Robert Bradshaw, David Harvey, Liang Xiao) and is implemented in the computer algebra system Sage.

3. Gauss-Manin connections

In this section, we introduce the notion of a Gauss-Manin connection for a smooth proper morphism (of algebraic, complex analytic, or rigid analytic varieties). We then describe how in some cases, such a connection carries a Frobenius action which can be used to compute zeta functions.

3.1. Connections in geometry and algebra.

DEFINITION 3.1.1. Let X be any of the following:
- a C^∞ real manifold;
- a complex analytic manifold;
- a smooth algebraic variety over a field of characteristic zero;
- a smooth rigid analytic variety over a field of characteristic zero.

Let Ω_X be the corresponding sheaf of differentials. Let V be a vector bundle (coherent locally free sheaf) on X. A *connection* on V is a bundle map $\nabla : V \to$

$V \otimes \Omega^1_X$ which is additive and satisfies the *Leibniz rule*: for any open set $U \subseteq X$, any $f \in \Gamma(U, \mathscr{O}_U)$ and $s \in \Gamma(U, V)$,

$$\nabla(fs) = f\nabla(s) + s \otimes df.$$

A section s is called *horizontal* if $\nabla(s) = 0$.

DEFINITION 3.1.2. Let $\nabla_1 : V \otimes \Omega^1_X \to V \otimes \Omega^2_X$ be the map

$$s \otimes \omega \mapsto \nabla(s) \wedge \omega + s \otimes d\omega,$$

where $\wedge : (V \otimes \Omega^1_X) \otimes \Omega^1_X \to V \otimes \Omega^2_X$ denotes the map given by wedging the second and third factors. The *curvature* is the map $\nabla_1 \circ \nabla : V \to V \otimes \Omega^2_X$; if it vanishes, we say ∇ is *integrable*. (This is automatic if $\dim(X) = 1$.)

REMARK 3.1.3. Here is another way to think about integrability of a connection ∇. Let $z_1, \ldots, z_n$ be local coordinates for X at a point x. Then $dz_1, \ldots, dz_n$ form a basis of Ω_X in some neighborhood of x. Form the dual basis $\frac{\partial}{\partial z_1}, \ldots, \frac{\partial}{\partial z_n}$ of tangent vector fields; we can contract ∇ with the vector field $\frac{\partial}{\partial z_i}$ to obtain a map $V \to V$ satisfying the Leibniz rule with respect to $\frac{\partial}{\partial z_i}$. If you think of this as an action of $\frac{\partial}{\partial z_i}$ on sections of V, then ∇ is integrable if and only if the $\frac{\partial}{\partial z_i}$ commute with each other.

REMARK 3.1.4. Here is the original differential-geometric interpretation of curvature. In real geometry, you can use a connection ∇ to tell you how to move between fibres of the bundle in a "horizontal" fashion, i.e., parallel to the base. Even in a small neighborhood of a point, moving parallel to different paths on the base leading to the same endpoint can give different results. But if the curvature vanishes, then this discrepancy does not arise; this means that on any contractible neighborhood of $x \in X$, we can write down a basis of V consisting of horizontal sections $s_1, \ldots, s_n$, and the connection is given in terms of this basis by

$$\nabla(f_1 s_1 + \cdots + f_n s_n) = s_1 \otimes df_1 + \cdots + s_n \otimes df_n.$$

It also means that given $x \in X$, parallel transport (the process of moving from one fibre to another via a horizontal path) gives a well-defined homomorphism $\rho : \pi_1(X, x) \to \mathrm{GL}(V_x)$, called the *monodromy*. (Differential geometers refer to integrable connections as *flat* connections, but for algebraic geometers this adjective is otherwise occupied.)

One can take the relationship between integrable connections and monodromy a step further.

DEFINITION 3.1.5. Let X be a connected complex manifold, and choose a point $x \in X$. A *local system* on X is a homomorphism $\rho : \pi_1(X, x) \to \mathrm{GL}_n(\mathbb{C})$. As noted above, there is a natural functor from vector bundles equipped with integrable connections to local systems; it turns out to be an equivalence of categories.

3.2. Connections and differential equations. It is worth pointing out that for purposes of explicit calculations, it is common to work with differential equations instead of connections. This discussion is formal, so it works in any of the categories we allowed when defining connections.

DEFINITION 3.2.1. Suppose that we are given a trivial vector bundle V of rank n over a subspace S of the t-line, together with a connection on V, or equivalently, with an action of $\frac{d}{dt}$. A *cyclic vector* is a section s of V such that

$$s, \frac{d}{dt}s, \dots, \frac{d^{n-1}}{dt^{n-1}}s$$

form a basis of V. Given a cyclic vector, we can describe horizontal sections as follows. We can write

$$\frac{d^n}{dt^n}s = a_0 s + a_1 \frac{d}{dt}s + \cdots + a_{n-1}\frac{d^{n-1}}{dt^{n-1}}s$$

for certain functions $a_0, \dots, a_{n-1}$ on S. Consider an undetermined section v of V, which must be given by

$$v = f_0 s + f_1 \frac{d}{dt}s + \cdots + f_{n-1}\frac{d^{n-1}}{dt^{n-1}}s$$

for certain functions $f_0, \dots, f_{n-1}$ in t. For v to be horizontal, we need

$$0 = f_0' + f_{n-1}a_0,$$
$$0 = f_1' + f_0 + f_{n-1}a_1,$$

$$\vdots$$

$$0 = f_{n-1}' + f_{n-2} + f_{n-1}a_{n-1}.$$

Eliminating $f_0, \dots, f_{n-2}$ leaves a differential equation of the form

$$f_{n-1}^{(n)} + b_1 f_{n-1}^{(n-1)} + \cdots + b_{n-2}f_{n-1}' + b_{n-1}f_{n-1} = 0.$$

Conversely, one can turn the differential equation into a first-order differential system in the usual fashion, and thus reconstruct ∇.

REMARK 3.2.2. One might imagine the above construction as being a differential analogue of the passage from a matrix to its characteristic polynomial, which can be reversed (up to similarity) by forming the companion matrix of a polynomial.

3.3. Gauss-Manin connections.

DEFINITION 3.3.1. Let $\pi : X \to S$ be a smooth proper morphism between objects in one of the categories we considered in Definition 3.1.1. Let $Z \subset X$ be a relative strict normal crossings divisor over S. Let $\Omega_{(X,Z)/S} = \Omega_{(X,Z)/K}/\pi^*\Omega_{S/K}$ be the sheaf of relative logarithmic differentials. Then the functor taking an open affine $U \subseteq S$ to the hypercohomology $\mathbb{H}^i(\Omega_{(\pi^{-1}(U),\pi^{-1}(U)\cap Z)/U})$ turns out to be a sheaf; we call this sheaf the *relative de Rham cohomology* $H_{dR}^i((X,Z)/S)$. It is a vector bundle on S whose fibre at a point b can be identified with $H_{dR}^i(X_b, Z_b)$, where $X_b = \pi^{-1}(b)$ and $Z_b = X_b \cap Z$.

REMARK 3.3.2. The formation of the relative de Rham cohomology throws away some information: it only uses the "vertical" part of the differential operator d. What this means is that given a relative i-form $\omega \in \Omega_{(X,Z)/S}^i$, if one lifts ω to an absolute i-form $\tilde{\omega} \in \Omega_{(X,Z)/K}^i$ and differentiates the result, one may get something nonzero even if ω was a relative cocycle. If one projects the result into $\Omega_{(X,Z)/S}^i \otimes \Omega_{S/K}^1$, one has essentially constructed the Gauss-Manin connection. We will give a more formal construction below, but the procedure just described is how one really computes the Gauss-Manin connection in practice; see Example 3.4.1.

DEFINITION 3.3.3. Equip the de Rham complex $\Omega^{\cdot}_{(X,Z)/K}$ with the decreasing filtration

$$F^i = \text{image}[\Omega^{-i}_{(X,Z)/K} \otimes_{\mathscr{O}_X} \pi^*(\Omega^i_{S/K}) \to \Omega^{\cdot}_{(X,Z)/K}],$$

then form the corresponding spectral sequence (as in Theorem 1.8.7). The E_1 term of the result has

$$E_1^{p,q} = \Omega^p_{S/K} \otimes_{\mathscr{O}_S} H^q_{\text{dR}}((X,Z)/S);$$

the *algebraic Gauss-Manin connection* is the differential $d_1 : E_1^{0,q} \to E_1^{1,q}$. This construction was introduced by Katz and Oda [29], who showed that this is an integrable connection, and also that it agrees with the more traditional analytic description for a real or complex manifold.

DEFINITION 3.3.4. Suppose we are working with real or complex manifolds, and that S is contractible. Then the fibrations $X \to S$ and $Z \to S$ are trivial in the category of real manifolds, so we get a notion of horizontality for sections of $H^i_{\text{dR}}((X,Z)/S)$. For general S, this gives a connection on $H^i_{\text{dR}}((X,Z)/S)$; this is the usual Gauss-Manin connection, and Katz and Oda showed that it agrees with their algebraic construction.

REMARK 3.3.5. The differential equations corresponding to Gauss-Manin connections (via the transformation in §3.2) were introduced long before anyone had defined a connection, and so they have their own name. They are known as *Picard-Fuchs equations*; they arise by taking a homology class across different fibres and integrating against a fixed differential form on the total space. A number of classical differential equations (e.g., hypergeometric equations) arise in this fashion.

3.4. Example: a family of elliptic curves.

EXAMPLE 3.4.1. Consider the family of smooth projective curves $\pi : X \to S$ with

$$X = \textbf{Proj } \mathscr{O}_S[X,Y,W]/(Y^2W - X^3 - a(t)XW^2 - b(t)W^3),$$

and S equal to the subscheme of the affine t-line on which $\Delta(t) = 4a^2 + 27b^3$ does not vanish. Then there exist $A, B \in \Gamma(S, \mathscr{O}_S)[x]$ such that

$$AP + BP_x = 1,$$

using subscripted x and t for the partial derivatives in x and t, respectively. Put

$$\omega = Ay\,dx + 2B\,dy,$$

so that a basis for $H^1_{\text{dR}}(X/S)$ is given by

$$\omega, x\omega.$$

On the other hand, if we let Z be the infinity section, then $H^1_{\text{dR}}((X,Z)/S) = H^1_{\text{dR}}(X,S)$, and it will be easier to compute the Gauss-Manin connection on the affine $U = X \setminus Z$.

In the relative module of differentials $\Omega^1_{U/S}$, i.e., modulo dt, we have as before the relation

$$2y\,dy = P_x\,dx,$$

and again

$$dx = y\omega, \qquad dy = \frac{1}{2}P_x\omega.$$

But in the full module $\Omega^1_{U/K}$, the relation lifts to

$$2y\,dy = P_x\,dx + P_t\,dt,$$

and it is this discrepancy that gives rise to the connection. It follows that

$$dx \wedge dt = y\omega \wedge dt,$$
$$dy \wedge dt = \frac{1}{2}P_x\omega \wedge dt,$$
$$dx \wedge dy = \frac{1}{2}P_t\omega \wedge dt.$$

To compute the connection, we are supposed to lift the basis of relative cohomology to a set of forms on the total space, then differentiate, then project onto $H^1_{\mathrm{dR}}((X,Z)/S) \otimes \Omega^1_{S/K}$. First,

$$\nabla(\omega) = A\,dy \wedge dx + A_t y\,dt \wedge dx + 2B_x\,dx \wedge dy + 2B_t\,dt \wedge dy$$
$$= (B_x P_t - \frac{1}{2}AP_t - A_t P - B_t P_x)\omega \wedge dt.$$

Second,

$$\nabla(x\omega) = Ax\,dy \wedge dx + A_t xy\,dt \wedge dx + 2(xB_x + B)\,dx \wedge dy + 2xB_t\,dt \wedge dy$$
$$= (xB_x P_t + BP_t - \frac{1}{2}xAP_t - xA_t P - xB_t P_x)\omega \wedge dt.$$

We then rewrite the quantities being wedged with dt as exact relative differentials plus a linear combination of $\omega, x\omega$ as in Example 2.3.1.

If you prefer, here is another way of describing essentially the same calculation.

EXAMPLE 3.4.2. Consider the same situation as in Example 3.4.1, but now let us redefine

$$\omega = \frac{dx}{y}.$$

(Remember that this agrees with $Ay\,dx + 2B\,dy$ only modulo dt.) Again, use $\omega, x\omega$ as the basis of relative differentials. These have poles along $y = 0$, but never mind that; we can still compute

$$\nabla(\omega) = d\left(\frac{dx}{y}\right)$$
$$= -\frac{dy \wedge dx}{y^2}$$
$$= \frac{P_t}{2y^2}\frac{dx}{y} \wedge dt.$$

To eliminate the pole, find the unique $C, D, E, F, G \in \Gamma(S, \mathscr{O}_S)$ (so these are functions of t alone) such that

$$P_t = a_t x + b_t = (Cx + D)P + (Ex^2 + Fx + G)P_x.$$

Then in relative de Rham cohomology,

$$\frac{P_t\,dx}{2y^3} = \frac{(Cx+D)\,dx}{2y} + \frac{(Ex^2+Fx+G)P_x\,dx}{2y^3}$$
$$\equiv \frac{(Cx+D)\,dx}{2y} + \frac{(2Ex+F)\,dx}{y}.$$

Similarly,

$$\nabla(x\omega) = d\left(\frac{x\,dx}{y}\right)$$

$$= \frac{xP_t}{2y^2}\frac{dx}{y} \wedge dt$$

and writing

$$xP_t = (Hx + I)P + (Jx^2 + Kx + L)P_x,$$

we get

$$\frac{xP_t\,dx}{2y^3} \equiv \frac{(Hx + I)\,dx}{2y} + \frac{(2Jx + K)\,dx}{y}.$$

3.5. Gauss-Manin connections and Frobenius. The reason why Gauss-Manin connections are relevant in rigid cohomology is that they can be used to compute Frobenius actions *en masse*.

EXERCISE 3.5.1. Let $\operatorname{Spec} A$ be a smooth affine $\mathbb{Z}_q$-scheme. Let $\hat{A}$ be the p-adic completion of A, and put $\overline{A} = A/pA$. Let $\phi : \hat{A} \to \hat{A}$ be a q-power Frobenius lift on $\hat{A}$; that is, ϕ acts on $\overline{A}$ by the q-power map. Prove that for each $\overline{x} \in (\operatorname{Spec}\overline{A})(\mathbb{F}_q)$, there is a unique $x \in (\operatorname{Spec}\hat{A})(\mathbb{Z}_q)$ which specializes to $\overline{x}$, such that $\phi(x) = x$.

DEFINITION 3.5.2. With notation as in Exercise 3.5.1, we call x the *Teichmüller lift* of $\overline{x}$ with respect to ϕ. If x is a Teichmüller lift of a point not specified, we say x is a *Teichmüller point*.

EXERCISE 3.5.3. State and prove a generalization of Exercise 3.5.1 to the case where the residue field is perfect but not necessarily finite, and/or the p-adic field is not necessarily unramified.

THEOREM 3.5.4 (Berthelot). *Let $S = \operatorname{Spec} A$ be a smooth affine $\mathbb{Z}_q$-scheme. Let ϕ be a q-power Frobenius lift on $\hat{A}$. Let $\pi : X \to S$ be a smooth proper morphism, and put $\mathscr{E} = H^i_{\mathrm{dR}}(X_{\mathbb{Q}}/S_{\mathbb{Q}})$ as a vector bundle equipped with the Gauss-Manin connection. Then there exists an isomorphism $F : \phi^*\mathscr{E} \cong \mathscr{E}$ of vector bundles with integrable connection on the affinoid space $\hat{S}^{\mathrm{an}}$ (the generic fibre of $\hat{S} = \operatorname{Spf}\hat{A}$), such that for any positive integer a, and any Teichmüller point $x \in \hat{S}^{\mathrm{an}}(\mathbb{Q}_{q^a})$, $F^a : (\phi^a)^*\mathscr{E} \cong \mathscr{E}$ induces the q^a-power Frobenius action on $H^i_{\mathrm{dR}}(X_x)$.*

REMARK 3.5.5. The key feature of Theorem 3.5.4 is that the Frobenius action commutes with the action of the connection; this constraint can be interpreted as a differential equation on the Frobenius action. Let's see what this looks like for S a subscheme of the affine t-line and $\mathscr{E}$ admitting a basis $s_1, \ldots, s_n$. Define the matrices A, N by

$$Fs_j = \sum_i A_{ij}s_i,$$

$$\frac{d}{dt}s_j = \sum_i N_{ij}s_i.$$

Then the compatibility between Frobenius and the connection is equivalent to the equation

$$(3.5.6) \qquad NA + \frac{d}{dt}A = \left(\frac{d\phi(t)}{dt}\right)A\phi(N);$$

given N, this constitutes a system of linear differential equations on the entries of A.

REMARK 3.5.7. Theorem 3.5.4 forms the basis of another method for using p-adic cohomology to compute zeta functions, originally proposed by Lauder [34]. Let us explain how this works in the example of a family of elliptic curves, as in Example 3.4.1. (A version of this has been worked out in detail by Hubrechts [25], and is implemented in Magma.)

In order to start this method, we must already have the Frobenius matrix on $H^i_{\mathrm{dR}}(X_x)$ for a single Teichmüller point x, to use as an initial condition. Say for simplicity that this point is at $t = 0$. In (3.5.6), we now have the value of $A(0)$. We then compute the unique matrix $U \in M_{n \times n}(\mathbb{Q}_q[\![t]\!])$ with $U(0) = I_n$ such that $NU + \frac{d}{dt}(U) = 0$; we then obtain a solution of (3.5.6) given by

$$N = -\frac{d}{dt}(U)U^{-1}, \qquad A = UA(0)\phi(U)^{-1}.$$

Although U only converges on the open unit disc (by a result of Dwork), the matrix A has better convergence properties: it has bounded coefficients, and modulo any power of p its entries are the power series representing certain rational functions with no poles in S. If one can control the number of zeroes and poles of these rational functions, one can reconstruct a p-adic approximation to A, and then evaluate at any other Teichmüller point.

REMARK 3.5.8. One way to gain control of the degrees of the rational functions appearing in Remark 3.5.7 is to understand how the Frobenius structure behaves in discs over which the fibration has one or more singular fibres. We will discuss a particularly simple instance of this in the next section.

4. Beyond good reduction

So far, we have only allowed consideration of varieties over $\mathbb{Q}_q$ with good reduction. In this section, we explore what happens when we relax this restriction slightly.

4.1. Semistable reduction and logarithmic connections. To keep things simple, we will work only over one-dimensional base spaces.

DEFINITION 4.1.1. Let (X, Z) be a smooth pair with X one-dimensional. A *logarithmic connection* on X is a bundle map $\nabla : V \to V \otimes \Omega_{(X,Z)}$ that satisfies the Leibniz rule.

DEFINITION 4.1.2. Let ∇ be a logarithmic connection on (X, Z). For each $z \in Z$, ∇ induces a linear map $V_z \to V_z$, called the *residue* of ∇ at z, as follows. For a section s specializing to a given point $\overline{s} \in V_z$, write $ds = f\frac{dt}{t}$ for f a section of V and t a local parameter for z. Then the residue map carries $\overline{s}$ to $\overline{f}$, the specialization of f to V_z.

DEFINITION 4.1.3. Let $\pi : X \to S$ be a proper, flat, generically smooth morphism (in any of the categories from Definition 3.1.1) with S one-dimensional. We say π is *semistable* at $z \in S$ if the fibre X_z is a reduced divisor with simple (but not necessarily strict) normal crossings. That is, étale locally, S looks like $\operatorname{Spec} k[t]$ and X looks like $\operatorname{Spec} k[x_1, \ldots, x_n]/(x_1 \cdots x_m - t)$ for some m, n (where m varies from point to point).

EXAMPLE 4.1.4. For example, the Legendre family of elliptic curves
$$y^2 = x(x-1)(x-\lambda)$$
is smooth over $\lambda \notin \{0, 1, \infty\}$, and semistable at $\lambda = 0$ and $\lambda = 1$. To check at ∞, we reparametrize $\mu = \lambda^{-1}$, $x = X/\mu$, $y = Y/\mu^2$ to get
$$Y^2 = \mu X(X - \mu)(X - 1),$$
which means that the fibre at $\lambda = \infty$ is not reduced.

If we instead consider the family
$$y^2 = x(x-1)(x-\lambda^2),$$
this family is indeed semistable everywhere.

By a suitable variation of the Katz-Oda arguments [29], one obtains the following.

THEOREM 4.1.5. *Let $\pi : X \to S$ be a semistable morphism, smooth over $U \subseteq S$. Then the Gauss-Manin connection on $H^i_{\mathrm{dR}}(\pi^{-1}(U)/U)$ extends to a logarithmic connection on S with nilpotent residue maps.*

EXERCISE 4.1.6. Check that the Gauss-Manin connection for the Legendre family extends to a logarithmic connection on all of $\mathbb{P}^1$, but the residue map at infinity cannot be made nilpotent. (Hint: you can cheat by looking up this calculation in [42, §7].)

4.2. Frobenius actions on singular connections. Berthelot gave the following refinement of Theorem 3.5.4.

THEOREM 4.2.1 (Berthelot). *Let $S = \operatorname{Spec} A$ be a smooth affine $\mathbb{Z}_q$-scheme. Let ϕ be a q-power Frobenius lift on $\hat{A}$. Let $\pi : X \to S$ be a proper morphism smooth over an open dense subscheme $U \subseteq S$. Put $\mathscr{E} = H^i_{\mathrm{dR}}(\pi^{-1}(U_\mathbb{Q})/U_\mathbb{Q})$ as a vector bundle equipped with the Gauss-Manin connection. Then the isomorphism $F : \phi^* \mathscr{E} \cong \mathscr{E}$ over $\hat{U}^{\mathrm{an}}$ given by Theorem 3.5.4 extends to a space of the form*
$$\{x \in U^{\mathrm{an}}_\mathbb{Q} : |f(x)| \geq \eta\}$$
for some $f \in A$ and some $\eta \in (0, 1)$.

REMARK 4.2.2. To clarify, let us suppose $S = \mathbb{A}^1_{\mathbb{Z}_q}$ and let us consider the open unit disc in S^{an}. Then Theorem 4.2.1 gives us an isomorphism $\phi^* \mathscr{E} \cong \mathscr{E}$ on some annulus $A_\eta = \{t \in \mathbb{A}^1_{\mathbb{Q}_q} : \eta < |t| < 1\}$. For applications to machine computations, one would like to be able to predict the value of η. This is complicated in general, but there is a particular case in which it is comparatively easy; see below.

LEMMA 4.2.3. *Let $\mathscr{E}$ be a vector bundle on the open unit t-disc over $\mathbb{Q}_q$, equipped with a logarithmic connection with nilpotent residue at $t = 0$ and no other poles. Suppose there exists an isomorphism $\phi^* \mathscr{E} \to \mathscr{E}$ on some A_η, where $\phi : \Gamma(A_{\eta'}, \mathscr{O}) \to \Gamma(A_\eta, \mathscr{O})$ for some $\eta' < \eta \in (0, 1)$ satisfies $\phi(t) - t^q \in p\widehat{\mathbb{Z}_q((t))}$. (Note that ϕ does not have to be defined on the entire open unit disc.) Then $\mathscr{E}$ is a successive extension of copies of the trivial bundle with connection (i.e., the bundle is $\mathscr{O}$ and the connection is just the exterior derivative).*

PROOF. The existence of the Frobenius structure ensures that the differential module $\mathscr{E}$ is solvable at 1, so that we may deduce the claim from [32, Lemma 3.6.2]. □

REMARK 4.2.4. Resuming now Remark 4.2.2, now suppose that π is semistable and that its only singular fibre over the open unit disc is at $t = 0$. Define the matrices N, A as in Remark 3.5.5. Then there is a unique matrix $U \in M_{n \times n}(\mathbb{Q}_q[\![t]\!][\log t])$ with $U \equiv I_n \pmod{(t, \log t)}$ such that $NU + \frac{d}{dt}(U) = 0$. From Lemma 4.2.3, one may deduce that each power series appearing in U and U^{-1} converges in the entire open unit disc. Define

$$\phi(\log t) = q \log t + \sum_{i=1}^{\infty} \frac{(-1)^{i-1}}{i} (\phi(t)/t^q - 1)^i$$

and put $B = U^{-1} A \phi(U)$. Then (3.5.6) implies $\frac{d}{dt}(B) = 0$, so $B \in M_{n \times n}(\mathbb{Q}_q)$.

By writing $A = U B \phi(U^{-1})$, we deduce the following. Suppose that $\eta \in [0, 1)$ has the property that $|\phi(t)/t^q - 1|_\eta < 1$; note that this implies $|\phi(t)|_\eta = \eta^q < 1$. (Here $|\cdot|_\eta$ is the η-Gauss norm, i.e., $|\sum_i c_i t^i|_\eta = \sup_i\{|c_i|\eta^i\}$.) Then A converges on $\eta \leq |t| < 1$.

For explicit computations, one needs effective convergence bounds in the above argument. These are provided by results of Christol and Dwork [7].

REMARK 4.2.5. Continuing as in Remark 4.2.4, we obtain on $H_{dR}^i(X_0)$ a Frobenius action Φ as well as a nilpotent operator N. It is natural to ask for a geometric interpretation of these; one expects them to coincide with the corresponding operators (Frobenius and monodromy) on the Hyodo-Kato cohomology $H_{HK}^i(\overline{X}_0)$ via the comparison isomorphism

$$H_{dR}^i(X_0) \cong H_{HK}^i(\overline{X}_0).$$

(These notions are introduced in [26]; a similar construction in topology is the "Milnor fibre".) We do not have a proof of this coincidence, though it is strongly suggested by results of Coleman and Iovita [8] and Grosse-Klönne [18].

Note that if we wish to transfer the Frobenius action from X_0 to another fibre X_t with $|t| < 1$, we must at some point evaluate $\log(t)$, which requires choosing a branch of the p-adic logarithm. This is consistent with the fact that for $t \neq 0$, the Hyodo-Kato isomorphism $H_{dR}^i(X_t) \cong H_{HK}^i(\overline{X}_0)$ depends on such a branch choice; the monodromy operator transfers canonically to $H_{dR}^i(X_t)$, but the Frobenius operator on $H_{dR}^i(X_t)$ does depend on the branch choice.

Bibliography

[1] T.G. Abbott, K.S. Kedlaya, and D. Roe, Bounding Picard numbers of surfaces using p-adic cohomology, arXiv:math.NT/0601508 (version of 18 Jan 2007), to appear in *Arithmetic, Geometry and Coding Theory (AGCT 2005)*, Societé Mathématique de France.

[2] F. Baldassarri and P. Berthelot, On Dwork cohomology for singular hypersurfaces, *Geometric Aspects of Dwork Theory, Vol. I*, de Gruyter, Berlin, 2004, 177–244.

[3] P. Berthelot, *Cohomologie Cristalline des Schémas de Caractéristique p > 0*, Lecture Notes in Math. 407, Springer-Verlag, Berlin, 1974.

[4] P. Berthelot, Géométrie rigide et cohomologie des variétés algébriques de caractéristique p, Introductions aux cohomologies p-adiques (Luminy, 1984), *Mém. Soc. Math. France* **23** (1986), 7–32.

[5] P. Berthelot and A. Ogus, *Notes on Crystalline Cohomology*, Princeton Univ. Press, Princeton, 1978.

[6] R. Bott and L.W. Tu, *Differential Forms in Algebraic Topology*, corrected third printing, Graduate Texts in Math. 82, Springer-Verlag, 1995.

[7] G. Christol and B. Dwork, Effective p-adic bounds at regular singular points, *Duke Math. J.* **62** (1991), 689–720.

[8] R. Coleman and A. Iovita, The Frobenius and monodromy operators for curves and abelian
     varieties, *Duke Math. J.* **97** (1999), 171–215.
[9] D.A. Cox and S. Katz, *Mirror Symmetry and Algebraic Geometry*, Math. Surveys and Mono-
     graphs 68, Amer. Math. Soc., 1999.
[10] P. Deligne, *Équations Différentielles à Points Singuliers Réguliers*, Lecture Notes in Math.
     163, Springer-Verlag, Berlin, 1970.
[11] P. Deligne and L. Illusie, Relèvements modulo p^2 et dècomposition du complexe de de Rham,
     Invent. Math. **89** (1987), 247–270.
[12] J. Denef and F. Vercauteren, An extension of Kedlaya's algorithm to hyperelliptic curves in
     characteristic 2, *J. Cryptology* **19** (2006), 1–25.
[13] B. Dwork, On the rationality of the zeta function of an algebraic variety, *Amer. J. Math.* **82**
     (1960), 631–648.
[14] B. Edixhoven, Point counting after Kedlaya, course notes at
     http://www.math.leidenuniv.nl/~edix/oww/mathofcrypt/.
[15] T. Ekedahl, On the multiplicative properties of the de Rham-Witt complex I, II, *Ark. Mat.*
     22(1984), 185–239; *ibid.* **23** (1985), 53–102.
[16] P. Griffiths, On the periods of certain rational integrals. I, II, *Annals of Math.* **90** (1969),
     460–495, 496–541.
[17] P. Griffiths and J. Harris, *Principles of Algebraic Geometry*, John Wiley & Sons, 1978.
[18] E. Grosse-Klönne, Frobenius and monodromy operators in rigid analysis, and Drinfel'd's
     symmetric space, *J. Alg. Geom.* **14** (2005), 391–437.
[19] A. Grothendieck, On the de Rham cohomology of algebraic varieties, *Publ. Math. IHÉS* **29**
     (1966), 95–103.
[20] A. Grothendieck, Élements de géométrie algébrique III: Étude cohomologique des faisceaux
     cohérents, première partie (EGA 3-1), *Publ. Math. IHÉS* **11** (1961), 5–167.
[21] A. Grothendieck et al, *Revêtements Étales et Groupe Fondamental (SGA 1)*, revised version,
     Société Mathématique de France, 2003.
[22] R. Hartshorne, On the De Rham cohomology of algebraic varieties, *Publ. Math. IHÉS* **45**
     (1975), 5–99.
[23] R. Hartshorne, *Algebraic Geometry*, Graduate Texts in Math. 52, Springer, 1977.
[24] D. Harvey, Kedlaya's algorithm in larger characteristic, *Int. Math. Res. Notices* (2007), article
     ID rnm095.
[25] H. Hubrechts, Point counting in families of hyperelliptic curves, to appear in *Found. Comp.
     Math.*
[26] O. Hyodo and K. Kato, Semi-stable reduction and crystalline cohomology with logarithmic
     poles, Périodes p-adiques (Bures-sur-Yvette, 1988), *Astérisque* 223 (1994), 221–268.
[27] K. Kato, Logarithmic structures of Fontaine-Illusie, *Algebraic Analysis, Geometry, and Num-
     ber Theory (Baltimore, MD, 1988)*, Johns Hopkins Univ. Press, Baltimore, 1989, 191–224.
[28] N.M. Katz and W. Messing, Some consequences of the Riemann hypothesis for varieties over
     finite fields, *Invent. Math.* **23** (1974), 73–77.
[29] N.M. Katz and T. Oda, On the differentiation of de Rham cohomology classes with respect
     to parameters. *J. Math. Kyoto Univ.* **8** (1968), 199–213.
[30] K.S. Kedlaya, Counting points on hyperelliptic curves using Monsky-Washnitzer cohomology,
     J. Ramanujan Math. Soc. **16** (2001), 323–336; errata, *ibid.* **18** (2003), 417–418.
[31] K.S. Kedlaya, Computing zeta functions via p-adic cohomology, *Algorithmic Number Theory*,
     Lecture Notes in Comp. Sci. 3076, Springer, 2004, 1–17.
[32] K.S. Kedlaya, Semistable reduction for overconvergent F-isocrystals, I: Unipotence and log-
     arithmic extensions, *Compos. Math.* **143** (2007), 1164–1212.
[33] U. Köpf, Über eigentliche Familien algebraischer Varietäten über affinoiden Räumen, *Schr.
     Math. Inst. Univ. Münster* Heft 7 (1974).
[34] A.G.B. Lauder, Deformation theory and the computation of zeta functions, *Proc. London
     Math. Soc.* **88** (2004), 565–602.
[35] B. le Stum, *Rigid Cohomology*, Cambridge Univ. Press, 2007.
[36] B. Mazur, W. Stein, and J. Tate, Computation of p-adic heights and log convergence, *Doc.
     Math.* Extra Vol. (2006), 577–614.
[37] P. Monsky, Formal cohomology. II: The cohomology sequence of a pair, *Annals of Math.* **88**
     (1968), 218–238.

[38] P. Monsky, Formal cohomology. III: Fixed point theorems, *Annals of Math.* **93** (1971), 315–343.

[39] P. Monsky and G. Washnitzer, Formal cohomology. I, *Annals of Math.* **88** (1968), 181–217.

[40] J.-P. Serre, Géométrie algébrique et géométrie analytique, *Ann. Inst. Fourier, Grenoble* **6** (1955–1956), 1–42.

[41] J. Tate, Residues of differentials on curves, *Ann. Scient. É.N.S.* **1** (1968), 149–159.

[42] M. van der Put, The cohomology of Monsky and Washnitzer, Introductions aux cohomologies p-adiques (Luminy, 1984), *Mém. Soc. Math. France* **23** (1986), 33–59.